ONCOGENE PROTEINS: NEW RESEARCH

ONCOGENE PROTEINS: NEW RESEARCH

ARTUR H. MALLOY

AND

EARL C. CARSON

EDITORS

Nova Science Publishers, Inc.

New York

For permission to use material from this book please contact us:
Telephone 631-231-7269; Fax 631-231-8175
Web Site: http://www.novapublishers.com

Library of Congress Cataloging-in-Publication Data

Oncogene proteins. New research / [edited by] Artur H. Malloy and Earl C. Carson.
p. ; cm.
Includes bibliographical references and index.
ISBN 978-1-60456-290-3 (hardcover)
1. Tumor proteins. 2. Oncogenes. I. Malloy, Artur H. II. Carson, Earl C. III. Title: New research.
[DNLM: 1. Oncogene Proteins. QZ 202 O577 2008]
RC268.42.O523 2008
362.196'9940072--dc22 2007050785

Published by Nova Science Publishers, Inc. ≃ New York

Contents

Preface

Oncogene proteins are proteins coded by oncogenes. They include proteins resulting from the fusion of an oncogene and another gene (oncogene proteins, fusion). An oncogene is a modified gene, or a set of nucleotides that codes for a protein and is believed to cause cancer. This book brings together the latest research in this field from around the world.

Short Communication A – Bmi1 is a polycomb group (PcG) transcription repressor initially identified as a cmyc cooperating oncogene in murine lymphomas. Bmi1 has been implicated in regulating diverse biological processes, including carcinogenesis, and stem cell self-renewal. Furthermore, this oncogene appears to play a role in gene regulation in both cancerous stem cells (CSCs) and normal stem cell (NSCs) proliferation through epigenetic mechanisms. Isolation and identification of CSCs have been performed in a number of glioma tissues utilising CD133 as a cell surface antigen marker

The current study used five cell lines (three gliomas, a normal astrocyte and an embryonic) and ten brain tissues (a recurrent anaplastic ependymona, seven glioblastoma and two normal) to assess the role of bmi1 in glioma by determining transcriptional relationship between stemness, cmyc and bmi1.

The oncogene bmi1 is highly expressed in all the cell lines tested at various levels, all the glioblastoma tissues and one of the normal brain tissues. No bmi1 mRNA was detected in the recurrent anaplastic ependymona and one of the normal tissues. However, cd133, the marker for stem cells was detected in one of the glioma cell lines and the embryonic cell lines, recurrent anaplastic ependymoma and six of seven glioblastoma and one of the normal tissues. The cooperating gene, cmyc was expressed in all cell lines tested and to a lesser extent in all the tissues. The activation of the oncogene bmi1 was found to be much higher in the cell lines than the tissues. There was no evidence to indicate that bmi1 is involved in carcinogenicity since it was expressed in both normal tissue and all the cell lines.

Bmi1 is required for self-renewal and for extending the replicative potential of stem cells through the suppression of the senescence pathway. Since CSCs are quiescent and can survive serum deprivation, the authors challenged the transcriptional relationship between cd133 and bmi1 using serum deprived media. Human brain astrocytoma/oligodendrogliomas GOS-3 cells starved of serum for 6 days were analysed for mRNA levels for cd133, bmi1 and cmyc. While serum deprivation (0% serum) enriched cd133 mRNA, it reduced the transcription of bmi1 and cmyc. There is a strong transcriptional association between bmi1 and cmyc, however, there is an inverse transcriptional association between these genes and cd133.

These results suggest that bmi1 may not be essential for the maintenance of glioma stem cells and therefore further work is necessary in order to explore the role of bmi1 in glioma.

Short Communication B – There is currently an increasing interest in the notion of modules in biological networks. In a very general way, modules can be defined as compartmentalized sets of molecular devices with interconnected functions that perfom complex tasks. The specific modularity of a network may then translate into cellular behaviors or phenotypes. The biological relevance of such a module seems thus evident, but its precise definition is less clear.

Here the authors comment on several aspects of modularity of the MAPK / Erk pathway in Xenopus oocytes, of which the basic cascade topology has been described more than a decade ago. Oocytes offer to study protein-protein interaction (PPI) networks and their functions during cell cycle transitions, independently of transcription. In these cells, the MAPK / Erk cascade may either be considered as one module among others in the MAPK family (vs JNK (Jun N-terminal Kinase) and p38MAPK pathways) or as composed of three modules reminding of domino toppling, where each module phosphorylates and activates the downstream one.

We demonstrate that the construction of modules and their interconnections are highly dependent on the biological questions addressed. Whereas two modules, (i) oncoprotein Mos (MAPK Kinase Kinase responsible for Erk cascade activation) and (ii) MAPK Kinase - MAPK - MAPK Activated Kinase, may be defined considering meiotic spindle morphogenesis in amphibian oocytes, these two modules are neither relevant nor sufficient to address both the regulation and the dynamics of this cascade within the same cellular model.

Chapter 1 – Human Papillomaviruses (HPVs) are involved in the etiology of at least 10% of all human cancers and are mainly implicated in the development of pre-malignant and malignant lesions of the ano-genital tract. In particular, genital High-Risk HPVs are responsible for nearly 100% of cases of cervical cancer, which represents the second most prevalent cancer in women worldwide.

The carcinogenic risk associated with HPV infection is primarily due to the activity of two viral oncoproteins, E6 and E7. Indeed, their expression is necessary for the maintenance of the malignant phenotype, and is always retained in HPV-positive cancer cells.

Although malignant transformation results from the synergistic and complementary effects of E6 and E7, the latter is reported to be the main oncoprotein of HPVs. One of the major aims of this review is to update the understanding of the role of E7 in HPV-associated carcinogenesis, focusing on the binding with its cellular targets and the effects induced by these associations. The interaction of E7 with its main target, the tumor suppressor protein retinoblastoma (pRb), will be described, with emphasis on the biological consequences and the domains involved. However, as E7 oncogenic properties are not only attributable to this association, the significance of other relevant interactions will be evaluated as well.

Due to the essential role of E7 in cervical carcinogenesis, the lack of homology with cellular proteins and the exclusive expression in cancer cells, this oncoprotein represents the main target for cervical cancer therapy. The therapeutic potential of strategies aimed at E7 inhibition for the treatment of HPV-associated lesions will be addressed. Among others, the use of small interfering RNA (siRNA), intracellular antibodies ("intrabodies"), ribozymes, anti-sense oligonucleotides and aptamers to knock out E7 at the gene/protein level will be covered. The results of these studies will be reviewed with the aim of highlighting E7 inhibition as a promising approach for the treatment of HPV-associated cancer.

Chapter 2 – Cancer cells are often physiologically dependent on activation and/or increased expression of some oncogenes, which are essential for growth and maintenance of the malignant phenotype. EGFR and HER2 are among those "oncogenes of addiction" in many solid cancers, including breast cancer. These oncogenes often cooperate with another oncogene, Src, which is also overexpressed or constitutively activated in many solid cancers, such as breast cancer. Many studies show a connection between diet and cancer, with cancer incidence inversely correlated with consumption of fruit and vegetables. Moreover, occurrence of breast cancers with particular genotypes of the EGFR gene is related to a diet poor in vegetables. Till recently, the molecular basis of a connection between cancer and diet has been obscure, however, recent studies have shown that dietary compounds found in some vegetables affect oncogenes in cancer cells and, thereby, reduce tumor growth. Dietary phytochemicals, such as indole-3-carbinol, resveratrol, genistein, curcumin and epigallocatechin gallate, can reduce expression and/or activation of EGFR or HER2. This review summarizes known inhibitory effects of these phytochemicals on the activation and expression of EGFR, HER2 and Src, eventually resulting in downregulated signaling from these oncogenes and reduced cancer cell viability in breast, colon and other solid cancers.

Chapter 3 – Cyclin dependent kinases (CDKs) and their activating cyclins are required for the ordered regulation of cell cycle progression throughout the cell cycle. Cyclin D1 forms active complexes with CDK4 or CDK6 that are essential for G1 to S phase progression. Together, these complexes phosphorylate the retinoblastoma (Rb) protein and facilitate E2F mediated gene expression. The E2F transcription factor regulates the expression of several genes required for progression into S phase. Cyclin D1 protein levels are elevated in a wide variety of cancers, including those of the breast, lung, oesophagus and prostate. The defective regulation of cyclin D1 degradation has been linked the elevated levels of cyclin D1 in many cancers. The ubiquitin dependent degradation pathway plays an important role in regulating cyclin D1 stability. Phosphorylation of threonine residue 286 and/ or 288 further enhances the rate of cyclin D1 degradation via this pathway. Additionally, cyclin D1 degradation can also occur independently of ubiquitin. Recent studies have identified the F-box proteins FBX4 and FBXW8 as E3-ligases that mediate cyclin D1 degradation. The loss of C-terminal regulatory sequences or expression of proteins that regulate cyclin D1 stability (and their cofactors) has been linked to deregulated cyclin D1 degradation. Numerous clinical and experimental pharmacological agents have been shown to induce cyclin D1 degradation. Generally, these agents induce the degradation of cyclin D1 via the ubiquitin dependent pathway. The functional characterization of FBX4 and FBXW8 is likely to increase current knowledge on the molecular pharmacology of these agents. Together, these findings suggest that cyclin D1 presents an important therapeutic target for cancer therapy. Currently, cyclin D1 ablative agents are in development and the utility of this treatment strategy has been demonstrated in clinical trials. In this chapter, recent advances in the authors understanding of the normal and pathological roles of cyclin D1 in cellular proliferation are discussed. The mechanisms that underlie the elevated levels of cyclin D1 in cancer cells are also explored. Finally, the development and current status of cyclin D1 as a therapeutic target in cancer therapy is discussed.

Chapter 4 – The RET (REarranged during Transfection) proto-oncogene plays a vital role in controlling cell proliferation, migration, differentiation, and continued cellular survival during development. It is central to the development of the kidney and enteric nervous systems as well as being chiefly responsible for at least four conditions associated with mutations at different sites.

Much progress has been made in identifying the functions of this gene and its ligands GDNF, GFRα and neurturin in Tyrosine kinase activation , particularly as it pertains to its influence in oncogenesis, in the thyroid and other endocrine glands (MEN syndromes) Up-regulation of the RAS-RAF-ERK and PI3K-PDK1-Akt sinaling pathways results. In addition, fusion to other genes (eg RET/PTC), activates the BRAF, RET, or RAS genes of the MAPK system.

This article examines the significance and functions of RET and its ligands reviewing its oncogenic potential as well as possible molecular targets in therapy.

Chapter 5 – In recent years the authors have witnessed the transition in studies of the insulin-like growth factor (IGF) system from its physiological functions to potential roles in cancer therapy. The two IGFs (I and II) and their main receptor, IGF-IR, have long been known to promote cell proliferation and tissue development, particularly prenatal growth. Their role in tumorigenesis is supported by studies from several aspects. IGF-IR function is essential for transformation of mouse embryonic fibroblasts by a variety of oncogenes. Epidemiological data indicate that the circulating IGF-I level is correlated with increased risk for development of colon, prostate, breast, lung and bladder cancers. Elevated plasma IGF-II level has also been found among colon and breast cancer patients. Various strategies have been designed to target components of the IGF system, namely the ligand (IGF-I and II), receptor, and ligand binding proteins, with the receptor and ligand being the main focus. Both small molecule inhibitors and protein-based therapeutics have been developed. Among them, antibodies against the receptor have been actively pursued by many academic groups and commercial entities. High affinity IGF-IR antibodies not only inhibit cell growth in vitro but also reduce tumor size in xenograft animal models. Phase I and II clinical trials of IGF-IR antibodies have been conducted by several companies with promising outcomes so far. The idea of inhibiting the signal from the pathway via targeting its ligand is relatively novel and development of those antibodies is at their early stage. The authors discuss in this review in detail these therapeutic antibodies currently under investigations and for comparison some small molecule inhibitors of the receptor kinase activity.

Chapter 6 – Among the EEF1A family, two constitutive and actively transcribed genes are known: EEF1A1 on chromosome 6q14 and EEF1A2 on chromosome 20q13.3. eEF1A carries aa-tRNA as a ternary complex eEF1A-GTP-aa-tRNA allowing to correct positioning of the aa-tRNA in the A site of the ribosome. In mammalian, eEF1A1 is ubiquitously expressed, whereas eEF1A2 is developmentally regulated, being its expression switched on in adult life in specialised tissues such as skeletal muscle, cardiomyocytes and nervous system . The expression of EEF1A2 in tissues other than skeletal muscle, heart and brain has been associated to cancer development and aggressiveness, playing a role in ovarian and in breast cancers. Analogously to EEF1A2, over-expression of EEF1A1 has been proposed to be related to cell proliferation and cancer development. Another transcript belonging to this family has been discovered and proposed for tumour transformation: the prostate tumour-inducing gene-1, alias PTI-1, a truncated form of eEF1A1, lacking of 67 aa at NH_2-end.

In the authors observations an isoform of eEF1A1 and a de-regulation of eEF1A1 and eEF1A2 expression has been found in human haematopoietic cancer cells, but not in the normal counterpart. In hepatocarcinoma cells the authors document the increased mRNA levels of EEF1A genes compared to normal liver and the higher nuclear-enriched/cytoplasm ratio of eEF1A protein in the less differentiated hepatoma cancer cells. Differently, in prostate cancer lines they observed a dramatic up-regulation of EEF1A2 and the concomitant down regulation of EEF1A1, compared to non cancer prostate cells. In prostate cancer cells the authors data are in favour of a minor role for PTI-1 to the onset of cancer, while a relation with the de-regulation of eEF1A genes was found. PTI-1 mRNA was indeed detected in some hematopoietic and colonadenocarcinoma cell lines, but its role in in vivo human tumors remains to be defined. These observations were discussed with the other available data about the involvement of eEF1A family components in human cancers. In conclusion, the up and/or down-regulation of the expression levels of the two eEF1A genes, as well as post-translational modifications of the eEF1A protein, may play a role in human tumorigenesis and cancer progression but in a different manner, depending on tumour type.

Chapter 7 – Thromboxane A2 (TXA_2) is a potent arachidonate metabolite in the cyclooxygenase-2 (COX-2) pathway. Thromboxane synthase (THXS) is a member of cytochrome P450 (CYP) superfamily which was first purified in human platelets and porcine lung. This enzyme catabolizes prostaglandin H2 (PGH_2) into 12-L-hydroxy-5,8,10-heptadecatrieonic acid (HTT), malondialdehyde (MDA) and TXA_2 in a ratio of 1:1:1. TXA_2 plays a role in platelet aggregation induction, vasocaonstriction and cell proliferation. Its dysfunction is associated with various pathological conditions including cancers. Although the expression of THXS has not been studied in lung cancer, the increase of THXS has been found in a number of other tumors. The level of TXA_2 in lung cancer has been shown to be higher than that in normal lung cells, but its role in lung cancer development is uncertain. TXA_2 can promote cell migration, angiogenesis, tumor metastasis and cancer proliferation. Therefore, the authors hypothesize that thromboxane plays a role in lung cancer progression. In this review, they will discuss the structure of THXS, the regulation of THXS expression, the effects of TXA_2 modulations and THXS inhibitors on the proliferation and growth of human lung cancer cells.

Chapter 8 – Programmed cell death (apoptosis) is essential for the development and survival of all metazoans. Apoptosis regulated by the Bcl-2 family of proteins involves a complex interplay between members that promote cell survival and those that promote cell death. However, despite extensive investigation, the molecular mechanisms underlying the control of apoptosis signalling via the Bcl-2 pathway remain controversial. Two major models have emerged: "indirect" and "direct" activation. These models differ significantly, both in terms of the role of the pro-survival proteins, and how the essential mediators of cell death, Bax and Bak, become activated following receipt of death stimuli. In this chapter, these opposing models are described in detail and the validity of each discussed. How these models impact upon strategies for the development of novel anti-cancer therapeutics will also be explored.

Chapter 9 – Tumor marker is a group of important laboratory diagnosis. In this review, the details on the general concepts of tumor marker, usefulness and limitation are collected and presented. Basically, a specific tumor marker is a fusion protein associated with a malignant process in which an oncogene is translocated. Oncogene proteins can be served as tumor markers. Here, important oncogene proteins as serum and tissue tumor markers will be presented and discussed. In addition, the clinical application of oncogene proteins as tumor markers for important tumors in medicine will be mentioned.

Chapter 10 – Lung cancer is the world's leading cause of cancer death. Currently there is no accepted adjuvant or palliative treatment modalities that have been conclusively shown to prolong survival in lung cancer. In order to improve outcomes with this disease, search for novel treatments is critical. Peroxisome proliferator activated receptors gamma (PPARγ) is a transcriptional factor belonging to the ligand-activated nuclear receptor superfamily. PPARγ is highly expressed in adipose tissue and has a dominant regulatory role in adipocyte differentiation. In addition to adipogenic and anti-inflammatory effects, PPARγ is expressed in multiple types of tissues, including the breast, colon, lung, ovary, and placenta in humans. The activation of PPARγ plays an inhibitory role in cell proliferation and growth and this property makes the activation of PPARγ by its natural and synthetic ligands an attractive tool in cancer treatment and prevention. A number of PPARγ ligands have been identified, such as natural prostaglandins and the synthetic anti-diabetic thiazolidinediones (TZDs). The discovery of PPARγ agonists enables the recognition of the mechanisms involved in PPARγ multiple effects such as inhibition of tumor cell growth. The importance of this transcription factor and its ligands in physiology and pathophysiology has instigated much research in this field. In this article, current situations in lung cancer therapy, structural features of PPARγ, mechanisms of PPARγ gene transcription and recent development in the discovery of their biological functions on growth inhibition and apoptosis of lung tumors are reviewed. Prospects for future research leading to new therapies for lung cancer are also discussed.

Chapter 11 – Hepatitis B virus (HBV) and hepatitis C virus (HCV) are global public health problems. The clinical courses of both HBV and HCV infection vary from acute hepatitis to chronic persistent infection that may progress to cirrhosis and carcinoma. Acute inflammation is a defense response and chronic inflammation can lead to cancer. Several proinflammatory gene products have been identified in both HBV- and HCV-infected patients. The expression of these genes is mainly regulated by the transcription factor NF-κB, which is constitutively active in most tumors and is induced by tumor promoters and carcinogenic viral proteins. Anti-inflammatory agents that suppress NF-κB may be useful in both the prevention and treatment of cancer. Tumor necrosis factor alpha (TNFα) is thought to be an important factor underlying the mechanisms of action of these types of viral hepatitis and carcinoma because TNFα plays an important role in the host immune response to HBV and HCV infection. Recent studies indicated that TNFα promoter polymorphisms are significantly associated with viral clearance. Neutralization of TNFα (anti-TNF therapy) has been shown to be associated with activation of HBV infection, but not HCV infection. However, TNF has been reported to induce clearance of HBV. One of the inhibitory mechanisms of anti-TNF therapy is that TNF blocks HBV replication by promoting destabilization of viral nucleocapsids. HCV proteins are able to trigger production of TNFα and modulate nuclear factor kappa B activation and apoptosis stimulated by this cytokine.

This review highlights the importance of TNFα production for hepatitis and carcinoma by HBV and HCV infection.

Chapter 12 – Thyroid cancer is the commonest endocrine malignancy and accounts for the majority of endocrine cancer deaths each year. It is notable that the incidence of thyroid cancer is roughly three times more frequent in woman than in men, and further that the incidence of thyroid cancer decreases after menopause in females. This gender difference suggests that the growth and progression of thyroid tumors may be influenced by female sex hormones, particularly estrogen. Estrogen induction of cell viability plays a crucial role in the carcinogenesis of cancers such as breast, colon and prostate. Estrogen also plays a role in thyroid cancer. The action of estrogen on target sites in the human body is mediated through related but distinct estrogen receptors designated ERα and ERβ, both of which are known to be expressed in thyroid cancer. The literature and data from the authors laboratory has shown that estrogen can promote thyroid cancer cell survival through an ERα-non-genomic signaling pathway and can promote cell apoptosis through an ERβ-non-genomic signaling pathway. The non-genomic signaling pathways are different to the classical ER-mediated genomic signal transduction pathway. Estrogen has opposite effects on cell proliferation, growth and apoptosis, depending on the balance between ERα and ERβ in the cell. This review will address the way in which estrogen interacts with apoptotic molecules and how such an interaction contributes to the development of thyroid cancer.

In: Oncogene Proteins: New Research
Editors: A. H. Malloy and E. C. Carson, pp.1-9

ISBN 978-1-60456-290-3

Short Communication A

The Transcriptional Correlation of *bmi1* Oncogene with *cd133* in Gliomas

A. Shervington*[1]*, C. Lu and R. Patel
Brain Tumour North West, Faculty of Science, University of Central Lancashire, Preston, United Kingdom

Abstract

Bmi1 is a polycomb group (PcG) transcription repressor initially identified as a cmyc cooperating oncogene in murine lymphomas. Bmi1 has been implicated in regulating diverse biological processes, including carcinogenesis, and stem cell self-renewal. Furthermore, this oncogene appears to play a role in gene regulation in both cancerous stem cells (CSCs) and normal stem cell (NSCs) proliferation through epigenetic mechanisms. Isolation and identification of CSCs have been performed in a number of glioma tissues utilising CD133 as a cell surface antigen marker

The current study used five cell lines (three gliomas, a normal astrocyte and an embryonic) and ten brain tissues (a recurrent anaplastic ependymona, seven glioblastoma and two normal) to assess the role of bmi1 in glioma by determining transcriptional relationship between stemness, cmyc and bmi1.

The oncogene *bmi1* is highly expressed in all the cell lines tested at various levels, all the glioblastoma tissues and one of the normal brain tissues. No *bmi1* mRNA was detected in the recurrent anaplastic ependymona and one of the normal tissues. However, *cd133*, the marker for stem cells was detected in one of the glioma cell lines and the embryonic cell lines, recurrent anaplastic ependymoma and six of seven glioblastoma and one of the normal tissues. The cooperating gene, cmyc was expressed in all cell lines tested and to a lesser extent in all the tissues. The activation of the oncogene *bmi1* was found to be much higher in the cell lines than the tissues. There was no evidence to

---

1 Corresponding author Telephone: +44 (0) 1772893598, Fax: +44 (0) 1772892929, Email: aashervington@uclan.ac.uk.

indicate that *bmi1* is involved in carcinogenicity since it was expressed in both normal tissue and all the cell lines.

Bmi1 is required for self-renewal and for extending the replicative potential of stem cells through the suppression of the senescence pathway. Since CSCs are quiescent and can survive serum deprivation, we challenged the transcriptional relationship between *cd133* and *bmi1* using serum deprived media. Human brain astrocytoma/oligodendrogliomas GOS-3 cells starved of serum for 6 days were analysed for mRNA levels for *cd133, bmi1* and *cmyc.* While serum deprivation (0% serum) enriched *cd133* mRNA, it reduced the transcription of *bmi1* and *cmyc.* There is a strong transcriptional association between *bmi1* and *cmyc*, however, there is an inverse transcriptional association between these genes and *cd133.*

These results suggest that *bmi1* may not be essential for the maintenance of glioma stem cells and therefore further work is necessary in order to explore the role of *bmi1* in glioma.

Keywords: cd133; bmi1; glioma tissues; stemness; serum deprivation; GOS-3 cells.

Introduction

Although cancer stem cells (CSCs) represent only a small fraction of a tumour, they give rise to a broad range of highly malignant cells [1-5]. CSCs have been identified in malignancies of haematopoietic origin and in some solid tumours using CD133 as a marker [1,6,7]. CD133 is a cell surface antigen, initially identified as a marker for haematopoietic stem cells and later observed as a marker of stem cells in other normal tissues [8-10]. Moreover, CSCs were sorted and purified from various cancer cells and tissues (1,11,12) to include neuroblastoma, breast, lung cancer tissues and glioblastoma cell lines [13,14]. CSCs were isolated from a C6 rat glioma cell line that had been maintained in culture for a prolong period of time through the use of serum-free medium supplemented with growth factors [12].

CSCs posses the capability of self renewal to maintain the stem cell number through a strict balance between cell proliferation, cell differentiation and cell death [2-5]. One mechanism that may underlie the capability of stem cells to replicate indefinitely is the expression of the putative oncogene *bmi1.*

bmi1 belongs to the *Polycomb* group which promotes cell cycle progression and controls cell senescence. *Bmi1* together with c*myc* cooperating oncogene is implicated in brain development, and may also prompt tumour development [15]. Extrinsic signals for self-renewal result in the elevation of the *bmi1* expression level, which allows the repression of various genes, and possibly activation, via indirect mechanisms of other genes [16].

Although, stem cells have self renewal properties, they generally quiescent, spending most of their time in G0 phase. Cell cycle analysis showed that more than 90% of the cells were in G0/G1 phase while the proportion of cells in the S and G2/M phase were approximately 5.56% and 2.08%, respectively [17]. The favourable environment for stem cells is in quiescent G0/G1 phase of the cell cycle, which can be used as a marker for stem cells.

CSCs subpopulation is by nature quiescent and remains in G0 phase and thus can survive serum deprivation which can be isolated using the stem cell marker CD133. Thus, a correlation between *bmi1*, *c-myc* transcription levels and stemness using serum deprivation were examined. Moreover, this study identified the presence of CSCs in glioma cell lines, normal and cancer tissues.

Materials and Methods

Tissues and Cells Culture

Ten human tissue samples (one recurrent anaplastic ependymoma, seven glioblastomas and two controls) were used for this study. Written consent was obtained from patients admitted to the Royal Preston Hospital, UK before using the tissues in the investigations. For each patient, one sample was surgically dissected and frozen immediately in liquid nitrogen until used for RT-PCR or formalin-fixed and paraffin-embedded (FFPE) for immunofluorescence. Human brain astrocytoma/oligodendrogliomas GOS-3 cells (DSMZ, Germany), astrocytoma 1321N1 (ECACC, UK) and glioblastoma U87-MG (ECCAC, UK) were cultured in Dulbecco's modified Eagle's medium (DMEM; Sigma, UK). Normal human astrocytes NHA cells (ScienCell, CA, USA) were cultured in astrocyte medium (ScienCell, CA, USA). The normal embryonic brain tissue FLOW3000 cells (JCRB, Japan) was cultured in EMEM. All culture media were supplemented with 2 mM L-glutamine (Sigma, UK) and 10% (v/v) foetal bovine serum (Gibco-BRL, UK), the exception to this was the addition of 4 mM of L-glutamine to the GOS-3 cell line medium. Cells were incubated at 37(C in a humidified 5% CO_2 incubator.

mRNA Isolation, Reverse Transcription and Qrt-PCR

An average of 1 ρg per cell of mRNA was extracted from the cells and tissues using mRNA Isolation Kit (Roche, West Sussex, UK) according to the manufacturer's instructions. Isolated mRNA (100 ng) was transcribed to cDNA using a First Strand cDNA Synthesis Kit (Roche, West Sussex, UK) which was then used as a template for PCR. Real time PCR containing cDNA was performed quantitatively using FastStart DNA master PLUS SYBR Green1 (Roche, West Sussex, UK). After an initial denaturation for 10 min at 95°C, the samples were run for 35 cycles at 95°C (10 s) with the recommended annealing temperature 64°C (*cd133*), 66°C (*bmi1* and *cmyc*) and 56°C (*GAPDH*) (15 s), 72°C (15 s). At the end of each cycle, the fluorescence was measured in a single step in channel F1 (gain 1) to obtain data for the melting curve analysis (Roche manual, 2004). PCR reactions were performed in triplicate and a negative control (primers with no DNA) was included. Primer (TIB MOLBIOL, Germany) sequence and length of amplified products were: *cd133* (162bp) sense: 5′TTGTGGCAAATCACCAGGTA, antisense: 5′TCAGATCTGTGAACGCCTTG; *bmi1* (186bp) sense: 5′TCGGTTGTTGCTGATCTGTC antisense: 5′CTCCTGGCAAAAGGTC AGAG; *cmyc* (206bp) Sense: 5′CTCCAGCGCCTTCTCTCCGT, antisense: 5′GAGCCTGCCTCTTTTCCACAG; *GAPDH* (226bp) 5′GAGTCAACGGATTTGGTCGT antisense: 5′TTGATTTTGGAGGGATCTCG.

Copy Number Calculation

Quantitative amplification was monitored by the level of fluorescence reflecting the cycle number at the detection threshold (crossing point). Crossing points were used for quantification of the copy number of genomic DNA normalized by using GAPDH as a reference gene. A standard curve was generated using the crossing points generated from different concentrations of genomic DNA with known copy numbers. Different crossing points of each gene were normalized using a standard curve to obtain the copy numbers for each of the amplified genes. All experiments were carried out using similar cells numbers.

Treatment of Human Brain Astrocytoma/Oligodendrogliomas Cells:

GOS-3 cells were cultured in Dulbecco's modified Eagle's medium (DMEM) (Sigma, Poole, UK) in 75 cm^2 tissue culture-treated polystyrene flasks (Sigma, Poole, UK) for the use of RT-PCR. DMEM culture medium was supplemented with 4 mM L-glutamine, and the cells were treated with different concentration of FBS at 0, 2.5, 5 and 10% (control) for 6 days.

Statistical Analysis:

Student's two-tailed t-test for studies on mRNA copy number was carried out using SPSS 13.0 for Windows. All values are expressed as the mean ± standard deviation, values of $P < 0.05$ were considered to be statistically significant.

Results

The expression profile of *cd133, bmi1, cmyc* and *GAPDH* were carried out in three glioma cell lines namely: 1321N1, GOS-3 and U87-MG together with the normal embryonic FLOW3000 and the normal astrocytes NHA cell lines. In addition, eight glioma tissues and two controls were also used in this study.

Although, *cd133* mRNA was detected in 1321N1, GOS-3 and the positive control FLOW3000 cell lines, it was not expressed in U87-MG (Figure 1). *cd133* expression was detected in the recurrent anaplastic ependymona, 6 of 7 gliomas and one of the control tissues (Figure 1).

While *bmi1* was expressed in all cell lines and tissues including the controls with the exception of the recurrent anaplastic ependymona, *cmyc* was detected in all cell lines and tissues. (Figure 1). *GAPDH,* the house keeping gene expression profile was consistent in all cell lines and tissues including the controls (Figure 1).

To correlate the expression profiles of the candidate genes, GOS-3 cells cultured for 6 days in varying concentrations of serum, were analysed for *cd133, bmi1, cymc* and *GAPDH* gene expressions.

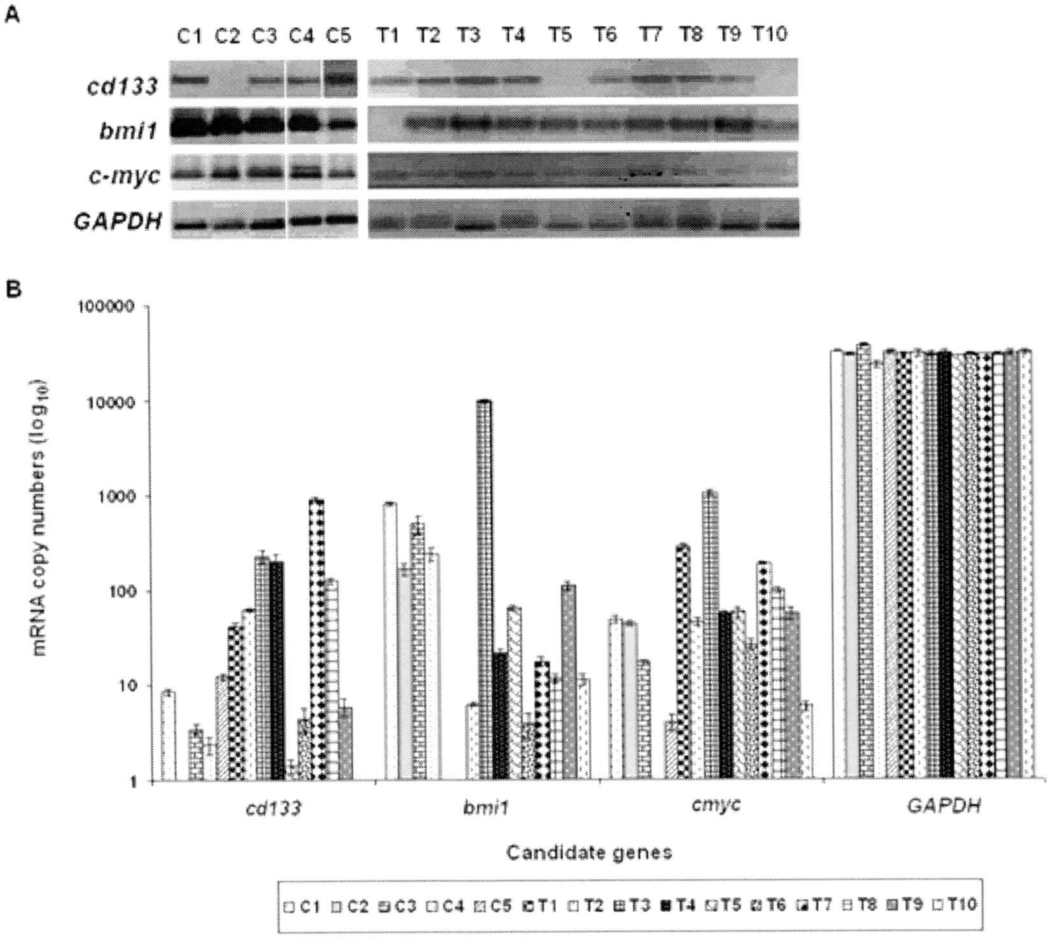

Figure 1: Differential expression of *cd133, bmi1, cmyc* and *GAPDH* in different cell lines and brain tissues: A) Lanes C1-C5 amplicons are cell lines 1321N1, U87-MG, GOS-3, FLOW3000, and NHA, respectively; Lanes T1, T2-T8, T9-T10 amplicons are tissues recurrent anaplastic ependymoma, glioblastomas and normal brain tissues, respectively. B) mRNA copy numbers of the candidate gene expression in each cell line and tissue (data values are mean ± standard deviation, n = 3).

RT-PCR quantitative analyses of *cd133* showed a decreased gene expression from 0-10% serum treated GOS-3 cells ($P < 0.001$). *cd133* mRNA level was increased 6.7 fold in stem-like cells (0% serum) compared to the same number of cancer cells (10%) (Figure 2).

While *bmi1* together with its cooperating oncogene c*myc* were expressed, the magnitude of expression was similar at 2.5, 5 and 10% serum treatment, however, their expression profiles were different at 0% serum treatment. *bmi1* expression was sharply down-regulated by 475 fold at 0% compared to 10% serum treatment (Figure 2). c*myc* mRNA down-regulation was gradual with serum deprivation (Figure 2). *GAPDH* was consistently expressed on treatment with all four of the serum concerntration with no significant difference between 0% and 10% serum treatment ($P = 0.057$, Figure 2).

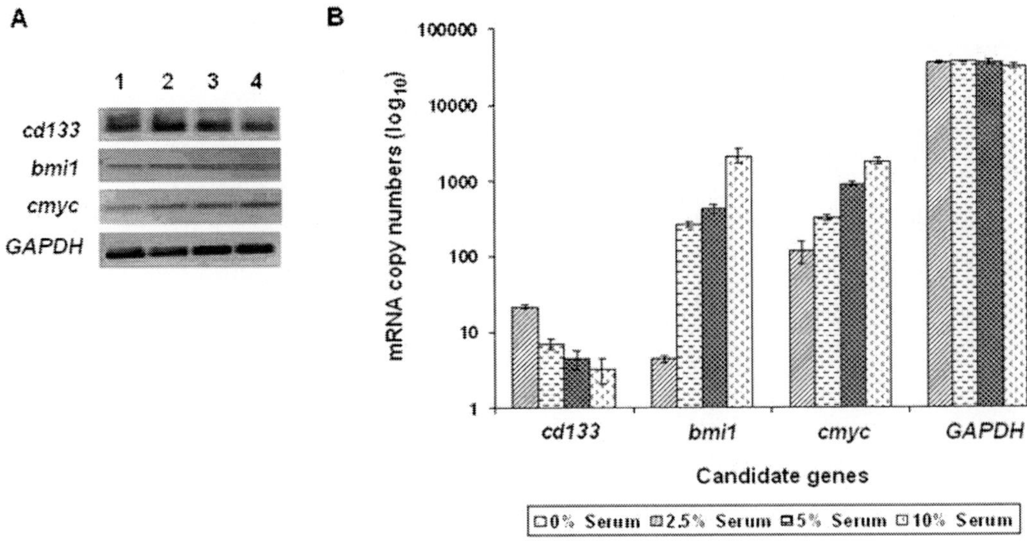

Figure 2: Differential expression of *cd133, bmi1, cmyc* and *GAPDH* under serum deprivation in GOS-3 cell line; A) Lanes 1-4 amplicons from COS-3 treated cells (0, 2.5, 5 and 10% FBS respectively. B) mRNA copy numbers of candidate gene expression in GOS-3 treated cells (data values are mean of ± standard deviation, n = 3).

Discussion

Several studies have demonstrated the existence of brain cancer stem cells in tissues and cell lines [12,14,18] but the key issue in normal and cancerous stem cell studies is the identification of specific molecular markers that can contribute to isolate and study 'pure' stem cell populations.

Normal neural stem cells as well as early progenitors express the CD133 antigen, whereas full mature cells do not [15]. The results of this study clearly demonstrate the existence of CD133 antigen as a stem marker in brain normal and cancerous tissues. The glioma sample size required for this independent uncorrected chi-square test is eight gliomas and eight controls. Due to the difficulties in obtaining brain control tissue, the study was limited to two control tissues, in addition to the negative and positive control cell lines. The importance of finding CD133 antigen in brain normal tissues need to be explored further using larger control samples.

Stem cells generate the differentiated cell types within many organs throughout the lifespan of an organism and are thus ultimately responsible for the longevity of multicellular organisms and therefore, senescence of stem cells need to be prevented. Bmi1 is required for the maintenance of adult stem cells in some tissues partly because it represses genes that induce cellular senescence and cell death [15]. Bmi1 does not appear to be crucial in maintaining cancer since it was expressed in glioma cell lines, tissues and controls. However, it may play a key role in other forms of cancer. *bmi1* mRNA was absent in the recurrent anaplastic ependymona, whilst cmyc was present, suggesting that Bmi1 appears to act independently of its oncoprotein cmyc. Bmi1 acts independently of cmyc in binding sequences in the promoter of specific genes [19].

CSCs subpopulation is by nature quiescent and remains in G0 phase and thus, can survive serum deprivation and be isolated using stem cell marker CD133. This suggests that culture conditions (serum deprivation) may initiate a response leading to differentiation-like suppression. Moreover, serum deprivation can introduce ectopic gene expression of *cd133, bmi1* and *cmyc*, and may establish a relationship(s) between these genes.

Cell viability determined after 6 days of treatment using varying concentrations of serum, showed the survival rate of the candidate stem cells to be approximately 1.2% compared with the 10% serum control culture. These results are in agreement with a previous investigation which showed that the isolated CD133+ SP from human prostate cancer xenografts were accounted for 0.1-1.0% of cells [18].

Cells survived serum deprivation expressed *cd133* mRNA showed down-regulation of *cmyc* expression. Moreover, serum deprivation led to cell death and a sharp downregulation of *bmi1.* The polycomb group protein BMI1 has been shown to support normal stem cell proliferation via its putative stem cell factor function, but there is no evidence to support the role of BMI1 as a cancer stem cell factor to promote the development of cancer [20-21]. *bmi1* maintains the stem cell pool by preventing premature senescence, either through repression of genes involved in senescence or possibly through the induction of telomerase to prevent telomere shortening. It is very likely that Bmi1 is important for maintenance of multiple types of somatic stem cells, since it is widely expressed in different tissues [22]. While Bmi1 is important for maintenance of leukemic stem cells and perhaps other tumorigenic stem cells [22], this may not be the case in glioma stem cells. While Bmi1 promotes cell cycle progression, serum deprivation leads to the G0 phase, thus the downregulation of Bmi1 is expected. The results of this study challenge the role of Bmi1 in the maintenance of these stem cells.

Conclusion

Cd133 mRNA was detected in normal and cancerous brain tissues and the cell lines examined. A pronounced upregulation of *cd133* and down-regulation of *cmyc* and *bmi1* was produced as a consequence of the decreased level of the serum supplement. These findings provide the first direct link between *cd133, cmyc* and *bmi1.* These genes are likely to play a role in stem cell fate decisions including self-renewal and differentiation.

Acknowledgements

The authors would like to thank the Brain Tumour North West (BTNW) for providing the tissue for this study. This work was supported by a grant from the University of Central Lancashire and Sydney Driscoll Neuroscience Foundation (SDNF).

References

[1] Dean M, Fojo T, Bates S (2005) Tumour stem cells and drug resistance. *Nature Reviews.* 5(4):275-284.

[2] Reynolds BA and Weiss S (1996) Clonal and population analyses demonstrate that an EGF-responsive mammalian embryonic CNS precursor is a stem cell. *Dev. Biol.* 175: 1–13.

[3] Gage FH (2000) Mammalian neural stem cells. *Science* 287: 1433–1438.

[4] Reynolds BA and Weiss S (1992) Generation of neurons and astrocytes from isolated cells of the adult mammalian central nervous system. *Science* 255: 1707–1710.

[5] Morrison SJ, Shah NM and Anderson DJ (1997) Regulatory mechanisms in stem cell biology. *Cell* 88: 287–298.

[6] Singh SK,, Hawkins C, Clarke ID, Squire JA, Bayani Jane, Hide T, Henkelman RM, Cusimano MD, and Dirks PB, Identification of Human Brain Tumour Initiating Cells. *Nature*, 2004, 432, 396-401.

[7] Hemmati HD, Nakano I, Lazareff JA, Masterman-Smith M, Geschwind DH, Bronner-Fraser M and Kornblum HI (2003) Cancerous stem cells can arise from pediatric brain tumors. *Proc. Natl. Acad. Sci. USA* 100: 15178–15183.

[8] Singh SK, Clarke ID, Terasaki M, Bonn VE, Hawkins C, Squire J and Dirks PB (2003) Identification of a cancer stem cell in human brain tumors. *Cancer Res.* 63: 5821–5828.

[9] Richardson GD, Robson CN, Lang SH, Neal DE, Maitland NJ, Collins (2004) CD133, a novel marker for human prostatic epithelial stem cells. *J. Cell Sci.* 117:3539-3545.

[10] Yu S, Zhang J, Zhao CL, Zhang HY , Xu O (2004) Isolation and characterization of the CD133+ precursors from the ventricular zone of human fetal brain by magnetic affinity cell sorting. *Biotechnol. Lett.,* 26:1131-1136.

[11] Hirschmann-Jax C, Foster AE, Wulf GG, Nuchtern JG, Jax TW, Gobel U, Goodell MA, Brenner MK (2004) A distinct 'side population' of cells with high drug efflux capacity in human tumour cells. *Proc, Natl. Acad. Sci.* 101:14226-14233.

[12] Kondo T, Setoguchi T, Taga T (2004) Persistence of a Smal Subpopulation of Cancer Stem-like cells in the C6 Glioma Cell Line. *Proc., Natl. Acad. Sc*i. 101, 781-786.

[13] Al-Hajj M, Wicha MS, Benito-Hernandez A, Morrison SJ and Clarke MF (2003) Prospective identification of tumorigenic breast cancer cells. *Proc. Natl. Acad. Sci. USA* 100: 3983–3988.

[14] Yuan X, Curtin J, Xiong Y, Liu G, Waschsmann-Hogiu S, Farkas DL, Black KL and Yu JS (2004) Isolation of cancer stem cells from adult glioblastoma multiforme. *Oncogene* 23: 9392–9400.

[15] Park I, Morrison S, Clarke M (2004). Bmi1, stem cells, and senescence regulation Clin. Invest. 113:175-179.

[16] Pelengaris S, Khan M, Evan G (2002). c-Myc: more than just a matter of life and death. Nature Review, *Cancer* 2:764-774.

[17] Cheong, H.T., Park, T.M., Ikeda, K. and Takahashi, Y. (2003) Cell cycle analysis of bovine cultured somatic cells by flow cytometry. *Jap. Vet. Res.* 51(2), 95-103.

[18] Patrawala L, Calhoun T, Schneider-Broussard R, Zhou J, Claypool K, Tang DG (2006) Side population is enriched in tumourigenic, stem-like cancer cells, whereas ABCG2+ and ABCG2- cancer cells are similarly tumourigenic . *Cancer Res.* 65(14):6207-6219.

[19] Dimri GP, Martinez JL, Jacobs JJ, Keblusek P, Itahana K, Van Lohuizen M, Campisi J, Wazer DE, Band V (2002). The bmi1 oncogene induces telomerase activity and immortalizes human mammary epithelial cells, Cancer Res. 62:4736-4745.

[20] Liu L, Andrews LG and Tollefsbol TO (2006) Loss of the human polycomb group protein BMI1 promotes cancer-specific cell death. *Oncogene*, 25, 4370–4375.

[21] Schmitt, C. 2003. Senescence, apoptosis and therapy: cutting the lifelines of cancer. *Nat. Rev. Cancer.* **3**:286-295.

[22] Molofsky AV, Pardal R, Iwashita T, Park IK, Clarke MF and Morrison SJ (2003) Bmi-1 dependence distinguishes neural stem cell self-renewal from progenitor proliferation. *Nature* 425: 962.

In: Oncogene Proteins: New Research
Editors: A H. Malloy and E.C. Carson, pp.11-17
ISBN 978-1-60456-238-5

Short Communication B

Modularity within the MAPK Cascade Network

Blossey Ralf [‡]***, Russo Christophe*** [‡,†] ***and Bodart Jean-François*** [‡,†2].
‡. Institut de Recherche Interdisciplinaire, CNRS USR 3078, Nanosystèmes Biologiques, Université des Sciences et Technologies de Lille, 59655 Villeneuve d'Ascq cedex, France
†. Laboratoire de Regulation des Signaux de Division, EA 4020, building SN3, Université des Sciences et Technologies de Lille, 59655 Villeneuve d'Ascq cedex, France

Abstract :

There is currently an increasing interest in the notion of modules in biological networks. In a very general way, modules can be defined as compartmentalized sets of molecular devices with interconnected functions that perfom complex tasks. The specific modularity of a network may then translate into cellular behaviors or phenotypes. The biological relevance of such a module seems thus evident, but its precise definition is less clear.

Here we comment on several aspects of modularity of the MAPK / Erk pathway in *Xenopus* oocytes, of which the basic cascade topology has been described more than a decade ago. Oocytes offer to study protein-protein interaction (PPI) networks and their functions during cell cycle transitions, independently of transcription. In these cells, the MAPK / Erk cascade may either be considered as one module among others in the MAPK family (vs JNK (Jun N-terminal Kinase) and p38MAPK pathways) or as composed of three modules reminding of domino toppling, where each module phosphorylates and activates the downstream one.

We demonstrate that the construction of modules and their interconnections are highly dependent on the biological questions addressed. Whereas two modules, (i)

2 Correspondence: Bodart Jean-François, Laboratoire de Regulation des Signaux de Division, EA 4020, building SN3, room 304, Université des Sciences et Technologies de Lille, 59655 Villeneuve d'Ascq cedex, France Tel: 33320436847; Fax: 33320434038; E-mail: jean-francois.bodart@univ-lille1.fr

oncoprotein Mos (MAPK Kinase Kinase responsible for Erk cascade activation) and (ii) MAPK Kinase - MAPK - MAPK Activated Kinase, may be defined considering meiotic spindle morphogenesis in amphibian oocytes, these two modules are neither relevant nor sufficient to address both the regulation and the dynamics of this cascade within the same cellular model.

Modularity: A Key Notion for Cell Biology?

The advent of high-throughput technologies in biology is accompanied by a paradigm shift in the conceptual understanding of biological systems : systems biologists believe the answers to the key biological questions cannot be uncovered from a knowledge of all molecular details, but rather be obtained from an understanding of the concerted action of the molecules within the dynamic interactive networks they constitute in the cell. The shift from a molecular perspective to a network perspective has brought with it the transfer of concepts from computational disciplines to biology ; one such notion is modularity. Within computer science, « a module can be defined variously, but generally must be a component of a larger system, and operate within that system independently from the operations of the other components. » (Wikipedia). The transferral of this concept to biological systems at a molecular level may be appealing, but it is in practice less than obvious to implement. In fact, systems biologists try to find a descriptive level below conventional cellular structure and function which cannot be easily delimited and identified with specific (spatial) structures within the cell. Without a generally applicable rule, the concept of modularity is therefore strongly context–dependent, as we illustrate for the MAPK cascade.

MAPK Cascade : Pathways

In many cellular models, MAPK (Mitogen Activated Protein Kinase) have been involved in cell cycle control, through the regulation of transcription (G1/S and early to mid G2 phases) or through non-genomic functions (late G2 and M phases). Among the numerous members of the MAPK family, different MAPK pathways can be distinguished and each such pathway may be regarded as a specific 'node' which integrates one or more stimuli (growth factors, inflammation, cytokines) and promotes different cellular responses (proliferation, differentiation, apoptosis, inflammation and / or angiogenesis).

Several MAPK pathways may then be distinguished through the involvement of specific molecules (although they are not necessarily independent): Erk 1/2 (Extracellular Regulated Kinases), JNK (Jun N-terminal Kinase), p38MAPK isoforms, Erk3 and Erk5. Within each one of these pathways, three essential functional components can be noted, with which is associated the concept of phosphorylation cascade: a MEKK, a MEK and a MAPK. These three components may also be regarded as individually independent and then remind of domino toppling, where each component phosphorylates and activates the downstream one.

The functional organization of MAPK/Erk signal transduction pathway greatly enhances the sensitivity of cellular targets to external stimuli and leads to ultrasensitivity of Erk to the input signal; the basis of the topological representation has been performed ten years ago by

Huang and Ferrell in *Xenopus* oocytes [1]. Levels of MAPK/Erk activity are not transitory or sustained but bistable in these oocytes. In somatic cells such as PC12 cells, the cellular effects of MAPK/Erk activation are dependent upon the configuration of the network the cascade is embedded in. On NGF stimulation, MAPK/Erk shows a sustain activation and leads to differentiation whereas EGF stimulation induced transitory response of MAPK/Erk and proliferation [2]. These dynamics originate from a configuration driven by the upstream processes activated by the receptors. EGF-dependent MAPK / Erk activation involves tyrosine-phosphorylation of EGF-receptor, SOS-dependent Ras activation, followed by the activation of Raf (MAPKKK / MEK Kinase). Transitory activation of MAPK / Erk results from a MAPK / Erk-dependent feedback inhibition of SOS. While NGF-dependent activation of MAPK / Erk also involves Ras, the Rap1 system mediates the sustain MAPK / Erk activation by activating B-Raf [3]. Depending upon the stimulation and the cellular context, the MAPK/Erk cascade may then be controlled by positive and/or negative feedback loops.

MAPK Cascade: From Pathways to Networks

Vertebrate oocytes offer opportunities to study protein-protein interaction networks and their functions during M phase progression, independently of transcription, with an amenability that is not offered by somatic cells or *in vitro* systems. The paradigm of the *Xenopus* oocyte has been advantageously used to analyze the role of the Mos/MEKK – MEK – MAPK/Erk as well as the role of each component of this cascade in morphological events of M-Phase. During meiosis, the stability of MAPK/Erk activity for several hours, depending upon the accumulation of the oncoprotein Mos, enables the cascade to perform or being involved in different successive tasks: spindle morphogenesis, replication inhibition and cytostatic activity establishment. The latter activity sets the arrest in metaphase II characteristic of most vertebrate oocytes [4, 5]. G2 arrested-synchronized oocytes in ovaries can easily be stimulated to enter M-Phase through MEK1/2 or Mos overexpression or insulin stimulation (oocytes express endogenous IGF1 receptor) or even through progesterone stimulation (Ras-independent pathway) [4, 6, 7]. The model also offers impressive opportunities for studies at the biochemical level : oocytes contain high level of proteins (each oocyte contains for example 50 to 70 ng of the catalytic subunit of PKA), and enable to perform anti-MAPK immunoblotting on one-tenth of a single cell, allowing to study feed-back regulation in one cell [8].

In these cells, the Mos - MAPK/Erk cascade may either be considered alone or as a part of a protein-protein interaction network, involving JNK and p38MAPK cascades, which are both activated during meiosis resumption [9, 10]. In the latter case, it is to note that: (i) JNK may reinforce MAPK/Erk cascade ultrasensibility through MEK-phosphorylated JNK, which phosphorylates in turn Raf, another MEKK [11], (ii) the Mos - MAPK/Erk and the p38MAPK pathways converge to regulate the activation of the M-Phase Promoting Factor (MPF). This heterodimer made up of a catalytic subunit, Cdk1, and a regulatory subunit, the Cyclin B, is negatively regulated by its phosphorylation *via* Myt1 or Wee1, and positively regulated by being dephosphorylated by the dual-phosphatase Cdc25 (4). p38MAPKγ [10] and MAPK/Erk [12] may phosphorylate Cdc25 and then promote MPF activation whereas

MAPK/Erk could also negatively regulate Myt1 activity, through its phosphorylation by Rsk, a MAPK activated protein kinase [13]. Moreover, the idea to consider the MAPK/Erk pathway itself as a network in vertebrate oocytes came from studies suggesting MAPK-independent roles or functions for Mos oncoprotein in amphibian and mice [14, 15].

Dealing with Modules within the MAPK Cascade Network

While Mos - MAPK/Erk cascade plays a role *in vitro* and *in vivo* in the establishment of a bipolar axis during meiotic spindle morphogenesis (Figure 1A, [7, 16]), the cascade may not be considered as one single functional module for spindle morphogenesis (Figure 1B) but has to be split into two different modules. If Rsk has been considered for a long time as the key-mediator or Mos effects, like in the establishment of the cytostatic activity [17, 18], Rsk is not sufficient on its own to promote spindle morphogenesis in absence of Mos : both Mos and Rsk appeared to be necessary and to play complementary roles in the establishment of the meiotic spindle bipolar axis [7]. From our experiments, two functional modules may be defined, as essentials for spindle morphogenesis in amphibian oocytes (Figure 1C). Nevertheless, these two modules are neither relevant nor sufficient to address both the regulation and the dynamics of this cascade within the same cellular model (Figure 1D).

Therefore, depending on context, the MAPK/Erk cascade may either be considered as a single functional module with pleiotropic cellular effects or as a network with pleiotropic cellular effects built from different functional modules.

As a final note we remark that the indicated modules in Figure 1D are not homogeneous and are to be considered more than three push-pull modules connected in series: the MAPK/Erk module exhibits one input (MEK phosphorylation) while it has several potential outputs (Mos and Cdc25 phosphorylations) in addition to its main one (Rsk phosphorylation and activation). In contrast to the MAPK/Erk module, while showing at least two outputs (MEK and Myt1 phosphorylation), the Mos module exhibits several potential inputs (phosphorylation by MAPK/Erk, phosphorylation by MPF), in addition to the one driven by progesterone through the stimulation of Mos synthesis. Additional complexity arises from the orchestration of protein synthesis upregulation, chaperones activity and impairment of degradation mechanisms which together sustain the accumulation and activity of the oncoprotein Mos. In order to capture and experimentally test the meaningful definition of modules we are currently developing a detailed computational model for the regulatory roles played by Mos which in particular allows to uncouple the feedbacks involving the active and inactive form of Mos [19].

Acknowledgements

We thank R. Beaujois and J.-P. Vilain for critical review of the manuscript, discussions and comments. Our work is partially supported by the 'Ligue Régionale contre le Cancer'. C.R. is supported by a fellowship from the french National Institute of Cancer (INCa).

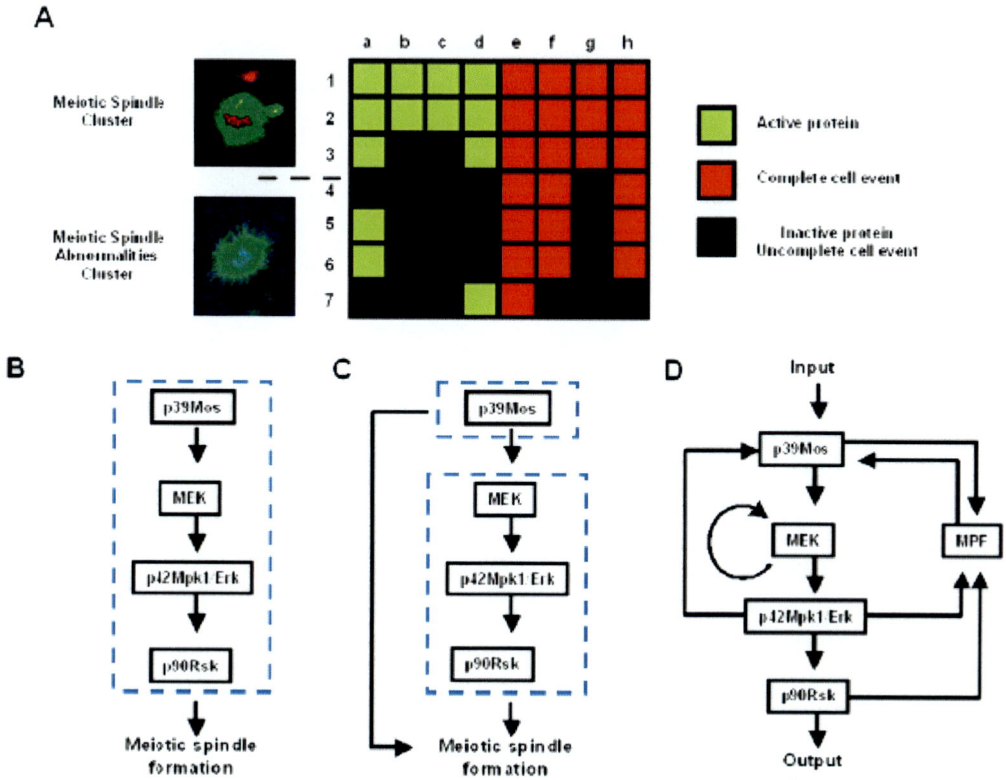

Figure 1. Building modules within the MAPK/Erk cascade network.

(A) Mos and Rsk are both required for meiotic spindle morphogenesis in *Xenopus laevis* oocytes. MEK 1 and 2 chemical inhibitor U0126 and oligonucleotides antisens strategies were employed. Left side : experimental procedures. 1 = Morpholinos sens oligonucleotides injection + DMSO (U0126 solvent) and progesterone coincubation ; 2 = Morpholinos antisens Mos oligonucleotides and murine Mos protein coinjection + progesterone incubation ; 3 = Rsk injection + U0126 and progesterone coincubation ; 4 = Morpholinos antisens Mos oligonucleotides injection + progesterone incubation; 5 = U0126 and progesterone coincubation; 6 = murine Mos protein injection + U0126 and progesterone coincubation ; 7 = Morpholinos antisens Mos oligonucleotides and constitutively active Rsk coinjection + progesterone incubation. On top of the diagram, proteins and cellular events considered: a = Mos oncoprotein, b = MEK, c = MAPK/Erk, d = Rsk, e = Germinal Vesicle Breakdown, f = microtubules nucleation, g = bipolar spindle organization, h = chromosomes condensation. Green: proteins are active, red : cellular events are complete, black : proteins are inactive or cellular events are not completed. Please report to [6, 7] for protocole details and data. Two clusters are defined according to treated oocytes exhibiting correct spindle formation (1-3) or showing spindle abnormalities (4-7). (B) Though the cascade Mos – MAPK/Erk itself may be considered as functional module, it may also be split into two functional modules (C). (D) Regulation motifs that have been reported for each molecule, drawing a complex pattern of interactions: Mos phosphorylates MEK and could promote MPF activation through Myt1 downregulation directly ; a short positive feedback loop has been described Adler and colleagues [11] at MEK level, involving JNK and Raf ; MAPK phosphorylates Mos directly and promotes Mos both stability and synthesis, but also may phosphorylate Cdc25 and then promote MPF activation ; Rsk activates MPF through Myt1 downregulation. Input = hormonal stimulation ; output = meiotic spindle formation, inhibition of DNA replication and establishment of cytostatic activity.

References

[1] Huang, C. Y., and Ferrell, J. E., Jr. (1996) Ultrasensitivity in the mitogen-activated protein kinase cascade. *Proc. Natl. Acad. Sci. U S A* 93, 10078-10083

[2] Santos, S. D., Verveer, P. J., and Bastiaens, P. I. (2007) Growth factor-induced MAPK network topology shapes Erk response determining PC-12 cell fate. *Nat. Cell Bio.l* 9, 324-330

[3] Sasagawa, S., Ozaki, Y., Fujita, K., and Kuroda, S. (2005) Prediction and validation of the distinct dynamics of transient and sustained ERK activation. *Nat. Cell Biol* .7, 365-373

[4] Haccard, O., and Jessus, C. (2006) Oocyte maturation, Mos and cyclins--a matter of synthesis: two functionally redundant ways to induce meiotic maturation. *Cell Cycle* 5, 1152-1159

[5] Bodart, J. F., Flament, S., and Vilain, J. P. (2002) Metaphase arrest in amphibian oocytes: interaction between CSF and MPF sets the equilibrium. *Mol. Reprod. Dev.* 61, 570-574

[6] Baert, F., Bodart, J. F., Bocquet-Muchembled, B., Lescuyer-Rousseau, A., and Vilain, J. P. (2003) Xp42(Mpk1) activation is not required for germinal vesicle breakdown but for Raf complete phosphorylation in insulin-stimulated Xenopus oocytes. *J. Biol. Chem.* 278, 49714-49720

[7] Bodart, J. F., Baert, F. Y., Sellier, C., Duesbery, N. S., Flament, S., and Vilain, J. P. (2005) Differential roles of p39Mos-Xp42Mpk1 cascade proteins on Raf1 phosphorylation and spindle morphogenesis in Xenopus oocytes. *Dev. Biol* 283, 373-383

[8] Ferrell, J. E., Jr., and Machleder, E. M. (1998) The biochemical basis of an all-or-none cell fate switch in Xenopus oocytes. *Science* 280, 895-898

[9] Bagowski, C. P., Xiong, W., and Ferrell, J. E., Jr. (2001) c-Jun N-terminal kinase activation in Xenopus laevis eggs and embryos. A possible non-genomic role for the JNK signaling pathway. *J. Biol. Chem.* 276, 1459-1465

[10] Perdiguero, E., Pillaire, M. J., Bodart, J. F., Hennersdorf, F., Frodin, M., Duesbery, N. S., Alonso, G., and Nebreda, A. R. (2003) Xp38gamma/SAPK3 promotes meiotic G(2)/M transition in Xenopus oocytes and activates Cdc25C. *Embo. J.* 22, 5746-5756

[11] Adler, V., Qu, Y., Smith, S. J., Izotova, L., Pestka, S., Kung, H. F., Lin, M., Friedman, F. K., Chie, L., Chung, D., Boutjdir, M., and Pincus, M. R. (2005) Functional interactions of Raf and MEK with Jun-N-terminal kinase (JNK) result in a positive feedback loop on the oncogenic Ras signaling pathway. *Biochemistry* 44, 10784-10795

[12] Wang, R., He, G., Nelman-Gonzalez, M., Ashorn, C. L., Gallick, G. E., Stukenberg, P. T., Kirschner, M. W., and Kuang, J. (2007) Regulation of Cdc25C by ERK-MAP kinases during the G2/M transition. *Cell* 128, 1119-1132

[13] Palmer, A., Gavin, A. C., and Nebreda, A. R. (1998) A link between MAP kinase and p34(cdc2)/cyclin B during oocyte maturation: p90(rsk) phosphorylates and inactivates the p34(cdc2) inhibitory kinase Myt1. *Embo J.* 17, 5037-5047

[14] Castro, A., Peter, M., Lorca, T., and Mandart, E. (2001) c-Mos and cyclin B/cdc2 connections during Xenopus oocyte maturation. *Biol. Cell* 93, 15-25

[15] Verlhac, M. H., Lefebvre, C., Kubiak, J. Z., Umbhauer, M., Rassinier, P., Colledge, W., and Maro, B. (2000) Mos activates MAP kinase in mouse oocytes through two opposite pathways. *Embo. J.* 19, 6065-6074

[16] Horne, M. M., and Guadagno, T. M. (2003) A requirement for MAP kinase in the assembly and maintenance of the mitotic spindle. *J. Cell Biol.* 161, 1021-1028

[17] Inoue, D., Ohe, M., Kanemori, Y., Nobui, T., and Sagata, N. (2007) A direct link of the Mos-MAPK pathway to Erp1/Emi2 in meiotic arrest of Xenopus laevis eggs. *Nature* 446, 1100-1104

[18] Nishiyama, T., Ohsumi, K., and Kishimoto, T. (2007) Phosphorylation of Erp1 by p90rsk is required for cytostatic factor arrest in Xenopus laevis eggs. *Nature* 446, 1096-1099

[19] Giuraniuc C., Russo, C., Bodart, J.-F., and Blossey R. (2007) On the Equilibria of the MAPK cascade, arXiv:q-bio/0702051, and in preparation.

In: Oncogene Proteins: New Research
Editors: A.H. Malloy and E.C. Carson, pp.19-63
ISBN 978-1-60456-290-3

Chapter 1

E7 Oncoprotein of Human Papillomavirus: Functions and Strategies of Inactivation for the Treatment of HPV-Associated Cancer

Maria Gabriella Donà
James Graham Brown Cancer Center, University of Louisville, KY, USA

Abstract

Human Papillomaviruses (HPVs) are involved in the etiology of at least 10% of all human cancers and are mainly implicated in the development of pre-malignant and malignant lesions of the ano-genital tract. In particular, genital High-Risk HPVs are responsible for nearly 100% of cases of cervical cancer, which represents the second most prevalent cancer in women worldwide.

The carcinogenic risk associated with HPV infection is primarily due to the activity of two viral oncoproteins, E6 and E7. Indeed, their expression is necessary for the maintenance of the malignant phenotype, and is always retained in HPV-positive cancer cells.

Although malignant transformation results from the synergistic and complementary effects of E6 and E7, the latter is reported to be the main oncoprotein of HPVs. One of the major aims of this review is to update the understanding of the role of E7 in HPV-associated carcinogenesis, focusing on the binding with its cellular targets and the effects induced by these associations. The interaction of E7 with its main target, the tumor suppressor protein retinoblastoma (pRb), will be described, with emphasis on the biological consequences and the domains involved. However, as E7 oncogenic properties are not only attributable to this association, the significance of other relevant interactions will be evaluated as well.

Due to the essential role of E7 in cervical carcinogenesis, the lack of homology with cellular proteins and the exclusive expression in cancer cells, this oncoprotein represents the main target for cervical cancer therapy. The therapeutic potential of strategies aimed at E7 inhibition for the treatment of HPV-associated lesions will be addressed. Among others, the use of small interfering RNA (siRNA), intracellular antibodies ("intrabodies"), ribozymes, anti-sense oligonucleotides and aptamers to knock out E7 at the gene/protein level will be covered. The results of these studies will be reviewed with the aim of highlighting E7 inhibition as a promising approach for the treatment of HPV-associated cancer.

Introduction

Human Papillomaviruses (HPVs) are species-specific viruses, which show strict tropism for the stratified squamous epithelium. Over 150 types have been identified so far that can infect either the skin (epithelial types) or the mucosal surface of the respiratory and ano-genital tract (mucosal types) [1]. Based on the pathology of the lesions they induce, mucosal HPVs can be distinguished as Low-Risk (LR) and High-Risk types (HR). Whereas the former group causes only benign lesions, such as genital warts (condilomata), the latter is involved in the development of pre-malignant and malignant lesions. HPV infection by HR-HPVs has been suggested to cause the majority of anal cancers, in addition to a subset of vulvar, vaginal, penile and head and neck cancers [2]. In particular, the association between HPV infection and cervical cancer (CC) has been uncontrovertibly established, biologically and epidemiologically. Almost 100% of CC cases are caused by HPV, with HPV16, 18, 31 and 45 being responsible for 80% of all cases [3]. According to a recent estimation, CC is the second most prevalent cancer in women, with 550,000 new cases/year and 270,000 deaths/year worldwide [4]. CC is the leading cause of cancer-related death for women in developing countries, where 80% of all cervical cancer cases occur.

The oncogenic potential of the HR-HPVs is mainly ascribed to the viral oncoproteins E6 and E7, which are essential for HPV-mediated carcinogenesis. Indeed, their oncogenic activity has been extensively documented in tissue culture and transgenic mouse model systems. E6 and E7 are necessary and sufficient to immortalize primary human epithelial cells [5, 6], although a fully transformed phenotype is acquired only after extended passages in culture or concomitant expression of activated oncogenes [7]. E6 and E7 continuous expression is required for the establishment and maintenance of the transformed phenotype [8, 9]. In fact, their expression is always retained in HPV-positive cancer cells [10, 11], whereas the expression of other early proteins, E2 and E4, and of the late proteins L1 and L2 (required for virus assembly) is usually lost [12]. This expression pattern is frequently related to the integration of the viral DNA into the cellular genome, which is observed in most invasive cancers but also in high-grade lesions of the cervix (HSILs) [13-15]. On the contrary, lesions induced by LR-HPVs do not contain integrated viral DNA, strongly suggesting that integration in the host genome is a key event in the acquisition of the malignant phenotype.

One of the major consequences of the integration is the loss of E2 expression. E2 has been reported to be a regulatory protein that, besides being involved in viral genome replication, can also repress expression of the viral oncogenes by displacing transcriptional activators from HPV long control region (LCR) [16]. Indeed, E2 expression suppresses E6 and E7 transcription and leads to growth arrest [9, 17, 18]. Upon integration, the E2 open reading frame (ORF) is often disrupted/deleted [19, 20]. Disruption of this regulatory gene results in loss of transcriptional repression, thus increasing expression of E6 and E7. The up-regulation of these oncoproteins confers a growth advantage to cells harboring an integrated viral genome versus cells containing episomal DNA. In fact, cervical cancer cells containing an integrated viral genome have little or no episomal DNA [15]. In addition, increased stability of E6-E7 mRNAs upon integration has also been reported [21]. Therefore, although the ultimate development of CC is a multistep process that relies on several events, the integration seems to be essential as it leads to over-expression of the viral oncogenes. Deregulated and prolonged expression of E6 and E7, associated with persistent infection, induces hyperproliferation and confers a genomic instability element that can cause the malignant conversion of the epithelial cells.

E7 in the Viral Life Cycle

Malignant transformation of HPV-positive cells is an unintended consequence of E6 and E7 activity upon a persistent viral infection. The primary function of these early proteins is not to transform the target cells, i.e. the oncogenic progression is not part of the normal viral life cycle, but to ensure the survival of the virus. Indeed, they play an essential role in the productive stage of the HPV life cycle [22, 23]. E7 expression has been demonstrated to be essential in the maintenance of viral episomes and extension of keratinocytes (KCs) life span [23]. In fact, HPV16-E7null genomes are deficient in inducing DNA synthesis and supporting viral DNA amplification [22]. Moreover, in cells containing E7-lacking HPV genomes, the expression of the main capsid protein, L1, is decreased [22].

During active infection, E7 is expressed at an early stage in order to establish a favourable environment for virus replication. At the beginning, the infection takes place at level of the dividing cells of the epithelial basal layer and the viral genome, maintained in episomal form, replicates along with the cellular DNA during the S-phase. As the basal cells proliferate, one of the daughter cells migrates toward the upper layers, undergoes differentiation and exits the cell cycle, losing its ability to divide. Therefore, these cells do not express anymore all the factors necessary for DNA synthesis. Nonetheless, HPV needs the target cells to provide these factors, since they are required for the replication of the viral genome. The infected cells of the suprabasal layers are then forced to proliferate, thus to be in a permanent replication-competent state, through the activity of E6 and E7. These viral oncoproteins profoundly deregulate the mechanisms of cell cycle control and apoptosis through the interactions established with their main targets, the tumor suppressors p53 and pRb respectively. The association with these key modulators of cell growth causes the inactivation of the pathways they control. E7 continuously stimulates the epithelial cells to support DNA synthesis by abrogating pRb function (see below), while p53-mediated pathways of DNA repair are lost due to E6 expression. Moreover, E7 delays the onset of differentiation, thus uncoupling proliferation and program of terminal differentiation, critical changes for virus survival and replication [24]. Ultimately, concomitant activity of E6 and E7 keeps HPV-positive KCs in a proliferating state as long as the virus needs the target cells to support the replication of its genome, until, at the final state of the productive cycle, the late genes are expressed and the virions are assembled.

It is worth noting that small DNA tumor viruses, such as Papillomaviruses, Adenoviruses and Polyomaviruses, have all independently evolved extremely similar mechanisms to subvert the same cellular pathways. In fact, these viruses have "developed" early proteins with remarkable functional similarity. Adenovirus (Ad) E1A and E1B, Polyomavirus Large Tumor antigen (T-Ag), HPV-E6 and E7 ultimately perform analogous tasks, in order to establish the most favourable environment for the viral replication. Some of these proteins share a large similarity in their amino acid sequence (see below) but, most interestingly, they also share the same privileged targets. HPV-E7, Ad E1A and Polyomaviruses (Simian Virus 40, JC virus and BK virus) T-Ag interact with pRb (and the other two members of the retinoblastoma family) via an LXCXE motif (see below), disrupting pRb-mediated control of cell proliferation [25-28]. On the other hand, Ad E1B, SV40 T-Ag and HPV-E6, even if structurally unrelated, show the ability of associating the tumor suppressor p53 [26, 27, 29]. The ultimate result of these interactions is the suppression of p53 function, though this is achieved through different mechanisms. In fact, E1B directly represses p53-dependent transcription by binding p53 transactivation domain [26], while T-Ag blocks p53 activity interacting with its DNA-binding domain [27]. Interestingly, p53 is stabilized (but still inactive) upon T-Ag binding, whereas in HPV-positive cells p53 is degraded as a consequence of E6-induced targeting of this tumor suppressor to the proteasome pathway [30, 31]. Recently, a novel mechanism to inactivate p53 activity has been reported for cutaenous HPVs associated with nonmelanoma skin cancer (NSSC), such as HPV38. In this case, p53 is not degraded but stabilized upon phosporylation, though it is kept functionally inactive through accumulation of its inhibitor ΔNp73 [32].

Even if the mechanisms that small DNA tumor viruses exploit to inactivate these key modulators of proliferation are not exactly the same, and this inactivation can be achieved by two viral products or simultaneously exerted by the same protein (as for T-Ag), the ultimate effect of the activity of the viral oncoproteins is the subversion of pRb and p53 pathways. This strengthens the notion that the concomitant inactivation of these tumor suppressors has a primary importance for the efficient replication of these viruses. In fact, inactivation of pRb is required to promote the proliferation of the infected cells in order to render available the cellular factors necessary for the replication of the viral genome. On the other hand, suppression of p53 pathway guarantees the survival of the infected cells, since it abrogates the cellular responses, such as apoptosis, that normally protect the cells in case of abnormal proliferation.

E7 as an Immortalizing and Transforming Protein

Several *in vitro* studies on E7 proteins of HR-HPVs have extensively demonstrated its transforming activity, in contrast with E7 proteins of LR-HPVs that do not show any oncogenic potential [5, 6, 33]. E7 is necessary and sufficient to transform established murine fibroblast cell lines [34]. The product of HPV16 and 18 shows the ability of transforming murine primary cells in cooperation with the activated *ras* oncogene [35, 36]. Continuous expression of E7 is required for the growth of the transformed cells and their tumorigenicity in nude mice. In addition, E7 is sufficient to extend the life span of human primary keratinocytes, although their immortalization also requires E6 or the entire HPV genome [5]. HPV16-E7 also immortalizes human foreskin keratinocytes (HFKs) when co-expressed either with Myc or with human telomerase (hTERT) [37]. As in the case of E6/E7-induced immortalization, co-expression of E7 and Myc induces telomerase activation [37]. In fact, HPV16-E6 (but not HPV16-E7) activates hTERT promoter through a c-myc-dependent mechanism [38, 39]. However, some studies report that E7 expression is sufficient to immortalize human epithelial cells [40, 41].

E7 oncogenic potential has been confirmed by additional studies on transgenic mice. In fact, mice expressing the HPV16-E7 oncoprotein in the basal layer of the squamous epithelium show hyperplasia, increased DNA synthesis in basal and suprabasal layers, loss of DNA damage response and lack of or delayed differentiation [42, 43]. E7 alone shows the capacity to induce both high-grade dysplasia and invasive cervical cancer in transgenic mice treated with estrogens [44].

E7 can abrogate several signals of growth arrest, inducing escape of quiescent cells from G0 or premature entrance of proliferating cells in S-phase [45-47]. Several studies have shown that E7 can overcome growth arrest signals provided by a number of cytokines with cytostatic effect on epithelial cells. For instance, resistance to TNFα, a cytokine involved in regulation of proliferation and differentiation of keratinocytes, has been reported [48]. Normal KCs respond to TNFα with arrest in G1, as a consequence of p53-independent $p21^{CIP1/WAF1}$ induction. On the contrary, E7-positive KCs continue to proliferate upon TNFα treatment. E7 ability to overcome the cytostatic signal provided by TNFα seems to be correlated with "pocket proteins" inactivation [49] (see below). E7 protein of HR-HPVs can also abolish growth suppression induced by TGFβ [50].

E7 Functional Domains

The E7 oncogene encodes a small acidic phosphoprotein of approximately 100 amino acid residues, in particular 98 for the HPV16 product and 105 for the HPV18 product.As already mentioned, E7 shows similarity with two other oncoproteins of DNA tumor viruses: Ad E1A and SV40 T-Ag [36]. Based on this similarity, E7 molecule can be divided in three functional regions, namely "conserved region 1" (CR1), CR2 and CR3, spanning the amino acids 1-15, 16-37 and 38-98 respectively for the product of HPV16 (Figure 1). The CR2 contains the high affinity pRb binding site (see below) and a consensus phosphorylation site for the casein kinase II (CKII) [51]. Additionally, the presence of another phosphorylation site has been reported at the C-terminus, though this represents the target of a not yet identified kinase [52].

Although the structural and functional homology between E7 and E1A is restricted to the N-terminus moiety, both oncoproteins display the ability to chelate zinc through the C-terminus. In E7 this property is due to the presence of two Cys-X-X-Cys motifs, usually spaced by 29 amino acids. The structure and length of E7 zinc-finger is unusual and unique and is not shared by any cellular protein with zinc finger DNA-binding domains [53, 54]. Changes in the length of the spacer between the motifs [55] as well as mutations in one or both cysteines in one or both Cys-X-X-Cys motifs [23, 53, 56] have been reported to affect

E7 intracellular stability. Furthermore, the zinc-binding domain seems to be responsible for E7 ability to dimerize/oligomerize [53, 57]. Most interestingly, it is involved in a large number of E7 biological activities.

Indeed, an intact zinc-finger is required to immortalize human keratinocytes [58] and mutations in this domain impair E7 ability of transforming rat primary embryo fibroblasts in cooperation with activated *ras* oncogene [53].

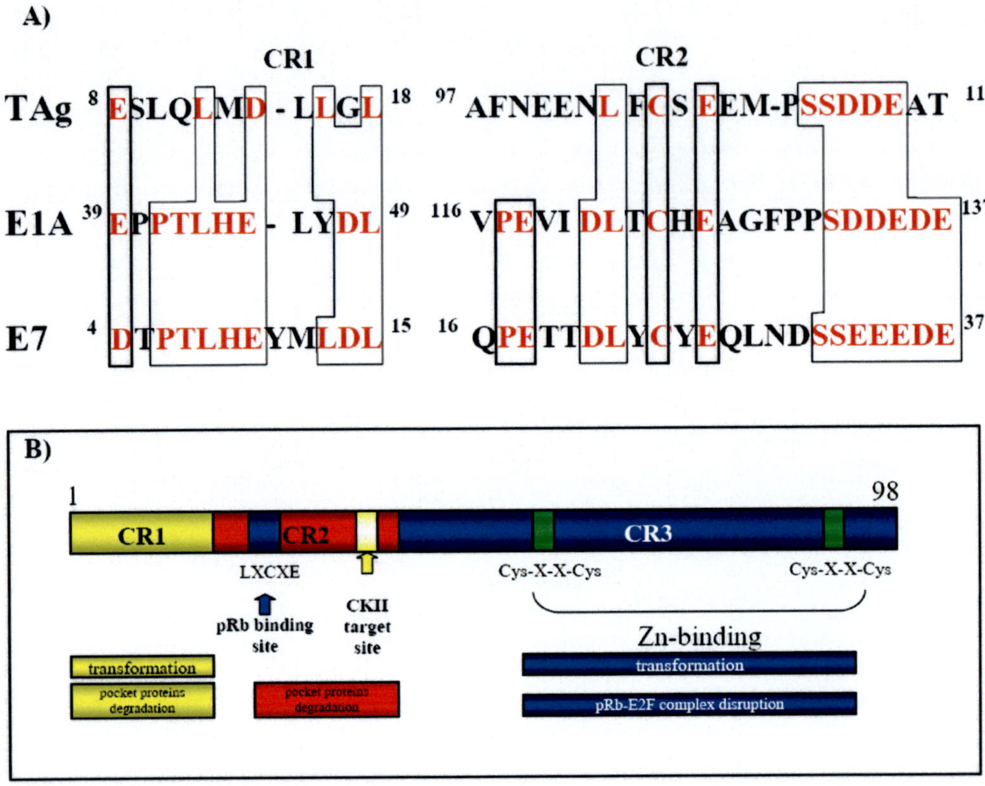

Figure 1. Sequence similarity among SV40 TAg, Ad E1A and HPV16-E7 and schematic representation of HPV-E7 domains. A) The primary sequence of Adenovirus 5 E1A oncoprotein conserved regions (CR) 1 and 2 is compared to SV40-TAg and HPV16-E7 homologous regions. The conserved amino acids are indicated in red. The position of the residues starting form the N-terminus is indicated. B) The three structural/functional domains of HPV-E7 are schematically illustrated. The LXCXE motif, representing pRb high-affinity binding site, is indicated. Moreover, the localization of the target site for the casein kinase II (CKII) is reported. The position of the Cys-X-X-Cys motifs involved in the formation of E7 Zn-finger is also shown.

In recent studies E7 has been described as an "intrinsically disordered" or "natively unfolded" protein. In fact, data collected by circular dichroism have suggested the presence of unfolded regions, together with some ordered secondary structure [59, 60]. More than 50% of the residues of E7 proteins from HR-HPVs have been identified as disordered [61]. In particular, recent results of the solution structure of E7 protein from HR-HPV45 have shown that the N-terminus is completely unfolded, while the C-terminus is well-structured [54]. This ordered region contains the zinc-binding domain [54, 61].

E7 has been reported to have a half-life between 1 and 2 hours in HPV16-positive CaSki and SiHa cells [62, 63]. This relative instability likely explains the relatively low steady-state level of the oncoprotein in cervical cancer cells. Indeed, it has been evidenced that E7 can be rapidly degraded by the proteasome pathway [64, 65]. However, this degradation does not involve classical ubiquitin conjugation to internal lysine residues, but depends on conjugation at the N-terminus [64].

Studies on E7 localization have been greatly impaired by the lack of sensitive antibodies and most of the available data have been obtained in transfected cells over-expressing tagged forms of the oncoprotein. However, some observations have also been accomplished in HPV-positive cervical carcinoma cell lines. The majority of the studies have indicated E7 as a mainly nuclear protein [66-68]. Interestingly, a preferential staining of nucleoli has been evidenced in HPV16-positive cells [66]. Localization in nucleolus has also been reported when E7 has been expressed in yeast cells, suggesting that a conserved mechanism is involved in its nucleolar transportation [66]. Since a canonical nuclear localization signal has not been identified, it is possible that E7 accumulation in the nucleus results from its association with cellular proteins that serve as a shuttle. However, since the fission yeast appears to lack the "pocket proteins", their role in serving as a shuttle for E7 has been excluded [66]. The mechanism of E7 nuclear import has not been clarified yet.

Since several cytoplasmic targets of E7 have also been described, E7 presence in the cytoplasm cannot be ruled out. Indeed, E7 cytoplasmic localization has been reported by some studies [62, 66].

E7 and the Retinoblastoma Family of "Pocket Proteins"

Retinoblastoma (pRb)

E7 is a multifunctional protein that can modulate the activity and/or stability of more than 20 cellular proteins, including cell cycle regulators, transcription factors and metabolic enzymes (Table 1). Through the interactions established with its numerous targets, E7 influences G1-S check-point transition, cellular differentiation, apoptosis, host immune response and cell metabolism.

Numerous observations have suggested that HPV-induced malignant transformation is intimately linked to E7 interference with cell cycle control pathways. Indeed, several cellular proteins involved in the control of cell proliferation represent the main targets of E7. The most relevant and well-characterized interactions are those established with the members of the retinoblastoma family of "pocket proteins": pRb [28], p107 and p130 [69]. The "pocket proteins" modulate the progression from G0 to G1 and the transition between G1 and S-phase, besides being implicated in regulation of differentiation, apoptosis and DNA damage response [70].

The "pocket proteins" function in cell cycle control is mainly accomplished through the interaction with the E2F family of transcription factors, which includes at least eight closely related proteins (reviewed in [71]). pRb is mainly found in complexes with E2F1 through

E2F3, which are typically associated with active promoters in S-phase. These transcription factors control the expression of genes necessary for nucleotide synthesis (dihydrofolate reductase and thymidine kinase), DNA replication (DNA polimerase α) and cell cycle progression (cyclin A and E, c-myc; [72]).

pRb represses these proliferation-associated genes by directly blocking E2F-mediated transactivation. Additionally, pRb actively represses E2F-responsive promoters recruiting chromatin remodelling enzymes, such as the histone deacetylases (HDACs) [73].

Table 1.

TARGET	FUNCTIONAL EFFECT	REFERENCES
pRb	Disruption of E2F complexes Degradation	[28, 75, 76]
p107	Disruption of E2F complexes Degradation	[69, 77]
p130	Disruption of E2F complexes Degradation	[69, 78]
p21^{WAF-1}	Inactivation of cdk inhibitory activity	[79, 80]
p27^{KIP1}	Inactivation of cdk inhibitory activity	[81]
AP-1	Activation of transcription	[82]
TBP	?	[83]
Mi2β (HDACs)	Modulation of transcription	[84]
M2-PK	?	[85]
α-glucosidase	?	[86]
S4 (26S proteasome)	?	[87]
IRF-1	Inhibition of IFN pathways	[88, 89]
TAF110	?	[90]
p48 (ISGF3)	Inhibition of IFNα stimulation	[91]
F-actin	?	[92]
MPP2	Activation of MPP2 transcriptional activity	[93]
hTid	?	[94]

pRb growth suppression function is regulated by a series of complex interactions, which also involve the tumor suppressor gene p53.

In fact, besides regulating pRb expression, p53 also modulates its activity by inducing the transcription of the inhibitor of cyclin-dependent kinases p21$^{CIP1/WAF1}$ that maintains pRb in the active form (see below) [74].

In turn, pRb modulates p53 activation since E2F also controls the expression of proteins that indirectly induce p53 pathway, such as the alternative reading frame (ARF) and ataxia-telangiectasia mutated (ATM) proteins. These proteins inhibit the activity of MDM2, which normally promotes p53 degradation. Importantly, p53 also represents a target of HPV. As already mentioned, the viral oncoprotein E6 suppresses p53 pathway by inducing degradation of this tumor suppressor via association with the cellular ubiquitin-protein ligase E6-associated protein (E6AP) [30, 31, 95].

The biological activity of retinoblastoma is regulated in response to growth factors through signalling pathways that lead to pRb phosphorylation, which represents the primary mechanism of control of pRb function in cell cycle regulation. In fact, pRb phosphorylation state determines pRb binding to its target regulatory proteins. Retinoblastoma is phosphorylated at several serine residues by G1 cyclin-dependent kinases (cdks) associated with their respective cyclins (cyc). In particular, cycD/cdk4-6 complexes first and cycE/cdk2 complexes later are mostly responsible for pRb phosphorylation, which begins at the G1/S boundary [96]. Dephosporylation by a specific phosphatase occurs at late M phase. In the hypophosphorylated form pRb binds E2F forming a transcriptional repressor complex; therefore the expression of E2F-responsive genes is inhibited. In late G1 phase, retinoblastoma phosphorylation by cdks determines the release of the E2F factor, now converted into a transcriptional activator.

A considerable amount of evidence strengthens the notion that association of E7 with pRb represents an essential event in E7-induced transformation, as the binding affinity of E7 for this target correlates with the transforming potential of the viral product [97].

E7 preferentially binds the hypophosphorylated form of pRb [69, 98], preventing its association with E2F and inducing E2F release from already established complexes [56]. In both cases, an increase in the level of free E2F is promoted, resulting in progression in S-phase (Figure 2).

Considering the new insights in E2F activity, it appears that induction of cell proliferation is not (or not only) a result of E2F-mediated transcription activation but also depends on E2F-mediated de-repression of its target genes [71].Aberrant expression of proliferation-associated genes, such as cycA, which is required during S-phase, and cycE, necessary for G1-S transition, has been detected in E7-positive cells, with concomitant activation of cdk2 [46, 99, 100]. Significantly, constitutive expression of these genes and simultaneous increase in associated kinase activities can be induced by E7 in the absence of external growth factors [46].

It is worth noting that the ability of E7 to transactivate E2F-responsive genes relies on complex and concomitant subversions of several regulatory pathways, through mechanisms that are not only linked to direct impairment of pRb activity. In fact, it has been shown that E7 can activate E2F1-driven transcription even in absence of pRb [101]. It has been demonstrated that E7 can interact with E2F1 both *in vitro* and *in vivo* and that, most interestingly, can form complexes with E2F1-DNA. These data suggest that, besides the interaction with pRb, the direct association with E2F might have a functional role in E7-induced transformation by promoting the expression of proliferation-associated genes.

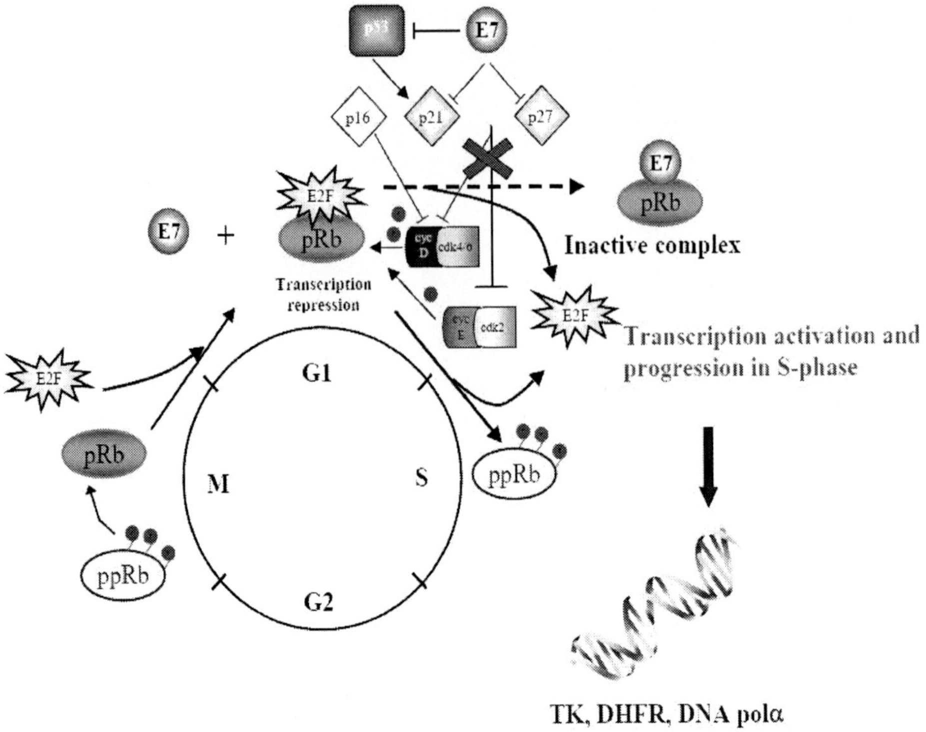

Figure 2. Schematic representation of E7-pRb interaction. In normal cells, hyposphorylated pRb binds the E2F transcription factor forming a repressor complex and repressing the expression of genes necessary for cell cycle progression. pRb phosphorylation by cyc/cdks complexes at the G1/S boundary induces E2F release and transcription of E2F-responsive genes, such as timidine kinase (TK), dihydrofalate reductase (DHFR) and DNA polymerase α (DNApolα). In HPV-positive cells, E7 binds the hyphosphorylated form of pRb and promotes E2F release, thus S-phase entry. In addition, E7 blocks the activity of the cdk inhibitors p21 and p27, which normally suppress cyc/cdks complexes.

Importantly, E7-mediated induction of E2F-responsive genes encoding cycA and cycE allows functional bypass of p16ink4a interference with cellular proliferation. This protein is a cdk-inhibitor (CKI) that suppresses the activity of cycD/cdk4-6 complexes, responsible for pRb phosphorylation in mid-late G1. Therefore, p16ink4a over-expression normally results in cell cycle arrest as it leads to accumulation of hypophosphorylated pRb and sequestration of E2F [102]. However, in the presence of E7 p16ink4a-mediated G1 arrest is bypassed, since pRb is inactivated/degraded as a result of the oncoprotein activity [103, 104]. Indeed, E7-mediated pRb degradation is required for the efficient suppression of G1 block imposed by p16ink4a [104].

Since E7-mediated impairment of pRb function causes substantial up-regulation of p16ink4a, E7-positive cells show aberrant p16ink4a accumulation, as frequently observed in HPV-associated neoplastic lesions and CC [105]. Interestingly, cervical biopsies from high-grade lesions of the cervix show strong staining for the serine/threonine kinase AKT as well [106]. Enhanced activity of AKT in E7-positive keratinocytes has been reported [106, 107] and also in this case a correlation with pRb impairment has been evidenced [106]. The biological relevance of promoting AKT function in cervical carcinogenesis has not been completely clarified. However, some observations have suggested its involvement in E7-induced HFKs migration upon $p27^{KIP1}$ mislocalization (see below) [107]. Moreover, the phosphoinositide 3 kinase (PI3K)-protein kinase B (PKB)/AKT pathway mediates resistance to anoikis, an apoptotic response that is activated in epithelial cells upon disruption of their interaction with the matrix [108]. Importantly, it has been shown that HPV16 E6/E7 can confer HaCaT cells resistance to anoikis and that this effect is enhanced when the viral oncogenes are co-expressed with a constitutively activated form of PKB/AKT [109].

E7 association with pRb is mediated by an LXCXE motif, which represents the canonical binding site for "pocket proteins". This motif is shared by many cellular proteins that interact with retinoblastoma, such as D-type cyclins, HDAC-1 and HDAC-2, besides the other viral oncoproteins able to functionally inactivate pRb, such as Ad5 E1A and SV40 T-Ag [25, 69, 110, 111]. The LXCXE motif maps in the CR2 region, therefore it is included in the unstructured and highly flexible N-terminal domain of E7 [54, 61]. Hence, this position might represent a way to keep the pRb-binding domain always accessible to pRb. The LXCXE binding site on pRb is localized in the pRb "pocket" domain, which is necessary, although not sufficient, for pRb-mediated growth inhibition [112, 113]. This "pocket" region has been co-crystallised with a nine-residue HPV16-E7 LXCXE peptide, which binds in an extended conformation to a hydrophobic groove included in the pRb-B box [114]. In particular, HPV16-E7 Leu 22, Cys 24, Glu 26 and Leu 28 make direct contact with highly conserved amino acid residues of pRb, evidencing that residues surrounding the LXCXE motif are also important for establishing a tight interaction.

The LXCXE motif is sufficient for pRb binding [115]. Nonetheless, several data suggest that other regions of the oncoprotein are involved in pRb association. A peptide containing the LXCXE sequence is 100-fold less efficient than the full-length E7 protein in inhibiting pRb binding in a competitive binding inhibition assay [115]. Moreover, it has been shown that E2F, which does not contain an LXCXE motif, and E7-LXCXE binding sites on pRb pocket domain do not overlap but map at two different sites, which are 30 Å apart [116]. Therefore, the LXCXE sequence alone cannot explain either the association or the functional inactivation of pRb. The C-terminus of E7 has been reported to contain a low-affinity binding site for pRb C-terminus [115, 117]. In addition, it seems that this domain harbors binding sites for the "marked box" region of E2F as well [117]. Indeed, the CR3 region of E7 is essential in inducing E2F displacement, whereas the LXCXE motif alone is not capable of promoting the release of transactivation-competent E2F [115, 117], suggesting the model depicted in Figure 3 for E7-mediated inactivation of pRb.

Besides the direct interaction and inactivation of pRb, there is evidence that E7 induces degradation of this tumor suppressor both *in vitro* [75] and *in vivo* [43, 77]. Most interestingly, retinoblastoma degradation seems to be necessary for efficient inactivation of pRb function by E7. Indeed, an E7 mutant able to bind but not to degrade pRb is partially defective in blocking pRb activity compared to the wild-type protein [77]. Notably, E7 capacity to induce pRb degradation strongly correlates with its transforming activity [118]. In fact, HPV1a-E7 that can bind pRb with almost the same efficiency of HR-HPV E7 but does not induce pRb degradation [77] cannot transform primary cells [119, 120]. Interestingly, E7 ability of promoting pRb degradation has been shown to correlate also with the E7 capacity of circumventing p16ink4a-induced cell cycle arrest [104]. Therefore, further dysregulation of the pRb pathway by E7-induced degradation strongly contributes to E7 transforming activity, as it may enhance the expression of E2F-responsive genes required for G1/S transition and DNA synthesis. Besides the impairment of pRb activity by direct binding, the ability to promote its degradation might represent a more efficient way of neutralizing the tumor suppressor function if E7 is not present in the cell at a sufficient stoichiometric level to sequester enough pRb.

Though the exact mechanism of E7-mediated pRb degradation has not been clarified yet, the ubiquitin-proteasome pathway seems to be involved [75, 77]. Although HPV16-E7 has been shown to recruit the proteasome 26S and E7 interaction with S4 subunit of the proteasome has been demonstrated [87], S4 does not appear to be responsible for pRb targeting to the proteasome pathway. In fact, an S4 binding deficient mutant of HPV16-E7 can still induce pRb degradation [77]. A recent study has reported that the cullin 2 ubiquitin ligase might be involved in HPV16E7-mediated pRb degradation [76]. Indeed, pRb/E7 complexes have been shown to contain cullin 2. In addition, silencing of this ubiquitin ligase determines an increase in pRb steady-state level.

E7 ability of inducing pRb degradation is impaired by mutations in CR1 and CKII phosphorylation site, suggesting that pRb binding, which occurs through CR2 and CR3 domains, is not sufficient for its degradation [77]. However, the amino acids located immediately after the LXCXE motif also seem to be essential for pRb destabilization [104].

p107 and p130

Several reports have shown that E7 targets two other factors that are functionally and structurally related to the retinoblastoma protein: p107 (retinoblastoma-like 1, RBL1) and p130 (RBL2) [69, 121]. The data available suggest importance of these interactions, even though they have not yet been well characterized. In comparison with pRb, p107 and p130 have not been conclusively linked to tumor suppressor activity and their contribution to E7 oncogenic potential is still debated.

These "pocket proteins" are differentially expressed during the cell cycle: p107 is mainly synthesized during the S-phase, while p130 level is abundant in G0, therefore in differentiated and quiescent cells [122]. These proteins do not associate the E2F1-3 members of the E2F family, which are exclusive targets of pRb. Instead, they form complexes with E2F4-5. These factors, opposed to E2F1-3, act as repressors and have a reduced ability to induce S-phase entry [72].

E7 association with p107 has been shown to occur independently of pRb binding and might contribute to E7-induced immortalization. In fact, HPV16-E7 mutants defective in pRb but not p107 binding are impaired in transformation but retain the ability of immortalizing human cells [58, 121]. However, in contrast with what has been evidenced for pRb, E7 is not able to induce E2F dissociation from p107 in S-phase [123, 124], suggesting that in this case an alternative mechanism is exploited in E2F activity regulation. On the other hand, E7 has been shown to disrupt E2F-p107 complexes in G1-phase [125], although the biological significance of this different behaviour is still unclear.

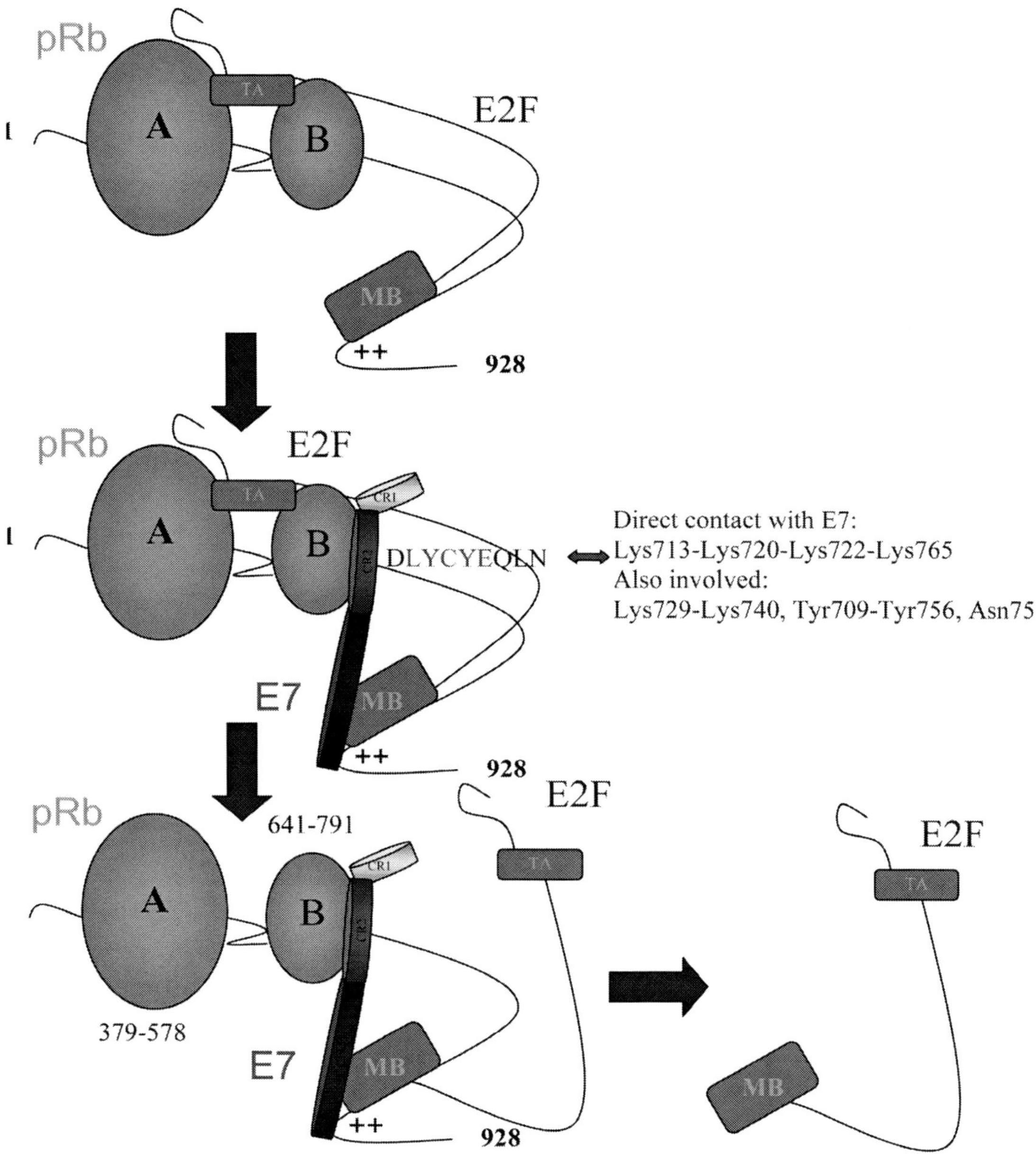

Figure 3. Schematic representation of E7-induced displacement of E2F from pRb. E2F transactivation domain (TA) binds at the interface between pRb A and B boxes (represented by the green ovals), while E2F "marked box" (MB) contacts a positively-charged region in pRb C-terminal (shown as ++). HPV-E7 CR2 domain interacts with the B box via an LXCXE domain (indicated in red). pRb amino acid residues involved in the interaction are illustrated. E7 C-terminus makes contacts both with pRb C-terminus and E2F-MB, displacing E2F probably through an allosteric mechanism. Modified from Liu *et al*, 2006 [117].

E7 displays the ability of interacting with p130 as well [69]. This "pocket protein" appears to be implicated in regulation of cellular differentiation and especially in the maintenance of terminal differentiated state [126]. Indeed, differentiated cells characteristically show accumulation of p130/E2F4 complexes. Therefore, it is likely that E7-mediated inhibition of differentiation is promoted by association with p130 and alteration of its function. In fact, the ability of E7 to bind and degrade p130, besides the other "pocket proteins", correlates with its capacity of disrupting keratinocyte differentiation [24]. Hence, this activity may be more relevant in the virus life cycle than in HPV-mediated carcinogenesis.

Although some experiments failed to prove that p107 and p130 levels are affected by E7 [127], other studies have evidenced their degradation consequent to E7 expression, as observed for pRb [77, 78]. Destabilization of these proteins might further contribute to E2F-dependent transcription. Complexes of p130, HDACs and E2F4-5 are prominent on E2F-promoters in quiescent and differentiating cells. Thus, E7-induced degradation of p130 might cause the replacement of these repressor complexes with the activators E2F1-3 and induce transcription activation. However, the correlation between the ability of promoting p107 and p130 degradation and E7-induced transformation is uncertain. Since E7 of LR-HPV6 also shows the capacity to destabilize p130, this activity might be a common feature of the Papillomaviruses, necessary to uncouple proliferation and differentiation during the viral life cycle [24, 78].

E7-mediated inactivation of all the "pocket proteins" suggests that this multiple knock out might be required in the context of the HPV life cycle, for several reasons. Impairment of all pRb family members may be necessary to achieve a complete deregulation of the cell cycle, since it may allow bypassing the compensation of pRb function loss by p107 and p130. In fact, the effects of inappropriate release of activators E2Fs upon E7-induced pRb loss could be compensated by their association with the other two "pocket proteins", i.e. these proteins could substitute for pRb in the regulation of E2F1-3. Substantially, in absence of functional pRb, p107 and p130 might act as tumor suppressors (though in pRb-deficient tissue/cells this circumstance depends on the availability of the other two "pocket proteins"). Additionally, the functional inactivation of all the three members of the retinoblastoma family might indicate that HPV needs to deregulate different processes, since the activities of the "pocket proteins" are only partially redundant. For instance, pRb is more involved in proliferation and early differentiation, while p130, as already mentioned, regulates later stages of differentiation. Therefore, pRb inactivation might be relevant early in the life cycle, whereas, once the cell cycle is deregulated and the early differentiation is impaired, disruption of p130 activity may become prominent to disturb terminal differentiation [24].

Other E7 Cellular Targets Potentially Contributing to E7 Oncogenic Activity

Although pRb is described as the main target of E7, association with this tumor suppressor protein alone is not sufficient to fully explain E7 oncogenic activity. In fact, mutations in the high-affinity binding site do not abrogate E7 ability to immortalize primary keratinocytes [121], whereas mutations in HPV16-E7 regions other than the pRb-binding site, such as the CR1, the CKII target site and the C-terminus, can impair E7 immortalizing and transforming properties [128]. Moreover, some non-transforming mutants of E7 show the ability of binding pRb and transactivating E2F-responsive promoters [58, 129, 130]. Studies conducted in transgenic mice with somatic *Rb* deletion have shown that E7 can still induce hyperplasia and displasia in pRb null tissues, suggesting that pRb-independent activities are also responsible for E7 oncogenic potential [43].

Association of E7 with numerous other cellular proteins has been reported and novel targets are continuously described, although the biological relevance of these interactions is often unclear.

Cyclins and Cyclin-Dependent Kinases (Cyc/Cdks)

Interaction of E7 with cyc/cdks complexes has been observed; in particular, E7 association with cycA/cdk2 [131, 132] as well as cycE/cdk2 [132] has been reported. It is conceivable that the "pocket proteins" operate as mediators in such interactions. Indeed, E7 binding to cycE does not appear to be direct but mediated by p107 [132]. However, other results suggest a direct interaction, as for the cycA/cdk2 complex [133]. HPV16-E7 shows the specific ability to promote the function of cdk2 complexed either with cycA or cycE. Although enhanced cdk2 activity might result from inhibition of $p21^{CIP1/WAF1}$ and $p27^{KIP1}$ (see below), at least a study has reported that the activation of cdk2 directly depends on its interaction with E7 and does not involve deregulation of the cdk inhibitors [133]. The ability of enhancing cdk2 activity seems to be a conserved feature among HPVs, as it has been demonstrated for LR-HPV E7 proteins as well [133]; therefore, whether the binding of E7 to this kinase has a functional role in E7-induced transformation remains uncertain. Induction of cdk2 function could be important to promote S-phase entry during the viral life cycle, since the histone H1 kinase activity is specifically enhanced and histone H1 phosphorylation models chromatin structure in preparation of DNA synthesis [133]. Additionally, since cdk2 is essentially involved in pRb phosphorylation, induction of its activity may have the significance of further enhancing pRb inactivation.

Finally, E7 might deregulate the activity of the cyc/cdks complexes modifying the time of activation during the cell cycle or the spectrum of targets phosphorylated by the kinases.

Cyclin-Dependent Kinase-Inhibitors (CKIs)

Modulation of cyc/cdks activity and abrogation of cell cycle inhibition occur also through E7 interaction with several other proteins involved in control of cell cycle progression. For instance, the function of the CIP/KIP family of cdk inhibitors (CKIs) $p21^{CIP1/WAF1}$ and $p27^{KIP1}$ has been reported to be blocked by E7 [79-81]. The interaction with both these CKIs has been ascribed to E7 C-terminus.

Inhibition of $p21^{CIP1/WAF1}$ ability of blocking cyc/cdk function, as well as of suppressing proliferating cell nuclear antigen (PCNA)-dependent DNA replication, has been demonstrated as a result of a direct interaction between the CKI C-terminus and HPV16-E7 [79]. Perturbation of $p21^{CIP1/WAF1}$ activity by E7 may have many relevant consequences due to the involvement of this CKI in numerous processes, including response to DNA-damage, senescence and differentiation. For instance, $p21^{CIP1/WAF1}$ is implicated in cell cycle withdrawal during keratinocyte differentiation; therefore, the abolishment of its function may have a role in inducing DNA synthesis in differentiated epithelial cells [79, 134]. Since HPV needs to uncouple cellular differentiation and proliferation, as already mentioned, deregulation of this CKI can be particularly relevant for the virus life cycle. Normally, as the level of $p21^{CIP1/WAF1}$ increases during the differentiation process, cdk2 is inactivated and growth arrest is induced. However, in E7-positive cells this CKI remains inactive, thus, cdk2 remains active and $p21^{CIP1/WAF1}$-mediated growth arrest is bypassed [79, 134]. Interestingly, these effects have been observed in spite of the elevated levels of $p21^{CIP1/WAF1}$, which are due to its stabilization [135, 136]. $p21^{CIP1/WAF1}$ inactivation seems to be required together with pRb deregulation to subvert cell cycle control in human epithelial cells [137]. Significantly, suppression of this CKI represents another mechanism to abolish pRb function, since $p21^{CIP1/WAF1}$ inhibits the activity of cdk2, which is responsible for pRb phosphorylation and inactivation (see above). Substantially, the repression of CKIs further contributes to guarantee that pRb is maintained in the inactive form and that pRb-regulated cellular processes are completely impaired. Moreover, direct inactivation of $p21^{CIP1/WAF1}$ also allows bypassing p53-mediated antiproliferative pathways that normally lead to $p21^{CIP1/WAF1}$ transcriptional activation, since this CKI represents a transcriptional target of p53. Obviously, since HPV-positive cells express also E6, its ability to promote p53 degradation fundamentally contributes to block p53-mediated responses as well.

Finally, since $p21^{CIP1/WAF1}$ is critically involved in G1-arrest upon DNA damage, $p21^{CIP1/WAF1}$ blockage might promote genetic instability during HPV-associated cancer progression.

E7 interaction with $p27^{KIP1}$ has also been reported, both *in vitro* and *in vivo* [81]. As for $p21^{CIP1/WAF1}$, E7 has been shown to inhibit $p27^{KIP1}$ activity. This ability appears to be responsible for E7 capacity of overcoming growth arrest in the presence of anti-proliferative signals. For instance, $p27^{KIP1}$ inhibition seems to be implicated in E7 ability of abrogating keratinocytes growth arrest in response to TGFβ [50]. In addition, a recent study has demonstrated that HPV16-E7 can cause $p27^{KIP1}$ mislocalization in HFKs, with cytoplasmic retention of this CKI [107]. Importantly, prominent cytoplasmic staining of $p27^{KIP1}$ has been shown in cervical cancer biopsies as well [107]. It has been postulated that E7-induced $p27^{KIP1}$ mislocalization might be an important determinant in HPV16-positive tumors invasiveness [107]. In fact, cytoplasmic $p27^{KIP1}$ appears to be a positive modulator of migration and tumor invasion [138-140]. In agreement with these data, E7-positive cells with cytoplasmic $p27^{KIP1}$ show increased migration capability. E7-induced cytoplasmic retention of $p27^{KIP1}$ is achieved through subversion of PI3K/AKT pathway, which seems to be involved in $p27^{KIP1}$ localization in primary KCs, since the use of specific inhibitors for these kinases determines increase of the CKI nuclear level [107].

E7 ability to subvert $p27^{KIP1}$ function and localization may therefore account for important events in the carcinogenic process, not only at level of tumor development but also tumor invasiveness.

Cdc25A

A further disruption of cell cycle control by E7 is achieved through modulation of Cdc25 proteins, which are tyrosine phosphatases involved in cyc/cdks activation upon removal of inhibitory phosphorylation [141]. In particular, Cdc25A controls the activity of cycA and cycE-dependent kinases, dephosphorylating and activating cycE/cdk2 and cycA/cdk2 complexes. Therefore, it largely contributes to progression in S-phase. HPV16-E7 has been reported to transactivate cdc25A promoter, whereas this activity has not been detected for the product of the LR-HPV11 [142]. Several data have suggested that E7 capacity of bypassing growth arrest signals is linked to the induction of cdc25A expression. For instance, a strict correlation has been evidenced between E7-induced Cdc25A overexpression and E7 ability to promote proliferation in conditions of serum starvation, normally associated with repression of cdc25A transcription [47]. Moreover, as Cdc25A level remains high during differentiation of E7-positive human KCs, E7-driven modulation of Cdc25A may be involved in E7-induced delay of primary keratinocyte differentiation [47]. It has been shown that increased expression of Cdc25A requires intact pRb and HDAC1-binding domains, suggesting that cdc25 modulation occurs in a HDAC-dependent manner through disruption/derepression of pRb/E2F/HDAC-1 complexes (see below).

Importantly, Cdc25 phosphatases have also a significant role in regulating G2/M transition upon DNA damage. Cdc25A is normally degraded or inactivated in response to DNA damaging agents because of the activation of the ATM and ATM-related (ATR) pathways [141]. Therefore, deregulation of this phosphatase may contribute to perturb cell response to DNA injury, rendering the cell unable to stop cell cycle progression either to repair DNA or to initiate apoptosis. Ultimately, defects in Cdc25A activity may result in genomic instability. All these data strongly support the idea that Cdc25A is an important mediator in E7-induced proliferation and tumorigenesis.

Histone Deacetylases (HDACs)

The association of HR-HPVs E7 proteins with the histone deacetylases 1 and 2 (HDAC-1 and HDAC-2) has been evidenced both *in vitro* and *in vivo* [23, 84]. HDACs are three classes of proteins that repress gene expression by removing acetyl groups from the N-terminal tails of the histones. Histone deacetylation leads to nucleosome condensation, which in turn blocks the access of transcription factors to the DNA. Several studies have demonstrated pRb's ability to recruit HDAC to E2F-regulated genes and an HDAC requirement for pRb to repress several cellular genes [143, 144]. However, if HDAC recruitment is strictly necessary to mediate pRb inhibition of E2F-responsive transcription is still controversial. The histone deacetylases might be involved in counteracting the activity of the histone acetyltransferases (HATs) p300/CBP and p/CAF that can interact with E2F1-3 [145]. It has also been reported that HDACs can deacetylate and thus directly inactivate E2F transcription factors [145].

The binding of HPV16-E7 to HDAC is mediated by the Mi2β component of the nucleosomal remodelling and deacetylation complex (NURD), which can directly associate both E7 and HDAC. As shown by mutational studies, the zinc-finger domain of E7 is involved in this association, whereas mutations in pRb binding site do not abolish E7-HDAC interaction [23, 84]. It has also been evidenced that, while mutations in the LXCXE-binding site of pRb abolish the interaction with E7, an intact LXCXE cleft is not required for pRb interaction with HDACs, even though they also contain an LXCXE motif [146]. However, it has been shown that the binding of pRb to the HDACs is suppressed in presence of E7 [143]. Modulation of histone deacetylase activity by E7 could further contribute to de-repress E2F-responsive genes. Moreover, as already described in this chapter, E7 binding to HDAC seems to be involved in cdc25A up-regulation [47]. Most interestingly, E7 inability to associate with HDAC correlates with lack of growth-promoting activity. In fact, E7 mutants defective in HDAC binding are also deficient in transforming activity, suggesting a tight link between the oncogenic potential of this viral product and the binding to HDAC.

The modulation of HDACs exerted by E7 might have a further functional relevance in HPV pathogenesis as it appears to be responsible for the transcriptional regulation of immune-modulating molecules. In fact, recruitment of HDAC is involved in E7-induced silencing of interferon regulatory factor-1 (IRF-1), which has a central role in interferon signalling and immune surveillance [88]. In addition, a recent study has shown that HPV16-E7 recruits and/or stabilizes the HDACs on the major histocompatibility complex (MHC) class I promoter [147]. Deacetylation of the chromatin in correspondence of this promoter has been demonstrated to be specifically linked to E7 presence, as abrogation of the oncoprotein expression by siRNA also eliminates the HDACs from the class I promoter and results in histone acetylation. E7-induced chromatin remodelling through modulation of the HDACs could play a key role in down-regulating MHC class I molecules expression on the surface of HPV-positive cells. Therefore, E7-HDAC association might contribute to the mechanisms of escape from immune surveillance evolved by the HPV, reducing the recognition and elimination of HPV-infected cells by the immune system and favouring the establishment of a persistent infection.

Finally, the binding to HDACs has also been demonstrated to play a key role in mediating E7 function in the viral life cycle, since it seems to be necessary to maintain HPV episomal genomes and to extend KCs life span [23].

600kda Retinoblastoma Protein-Associated Factor (P600)

Identification of novel E7 targets has recently led to isolation of p600 as a cellular target of HPV16-E7 [148]. E7 N-terminus is required for the interaction with p600, though the binding appears to be pRb-independent. Data obtained in this study strongly suggest a role of p600 in E7 transforming activities. In fact, p600 depletion from HPV16-positive CaSki cells, as well from E6/E7-transformed NIH3T3 cells, partially suppresses the transformed phenotype. However, as LR-HPV E7 proteins also show the ability of associating p600, the relevance of this factor in the HPV-mediated transformation process must be assessed.

Transcription Factors (AP-1, TBP) and Transcriptional Coactivators (P300, P/CAF)

E7 has been reported to interfere with the activity of several transcription factors, such as AP-1 [82], TATA-box binding protein (TBP) [83], IRF-1 [88] and the forkhead transcription factor MPP2 [93].

The interaction of E7 with several components of the AP-1 transcription complex, including c-Jun, JunB, JunD and c-Fos, has been proven both *in vitro* and *in vivo* [82]. E7 ability of up-regulating AP-1 activity, which requires an intact zinc-finger domain, has been demonstrated and suggests relevance for E7 transforming properties. In fact, E7 inability of activating c-Jun-mediated transcription correlates to its inability to transform rat embryo fibroblasts [82]. In addition, a correlation has been noted between the ability of E7 mutants of transactivating c-Jun-dependent transcription and immortalizing primary keratinocytes, strongly suggesting a direct link between E7-driven transformation and modulation of AP-1 components. Therefore, the biological significance of this interaction is clear, although its exact contribution to E7 oncogenic potential is still at hypothesis level. It has been argued that the formation of E7-AP1 complexes may be important in the viral life cycle since AP-1 is also involved in regulating the transcription of E6/E7 from the viral genome and in keratinocyte differentiation [82].

E7 association with TBP has also been reported [83]. Interestingly, this interaction is influenced by the state of the E7-CKII target site, as phosphorylation at Ser 31 and Ser 32 of HPV16-E7 greatly enhances the binding to TBP. Data have been obtained regarding E7 ability of inhibiting TBP's DNA binding [149]. Association between the two proteins seems to be required for this inhibitory effect, which is increased upon E7 phosphorylation as well. The significance of TBP inhibition is still not clear, as a broad regulation of all TATA box-containing promoters is unlikely.

Further data have confirmed E7 capacity of transcriptional control, since its interaction with the transcriptional coactivators p300 and p300/CBP-associated factor (P/CAF) has also been evidenced [150, 151]. p300 displays a number of functions: it is an essential component of the transcription apparatus, possesses intrinsic HAT activity and can recruit other HATs, such as P/CAF. Numerous studies have demonstrated that this protein is essentially a tumor suppressor, involved in cell cycle control, differentiation and development. HPV16-E7 has been shown to interact with p300 and modulate its activity [150], which is, interestingly, a common feature of other viral oncoproteins, such as Ad E1A [152], SV40 T-Ag [27] and HPV-E6 [152]. Most importantly, interaction with p300 seems to be required for the transformation induced by these proteins [27, 152, 154]. However, if E7-p300 association plays a role in E7-mediated transformation is still debated, since also the product of LR-HPVs shows the ability of associating this HAT [150]. Both HPV16- and HPV11-E7 abolish p300 transcriptional function, even if to a different extent. In particular, E7 represses p300 ability to coactivate E2-dependent transcription [150]. Due to the role of E2 in regulating the expression of HPV oncoproteins, these data suggest a possible contribution of E7 to the regulation of its own transcription.

The association of E7 with P/CAF has also been described [151]. Importantly, E7 binding to this coactivator has been shown to be responsible for E7-induced repression of the interleukin-8 (IL-8) promoter [151], further confirming E7 ability of interfering with the expression of immuno-modulating molecules at transcriptional level. Subversion of the immune response may be important in evasion of HPV-infected cells from the immune surveillance, favoring a persistent infection and subsequent tumorigenesis (see below).

M2 Pyruvate Kinase (MP-K2)

Although most E7 targets are nuclear proteins, the cytolpasmic enzyme M2 pyruvate kinase (M2-PK) has also been recognized as a target of E7 [85]. This enzyme catalyses one of the limiting passages of glycolysis, i.e. the conversion of phosphoenolpyruvate (PEP) into pyruvate. E7 can influence the equilibrium between the two multimeric forms of the enzyme, which are characterized by different affinity for the substrate PEP [85]. As a consequence of this perturbed equilibrium, pyruvate production is decreased and the pool of phosphometabolytes upstream of PEP is expanded. The level of these metabolites, necessary for the biosynthesis of nucleotides, is usually low in quiescent cells. Therefore the interaction of E7 and M2-PK may have role in inducing the proliferation of the infected cells.

E7 has also been reported to affect the cellular carbohydrate metabolism through the direct interaction with the glycogen-degrading enzyme α-glucosidase [86]. This association promotes the activation of the enzyme catalytic activity *in vivo*, resulting in depletion of intracellular glycogen. Changes in carbohydrate metabolism are typical during malignant transformation. Significantly, they are frequently observed in early lesions of the cervix. Therefore, a potential link between E7 binding to α-glucosidase and E7-induced transformation may be hypothesized, though the precise role of the modulation of this enzyme is still not clear.

Subversion of p53 Function and the Apoptotic Pathway

Although in HPV-positive cells subversion of p53 functions is mainly ascribed to E6, interference with p53 activity by E7 has also been reported. Several studies have evidenced that E7 can overcome p53-induced growth arrest in response to cell damage [155, 156]. This phenomenon could depend on E7 capacity of inducing aberrant expression of cycA and cycE, as well as decreasing pRb levels and inactivating p53- responsive $p21^{CIP1/WAF1}$. Interestingly, E7-expressing cells show increased levels of p53 [157], suggesting that p53 degradation is perturbed. Indeed, in E7-positive cells p53 is not efficiently bound by the ubiquitine ligase MDM2, which normally regulates p53 steady-state level by targeting the tumor suppressor for proteasomal degradation [158]. As a result of p53 stabilization, the expression of E7 alone could induce apoptotic response [159]. However, as a consequence of E6/E7 synergistic and complementary effect, E7-induced apoptotic death of HPV-positive cells is bypassed by E6-mediated degradation of p53 and Bak [160].

E7 ability to decrease p53-dependent transcription in a transient transfection system has also been reported [161]. The binding to pRb seems not to be required for this activity while phosphorylation at the E7-CKII recognition site appears to be relevant. In fact, E7 phosphorylation by CKII increases E7 binding to p53, possibly mediated by TBP. Repression of p53 transcriptional activity is also induced by LR-HPV E7 proteins, suggesting the lack of a direct correlation between this property and HR-HPVs E7 transforming potential.

In spite of the described effects of E7 on p53 and the consequent apoptotic death E7 could induce in absence of E6, it is worth noting that E7 is a pleiomorphic protein as for apoptotic pathways. In fact, even if the majority of studies have evidenced E7 pro-apoptotic activity [162, 163], several data have indicated E7 as an anti-apoptotic protein. In a recent study, E6/E7 have been shown to inhibit TNF-mediated apoptosis in keratinocytes through up-regulation of the cellular inhibitor of apoptosis protein c-IAP2, though this activity seems to be mainly due to E6 [164]. Moreover, E7 has been reported to prevent TNF- and Fas-induced apoptosis in fibroblasts [165]. These contradictory results may be ascribed to different genetic backgrounds, cellular environments and experimental systems used, but also underline the complexity of E7 activities and effects.

E7 as an Inducer of Genomic Instability

Numerous recent studies have evidenced E7 ability to induce abnormal centrosomal duplication, which is a unique property of the HR-HPV protein [166]. The data obtained are consistent with the previous observations that E7 expression rapidly leads to aneuploidy and genomic instability [167]. In fact, as a result of defects in centriole assembly, multipolar mitoses occur and chromosomes do not symmetrically segregate during cell division. Indeed, tripolar mitotic figures are histopathological marks of HR-HPVs positive lesions, which typically contain irregular centrosome number [44, 166].

Abnormal centrosomal synthesis in HPV-positive cells seems to be independent of E7 ability to bind and degrade the "pocket proteins". In fact, pRb-, p107- and p130-deficient cells, which already show centrioles numeric irregularity, display a significant increase in this phenomenon upon HPV16-E7 expression [168]. Moreover, this activity is not associated with E7 ability to deregulate the expression of Cdc25, though its homologue in *Drosophila* is involved in centriole synthesis and maturation. The inactivation of $p21^{CIP1/WAF1}$ and $p27^{KIP1}$ has been suggested to play an important role in the subversion of centrosome homeostasis, since loss of $p21^{CIP1/WAF1}$ results in alteration of centrosome number [168]. Besides, the complexes cycE/cdk2 and cycA/cdk2 have been implicated in centrosome duplication in mammalian cells [170], thus it is conceivable that deregulation of this kinase contributes to the presence of supernumerary centrosomes. Indeed, cycE/cdk2 has been recently shown to be required for centriole overduplication from a single maternal template [171]. A recent study has demonstrated that HPV16-E7 possesses the ability to associate the gamma-tubulin component of centrosomes and that this interaction alters its recruitment to the centrosomes [172]. This study has also confirmed that E7-induced abnormal centrosomal synthesis relies on a "pocket proteins"-independent mechanism, since E7 association to gamma-tubulin does not involve the members of the retinoblastoma family [172].

As E7 proteins of LR-HPVs do not show the capacity of inducing irregular centrosome synthesis [166], it seems clear that this ability is greatly relevant in HR-HPV E7-driven transformation, as it confers an intrinsic element of genomic instability.

HPV-E7 also induces structural chromosome instability *in vitro* [166, 173]. Significantly, HPV-positive cancer cells frequently show structural chromosomal changes, such as translocations, deletions and amplifications [174]. Chromosome breaks could be caused by increased susceptibility to DNA damage [175] or defective DNA repair [173], presumably linked to loss of p53 and pRb function. Moreover, while cdc25A does not appear to be involved in E7-induced abnormal duplication of centrosomes, as already mentioned [168], its deregulation may be an important determinant of genome instability in E7-positive cells (see above). Importantly, disruption of chromosomal stability may ultimately promote integration of viral DNA in the host genome, an essential step in HPV-induced malignant transformation.

E7 in the Evasion of the Host Immune Response

Besides the numerous pathways of cell cycle control E7 is able to subvert, this oncoprotein appears to be essentially involved in HPV-induced immune suppression, which may partially account for the cases of persistent viral infection. In fact, even though most of the HPV infections in immunocompetent hosts are effectively controlled by the immune system, in a limited number of cases the virus establishes a persistent infection. A thorough understanding of the determinants that enable the HPV persistance is fundamental, since this persistance represents the main risk factor for the development of CC.

HPV has evolved specific mechanisms to escape from the host defences. Some of them are intrinsically linked to the virus life cycle. For instance, HPV reproduction in the upper layers of the epithelium, where the cells are rapidly and frequently eliminated by shedding, may determine a delayed/inefficient immune response. The superficial squamous epithelial cells are the only site where the viral gene products are expressed at high level; therefore, they are scarcely exposed to recognition by the cells of the immune system. Moreover, the ability of HPV proteins to modulate directly several components of the host defenses essentially contributes to determine that HPV-infected cells are not readily detected, and then eliminated, from the host immunity.

Several immune suppressor strategies have been developed by the HPV to prevent recognition of its antigens from the cells of the immune system. Among these, numerous have been ascribed to E7, which shows the ability of interfering with both innate and adaptive immune response.

A recent study has evidenced as HPV16 E6 and E7 are able to suppress the expression of the Toll-like receptor 9 (TLR9) [176]. This protein belongs to TLR family, which plays a key role at the interface between innate and adaptive immune response, recognizing and promoting the elimination of bacterial and viral pathogens. TLR9, in particular, is activated by dsDNA-derived CpG motifs. HPV16 and, at a lower extent, HPV18 oncoproteins have been shown to induce a significant decrease in TLR9 mRNA level, greatly compromising the functionality of TLR9 pathway [176]. Significantly, it has been reported that TLR9-specific staining is absent in cervical cancer biopsies in comparison with normal cervical tissue [176].

As already mentioned, E7 is involved in the deregulation of the expression and activity of immuno-modulating molecules, such as IL-8 [151]. In addition, E7 also subverts interferon (IFN) signalling pathways, by both affecting the expression and the function of type I interferons. These cytokines are essentially involved in host immune response since they stimulate the expression of MHC class I and II molecules, which in turn mediate the antigen recognition by cells of the immune system. Besides, IFNs also promote the expression of factors part of the machinery used for processing and presentation of the antigens on the cell surface. HPV16-E7 has been shown to inhibit multiple IFNα-inducible genes, due to its direct association with and sequestration of IRF-9 [91]. This interaction leads to the loss of IFN stimulatory gene factor-3 (ISGF3) transcription complex, which normally binds IFN responsive elements (ISRE) promoting the initiation of transcription of IFN-inducible genes.

As mentioned above, E7 has also been reported to impair IFNγ-induced IRF-1 transactivation activity, which is in turn involved in IFNβ gene expression [88, 89]. It is worth noting that IRF-1 also contributes to transcriptional activation of $p21^{CIP1/WAF1}$ [177], evidencing E7 potential ability of interfering with different processes, i.e. cell cycle control and immune response, inactivating a single multifunctional target.

HPV18-E7 has been shown to repress the transcription of the genes encoding the LMP2 subunit of the proteasomes involved in the production of the antigenic peptides, and TAP1, one of the components of the transporter associated with antigen presentation (TAP), that allows the peptides to be accumulated in the endoplasmic reticulum and loaded on MHC I molecules [178].

Several studies have also evidenced that, besides the effects achieved by inhibiting IFN pathways, E7 directly interferes with MHC expression, further decreasing the chance of HPV-positive cells to be recognized thus eliminated by T-cells. HPV-positive cervical carcinoma cells typically show low levels of MHC I molecules [179, 180]. As already mentioned, E7 down-regulates MHC class I genes in HPV16-positive cells through interaction with HDAC at level of MHC I heavy chain promoter [146]. E7 capacity of modulating MHC I expression has been confirmed by transfecting E7-siRNA in HPV-positive cells. In this system E7 knock out induces a significant increase in cell-surface MHC I molecules [181]. Moreover, overexpression of E7 in HPV-negative cells determines a marked reduction in MHC I molecules [181].

E7 has also been shown to influence the functionality of the adaptive immune system. For instance, E7 appears to repress the antigen-presenting function of dendritic cells (DCs) [182]. Indeed, DCs expressing E7 and pulsed with E7 are poor stimulators of E7-specific immunity [182].

Taken together, all these data demonstrate that E7 actively and directly inactivates the host immunity, partially but significantly contributing to HPV ability of escaping the immune surveillance. E7-mediated strategies of immune evasion may eventually lead to the virus persistence and ultimately contribute to the carcinogenic progression.

E7 as a Target for Cervical Cancer Therapy

HPV-induced cancers represent up to up to 15% of all cancer cases in women and up to 5% in men [183]. In particular, as previously mentioned, cervical cancer is the second most common cancer among women, accounting for one-fifth of all their cancer-related deaths [4].

Early detection of pre-cancerous lesions based on regular PAP-testing can markedly reduce cervical cancer death rates [184]. However, neither effective population screening programs nor optimal treatment are available in developing countries, where CC may represent the most prevalent cancer in women. Research on the biology of cervical cancer together with its viral etiology has recently resulted in the approval by the Food and Drug Administration (FDA) of an anti-HPV prophylactic vaccine based on the major capsid protein L1 of the High-Risk HPVs 16 and 18. This vaccine has been demonstrated to be effective in reducing the incidence of persistent HPV infections and HPV-associated lesions [185, 186]. Nevertheless, a significant impact of the vaccination on CC incidence will not be observed for at least a couple of decades. Moreover, despite the availability of an effective vaccine against HPV infection, a disparity in vaccination programs will probably occur due to the high cost of this new HPV vaccination for developing countries. Unfortunately, there is little evidence that the vaccine has therapeutic activity in CC patients and the survival benefits of the currently available anti-cancer therapies are not satisfying, especially in cases of advanced stage CC. The 5-year survival rate for CC has improved only by 4% over the last three decades; for patients with CC in advanced stage it is around 15-20%. The main therapeutic options, represented by surgery and radiation, are successful only if the tumor is detected at an early stage. In advanced stage, the efficacy of these therapies is less successful and the available chemotherapeutic agents are often only a palliative [187]. Radiotherapy and chemotherapy alone or in combination are used to eliminate potential residual cancer cells. However, these approaches can severely diminish the quality of life of the patients and are limited by the need of achieving therapeutically relevant drug concentrations. The toxicity associated with the high drug dose required to maintain a state of remission thus represents a major problem. Therefore, the pre-clinical and clinical studies focused on the development of strategies capable of providing an effective therapy for HPV-induced lesions while avoiding significant side effects are numerous. Since the results in terms of tumor regression are still not encouraging, a continuous effort to develop effective and less invasive therapies continues to be made.

The viral etiology of cervical cancer allows for choosing viral antigens as specific targets. HPV oncoproteins represent the most attractive candidate for a targeted cancer therapy, as they are the primary contributors to the development of malignant transformation. Due to E7 profound involvement in the oncogenesis, the inhibition of the function of this protein might be an effective approach to block the process of malignant transformation. E7 offers the unique advantage of being constitutively expressed only in HPV-positive cells and of being unrelated to the host proteins. Therefore E7 is a specific target, allowing selective targeting of only the pre-malignant and malignant cells, without any effect on normal cells.

As already mentioned in this chapter, E2 gene product can directly bind the HPV early promoter and repress the transcription of HPV oncogenes [16]. In addition, a recent study has reported that E2 can also regulate E7 activity [188]. In fact, E2 can suppress E7-induced transformation and this phenomenon appears to be independent from its ability to suppress E7 expression. Due to its function as a transcriptional repressor, introduction of E2 gene in HPV-positive cells has been evaluated as an approach to suppress the oncoproteins expression, thus to revert the transformed phenotype. An SV40-derived viral vector has been exploited to deliver the E2 ORF in HPV18-positive HeLa cells [18, 189]. Adenovirus-mediated delivery of BPV-E2 to CaSki cells has also been reported [190]. In both HPV16- and HPV18-positive cells, E2-mediated abrogation of the endogenous E6/E7 has been successfully achieved. Most interestingly, profound effects on DNA synthesis with rapid growth arrest and acquisition of the senescent phenotype have been observed [189, 190]. Upon ectopical E2 expression in HPV-positive cells, apoptosis induction has also been shown [191, 192], though this effect has been detected in HPV-negative cells as well [191]. Effects elicited by E2 have been partially reversed by constitutive expression of E6/E7, confirming that growth inhibition is due to the suppression of the oncogenes transcription mediated by E2 [9]. Importantly, the delivery of BPV-E2 into HeLa cells stably expressing either HPV16-E6 or E7 has allowed elucidation of the individual effects of the suppression of each oncoprotein [192]. This study has shown that E7 repression does not reactivate p53 pathway but does restore pRb pathway. Indeed, an increase in the levels of the hypophosphorylated form of pRb, p107 and p130 has been observed and suppression of the E2F-responsive genes for cycA and cdc25A has been detected. Moreover, E7 abrogation alone has been shown to be sufficient to inhibit telomerase activity. In this system, E7 repression efficiently triggered senescence, while appearance of apoptotic cells was limited. Conversely, E6 suppression has been demonstrated to reactivate p53 pathway and to induce both senescence and apoptosis [192].

Abundant data in the literature have proven that E7 inactivation/down-regulation at the post-transcriptional level is an effective approach for reversal of the malignant phenotype and/or killing of cervical cancer cells as well. Numerous strategies to suppress E7 activity by blocking the oncogene mRNA translation have been explored. For instance, HPV16-E7 expression knock out has been obtained by delivery of antisense (AS) RNA molecules, that can form a duplex with the RNA of the opposite strand leading to its degradation. ASRNA for HPV oncoproteins have been delivered to tumor cells by using adeno-associated virus [193, 194], herpes simplex virus-derived vectors [195] and replication-defective retrovirus [196]. All these delivery systems have been proven to be effective in specific suppression of the E6/E7 mRNA and the relative proteins. Up-regulation of pRb and down-regulation of E2F1-mediated transcription have been reported as a result of E7 knock-down by ASRNA [194, 196]. Most importantly, delivery of HPV16E7-ASRNA has been demonstrated to induce partial reversal of the malignant features of HPV16-positive CaSki cell line. Reduced migration and invasion capability of these cells *in vitro* have been reported [193]. Furthermore, antisense HPV16E7 constructs have been able to inhibit and/or retard the tumorigenicity of CaSki cells in nude mice [193, 194, 196]. ASRNA-mediated repression of both E6 and E7 products of HPV16 has also been reported to induce senescence and apoptosis of the human cancer SiHa cells [194, 197].

Successful silencing of HR-HPVE7 by small interfering RNA (siRNA) has also been documented [198-201]. In fact, siRNA-induced specific degradation of the target mRNA enables knock-down of the relative protein. Due to the bicistronic nature of the mRNA encodying E6 and E7, expression of E7 siRNA actually results in suppression of both the oncoproteins. However, some studies have reported the design of siRNA specifically targeting only E7 [198]. Growth inhibition and induction of senescence in HPV18-positive HeLa cells have been achieved by delivery of E6-E7siRNA [199, 200]. On the other hand, siRNA-mediated silencing of E7 in SiHa cells has been shown to induce apoptotic cell death [198]. Importantly, intra-tumor injection of specific siRNA has resulted in successful suppression of tumor growth *in vivo* [200]. As expected, E7 repression has also been described to rescue pRb [200].

Reduction of E7 protein level has also been achieved by using a lentiviral vector to deliver a short hairpin RNA (shRNA) against HPV18-E6 [202]. Induction of HPV18-positive cells senescence or specific cell death via apoptosis have been reported in this study. Additionally, inhibition of tumor growth *in vivo* has also been obtained upon systemic delivery of the shRNA in mice.

Among the strategies exploiting nucleic acids as potential therapeutic agents for HPV-associated cancer, the use of ribozymes (Rz) has also been documented, although it is mostly restricted to *in vitro* studies with very limited practical applications. Ribozymes possess a secondary structure that provides the molecule with enzymatic activity. Rz specifically designed for the transcripts of either HPV16- or HPV18-E6/E7 have been tested, showing the ability of effectively cleaving the target HPV-RNA *in vitro* [203-205]. These results have provided the basis to evaluate the validity of Rz-based therapeutic approaches. Indeed, some studies have reported the ribozymes to be effective in preventing E6/E7-mediated immortalization of human keratinocytes [205] and in inhibiting CaSki cells growth [206]. Interestingly, CaSki cells expressing specific ribozymes have revealed higher sensitivity to cisplatin and radiation, offering the possibility of using Rz to increase the efficacy of the current available therapies. However, a pre-clinical study failed to detect any anti-tumor effect in nude mice upon delivery of ribozyme genes by adeno-associated virus [207].

Since strategies based on RNA are limited by mRNA turnover and can induce non-specific effects (siRNA), other approaches aimed at directly impairing the activity of E7 at protein level have been explored. For instance, the use of peptide aptamers to block E7 transforming potential has been assessed [208]. This study has demonstrated that 20-mer peptides selected against the CR3 of HPV16-E7 can modulate the oncoprotein activity in living cells. In particular, some of the isolated peptides have shown the ability of inhibiting the growth of NIH3T3E7 and CaSki cells, through an increase in the apoptotic rate.

Suppression of E7-mediated proliferation has also been achieved by intracellular expression of specific antibodies in single-chain format (scFv) [209]. Anti-E7 antibodies selected by phage display technology have been shown to reduce DNA synthesis when expressed in the nucleus and endoplasmic reticulum of HPV16-positive SiHa cells. In comparison with other approaches, therapeutic strategies based on intracellular antibodies ("intrabodies") offer the advantage of potentially blocking intermolecular interactions, which are fundamental for oncoproteins transforming activity. Indeed, in a recent paper specific "intrabodies" have demonstrated the ability to suppress E6-mediated degradation of p53, which occurs upon the formation of a trimeric complex involving E6, p53 and the cellular protein E6AP [210].

Finally, E6 and E7 might be exploited as targets for anti-cancer drug discovery since their domains as well as many of the intermolecular interactions that contribute to their oncogenic potential are identified. In fact, the development of new chemotherapeutics to manage CC should be directed by the knowledge of the mechanisms underlying the development of this tumor, even if partial. Therefore, the effort to develop new anti-HPV chemotherapeutics designed to affect the structure and/or the function of the oncoproteins should be pursued.

Several compounds, such as sulindac, cidofovir, curcumin and ursolic acid, among others, have been reported to be effective in suppressing the level of E6/E7 in HPV-positive cells [211-215]. These drugs have also shown a certain ability to inhibit the growth of cancer cells and to promote apoptosis. However, none of these compounds has been specifically designed for binding to the HPV oncoproteins and most of the time their effects are a mere result of serendipity.

Discovery of new anti-HPV compounds could greatly benefit from the knowledge of the specific functional relationship between the target domains and activity. Indeed, studies aimed to identify and characterize compounds that interfere with E6 binding to zinc have been reported [216, 217]. Importantly, several zinc-ejecting compounds have shown the ability to inhibit E6 binding to some of its target, such as E6AP and E6-binding protein (E6BP). Most significantly, they have been able to promote specific death of HPV-positive cells [217]. These studies have demonstrated that targeting a specific domain associated with a known and relevant biological function can be effective in suppressing the target activity. Moreover, they have shown that small molecules represent a valuable tool in inhibiting intermolecular interactions, which are often the basis of tumorigenic processes. However, the approach of screening compounds for their potential ability of interacting with E7 and interfering with its activity has not been reported yet. Computational biology may have an important role in this field, as the use of computational methods has been proven to be effective for isolating specific inhibitors of the molecules of interest [218-221] and potential anti-cancer drugs [222, 223].

Conclusion

The oncogenic potential of E7 protein of High-Risk Human Papillomaviruses has been extensively documented over the years. Its ability to deregulate numerous cellular modulators has evidenced how this oncoprotein can affect so many fundamental processes, such as cell cycle control, differentiation, apoptosis and immune response. Since the malignant progression represents an unintended consequence of the viral infection, it is obvious as the deregulation of these processes has the main significance of allowing the virus replication, which is, ultimately, the central function of E7. However, it is also clear as the extensive deregulation of E7 expression and the continuous subversion of crucial cellular pathways, for instance following the integration of the viral DNA in the host genome, essentially contribute to HPV-mediated carcinogenesis.

Although the list of E7 cellular targets keeps growing and new insights in the mechanisms of E7-induced transformation are emerging, the critical contribution of pRb inactivation and destabilization to E7 transforming properties is unquestionable. The loss of pRb negative modulation on the cell cycle certainly has a role in promoting deregulated proliferation of the infected cells through release of E2F-mediated transcription. Nevertheless, modulation of other cellular proteins has a biological relevance in the carcinogenic process, as evidenced by the capability of pRb binding-defective mutants to induce immortalization and of non-transforming mutants of E7 to still bind pRb. Numerous cellular factors targeted by E7, including the other "pocket proteins" p107 and p130, as well as cyclins and cyclin-dependent kinases, inhibitors of cdks, histone deacetylases, transcription factors and glycolytic enzymes, might be important determinants for E7 oncogenic potential. For instance, the property of E7 protein of the High-Risk HPVs to induce structural and numerical alterations of the cellular genome seems to be dependent on the deregulation of proteins other than pRb, though their exact contribution is still unclear. A thorough understanding of the effects induced by the additional associations established by E7, as well as of the underlying mechanisms, will clarify how extensively they contribute to the transformation. Regardless, it is evident that E7-induced carcinogenesis results from complex and mutually related events. It appears that E7-induced subversion of the host defenses, which essentially contributes to HPV ability to escape from the immune surveillance, ultimately leads to a condition that promotes the carcinogenic progression. In fact, together with other factors, it determines a persistent infection, which allows a prolonged expression of E7 and thus a situation of sustained oncogenic stimuli. Notably, E7-induced genetic changes clearly play a key role in the carcinogenesis, since they confer an endogenous element of instability that likely promotes the malignant conversion of the infected cells.

The concomitant expression of E6 and E7 in HPV-infected cells obviously determines a more elaborate scenario, in which the interplay between the two oncoproteins through a refined and specific "choice" of the respective targets ultimately establishes the conditions for the malignant progression.

It is worth noting that E7 ability to interact with multiple partners is really remarkable, considering that this protein contains just about 100 amino acids. However, as new data on E7 are emerging, this property seems to be less surprising. In fact, E7 has been described as an "intrinsically disordered" or "natively unfolded" protein. This "unstructured" aspect may determine a flexible and dynamic conformation that in turns may account for the extreme functional plasticity of this oncoprotein. Interestingly, in a recent study E7 proteins of High-Risk HPVs have been reported to be more disordered than E7 proteins of Low-Risk HPVs both in the N- and C-terminus, suggesting that this element of disorder might have a role with respect to the different biological function and oncogenic potential [61].

Finally, the essential role of E7 in HPV-mediated transformation and its features as a tumor-specific antigen make this protein an ideal target for the therapy of HPV-associated lesions. All the approaches to knock out E7 both at transcriptional and post-transcriptional level, mentioned in this chapter, have major therapeutic potential for the treatment of cervical cancer. However, it must be emphasized that most of the successful results obtained by these strategies are restricted to *in vitro* applications and pre-clinical models and that there are still several challenges to their clinical potential. For instance, several important issues remain unsolved in gene therapy, such as the development of a safe and effective delivery system.

The validation of new potential chemotherapeutics also presents important barriers. A great effort is required to identify cancer-specific drugs, to obtain their proper delivery to the targets cells and to ensure the administration of the therapeutic agent in relevant and effective dose without toxicity. Nonetheless, the scenario of therapeutic options for cervical cancer will benefit from a complete understanding of the mechanisms of E7-mediated carcinogenesis, together with the development of new technologies for the isolation, pre-clinical development and safe delivery of therapeutic molecules.

Reviewed by Prof. Alfred B. Jenson, MD (James Graham Brown Cancer Center, University of Louisville), Dr. Douglas C. Dean, PhD (Department of Ophthalmology, University of Louisville) Dr. Kenneth E. Palmer, PhD (Department of Pharmacology and Toxicology, University of Louisville).

Acknowledgments

The author gratefully acknowledges Miss Amanda B. Lasnik for reviewing the English of the manuscript.

References

[1] de Villiers EM, Gunst K, Stein H, Scherubl H. Esophageal squamous cell cancer in patients with head and neck cancer: Prevalence of human papillomavirus DNA sequences. *International journal of cancer*. 2004 Mar 20;109(2):253-8.

[2] Bosch FX, de Sanjose S. Human papillomavirus in cervical cancer. *Current oncology reports*. 2002 Mar;4(2):175-83.

[3] Walboomers JM, Jacobs MV, Manos MM, Bosch FX, Kummer JA, Shah KV, et al. Human papillomavirus is a necessary cause of invasive cervical cancer worldwide. The *Journal of pathology*. 1999 Sep;189(1):12-9.

[4] Parkin DM, Bray F, Ferlay J, Pisani P. Global cancer statistics, 2002. CA: a *cancer journal for clinicians*. 2005 Mar-Apr;55(2):74-108.

[5] Hawley-Nelson P, Vousden KH, Hubbert NL, Lowy DR, Schiller JT. HPV16 E6 and E7 proteins cooperate to immortalize human foreskin keratinocytes. *The EMBO journal*. 1989 Dec 1;8(12):3905-10.

[6] Munger K, Phelps WC, Bubb V, Howley PM, Schlegel R. The E6 and E7 genes of the human papillomavirus type 16 together are necessary and sufficient for transformation of primary human keratinocytes. *Journal of virology*. 1989 Oct;63(10):4417-21.

[7] [7] Durst M, Gallahan D, Jay G, Rhim JS. Glucocorticoid-enhanced neoplastic transformation of human keratinocytes by human papillomavirus type 16 and an activated ras oncogene. *Virology*. 1989 Dec;173(2):767-71.

[8] Goodwin EC, DiMaio D. Repression of human papillomavirus oncogenes in HeLa cervical carcinoma cells causes the orderly reactivation of dormant tumor suppressor

pathways. *Proceedings of the National Academy of Sciences of the United States of America.* 2000 Nov 7;97(23):12513-8.

[9] Wells SI, Francis DA, Karpova AY, Dowhanick JJ, Benson JD, Howley PM. Papillomavirus E2 induces senescence in HPV-positive cells via pRB- and p21(CIP)-dependent pathways. *The EMBO journal.* 2000 Nov 1;19(21):5762-71.

[10] Smotkin D, Wettstein FO. Transcription of human papillomavirus type 16 early genes in a cervical cancer and a cancer-derived cell line and identification of the E7 protein. *Proceedings of the National Academy of Sciences of the United States of America.* 1986 Jul;83(13):4680-4.

[11] Androphy EJ, Lowy DR, Schiller JT. Bovine papillomavirus E2 trans-activating gene product binds to specific sites in papillomavirus DNA. *Nature.* 1987 Jan 1-7;325(6099):70-3.

[12] Baker CC, Phelps WC, Lindgren V, Braun MJ, Gonda MA, Howley PM. Structural and transcriptional analysis of human papillomavirus type 16 sequences in cervical carcinoma cell lines. *Journal of virology.* 1987 Apr;61(4):962-71.

[13] Cullen AP, Reid R, Campion M, Lorincz AT. Analysis of the physical state of different human papillomavirus DNAs in intraepithelial and invasive cervical neoplasm. *Journal of virology.* 1991 Feb;65(2):606-12.

[14] Fujii T, Masumoto N, Saito M, Hirao N, Niimi S, Mukai M, et al. Comparison between in situ hybridization and real-time PCR technique as a means of detecting the integrated form of human papillomavirus 16 in cervical neoplasia. *Diagn.Mol. Pathol.* 2005 Jun;14(2):103-8.

[15] Hudelist G, Manavi M, Pischinger KI, Watkins-Riedel T, Singer CF, Kubista E, et al. Physical state and expression of HPV DNA in benign and dysplastic cervical tissue: different levels of viral integration are correlated with lesion grade. *Gynecologic oncology.* 2004 Mar;92(3):873-80.

[16] McBride AA, Romanczuk H, Howley PM. The papillomavirus E2 regulatory proteins. *The Journal of biological chemistry.* 1991 Oct 5;266(28):18411-4.

[17] Dowhanick JJ, McBride AA, Howley PM. Suppression of cellular proliferation by the papillomavirus E2 protein. *Journal of virology.* 1995 Dec;69(12):7791-9.

[18] Hwang ES, Riese DJ, 2nd, Settleman J, Nilson LA, Honig J, Flynn S, et al. Inhibition of cervical carcinoma cell line proliferation by the introduction of a bovine papillomavirus regulatory gene. *Journal of virology.* 1993 Jul;67(7):3720-9.

[19] Corden SA, Sant-Cassia LJ, Easton AJ, Morris AG. The integration of HPV-18 DNA in cervical carcinoma. *Mol.Pathol.* 1999 Oct;52(5):275-82.

[20] Choo KB, Pan CC, Han SH. Integration of human papillomavirus type 16 into cellular DNA of cervical carcinoma: preferential deletion of the E2 gene and invariable retention of the long control region and the E6/E7 open reading frames. *Virology.* 1987 Nov;161(1):259-61.

[21] Jeon S, Lambert PF. Integration of human papillomavirus type 16 DNA into the human genome leads to increased stability of E6 and E7 mRNAs: implications for cervical carcinogenesis. *Proceedings of the National Academy of Sciences of the United States of America.* 1995 Feb 28;92(5):1654-8.

[22] Flores ER, Allen-Hoffmann BL, Lee D, Lambert PF. The human papillomavirus type 16 E7 oncogene is required for the productive stage of the viral life cycle. *Journal of virology*. 2000 Jul;74(14):6622-31.

[23] Longworth MS, Laimins LA. The binding of histone deacetylases and the integrity of zinc finger-like motifs of the E7 protein are essential for the life cycle of human papillomavirus type 31. *Journal of virology*. 2004 Apr;78(7):3533-41.

[24] Collins AS, Nakahara T, Do A, Lambert PF. Interactions with pocket proteins contribute to the role of human papillomavirus type 16 E7 in the papillomavirus life cycle. *Journal of virology*. 2005 Dec;79(23):14769-80.

[25] Dyson N, Guida P, McCall C, Harlow E. Adenovirus E1A makes two distinct contacts with the retinoblastoma protein. *Journal of virology*. 1992 Jul;66(7):4606-11.

[26] Ben-Israel H, Kleinberger T. Adenovirus and cell cycle control. *Front Biosci*. 2002 May 1;7:d1369-95.

[27] Ali SH, DeCaprio JA. Cellular transformation by SV40 large T antigen: interaction with host proteins. Seminars in cancer biology. 2001 Feb;11(1):15-23.

[28] Dyson N, Howley PM, Munger K, Harlow E. The human papilloma virus-16 E7 oncoprotein is able to bind to the retinoblastoma gene product. *Scien*ce (New York, NY. 1989 Feb 17;243(4893):934-7.

[29] Werness BA, Levine AJ, Howley PM. Association of human papillomavirus types 16 and 18 E6 proteins with p53. Science (New York, NY. 1990 Apr 6;248(4951):76-9.

[30] Scheffner M, Werness BA, Huibregtse JM, Levine AJ, Howley PM. The E6 oncoprotein encoded by human papillomavirus types 16 and 18 promotes the degradation of p53. *Cell*. 1990 Dec 21;63(6):1129-36.

[31] Huibregtse JM, Scheffner M, Howley PM. A cellular protein mediates association of p53 with the E6 oncoprotein of human papillomavirus types 16 or 18. *The EMBO journal*. 1991 Dec;10(13):4129-35.

[32] Accardi R, Dong W, Smet A, Cui R, Hautefeuille A, Gabet AS, et al. Skin human papillomavirus type 38 alters p53 functions by accumulation of deltaNp73. *EMBO reports*. 2006 Mar;7(3):334-40.

[33] Halbert CL, Demers GW, Galloway DA. The E6 and E7 genes of human papillomavirus type 6 have weak immortalizing activity in human epithelial cells. *Journal of virology*. 1992 Apr;66(4):2125-34.

[34] Vousden KH, Doniger J, DiPaolo JA, Lowy DR. The E7 open reading frame of human papillomavirus type 16 encodes a transforming gene. *Oncogene research*. 1988 Sep;3(2):167-75.

[35] Matlashewski G, Schneider J, Banks L, Jones N, Murray A, Crawford L. Human papillomavirus type 16 DNA cooperates with activated ras in transforming primary cells. *The EMBO journal*. 1987 Jun;6(6):1741-6.

[36] Phelps WC, Yee CL, Munger K, Howley PM. The human papillomavirus type 16 E7 gene encodes transactivation and transformation functions similar to those of adenovirus E1A. *Cell*. 1988 May 20;53(4):539-47.

[37] Liu X, Disbrow GL, Yuan H, Tomaic V, Schlegel R. Myc and HPV-16 E7 genes cooperate to immortalize human keratinocytes. *Journal of virology*. 2007 Sep 5.

[38] Oh ST, Kyo S, Laimins LA. Telomerase activation by human papillomavirus type 16 E6 protein: induction of human telomerase reverse transcriptase expression through Myc and GC-rich Sp1 binding sites. *Journal of virology.* 2001 Jun;75(12):5559-66.

[39] Veldman T, Liu X, Yuan H, Schlegel R. Human papillomavirus E6 and Myc proteins associate in vivo and bind to and cooperatively activate the telomerase reverse transcriptase promoter. *Proceedings of the National Academy of Sciences of the United States of America.* 2003 Jul 8;100(14):8211-6.

[40] Halbert CL, Demers GW, Galloway DA. The E7 gene of human papillomavirus type 16 is sufficient for immortalization of human epithelial cells. *Journal of virology.* 1991 Jan;65(1):473-8.

[41] Wazer DE, Liu XL, Chu Q, Gao Q, Band V. Immortalization of distinct human mammary epithelial cell types by human papilloma virus 16 E6 or E7. *Proceedings of the National Academy of Sciences of the United States of America.* 1995 Apr 25;92(9):3687-91.

[42] Herber R, Liem A, Pitot H, Lambert PF. Squamous epithelial hyperplasia and carcinoma in mice transgenic for the human papillomavirus type 16 E7 oncogene. *Journal of virology.* 1996 Mar;70(3):1873-81.

[43] Balsitis SJ, Sage J, Duensing S, Munger K, Jacks T, Lambert PF. Recapitulation of the effects of the human papillomavirus type 16 E7 oncogene on mouse epithelium by somatic Rb deletion and detection of pRb-independent effects of E7 in vivo. *Molecular and cellular biology.* 2003 Dec;23(24):9094-103.

[44] Riley RR, Duensing S, Brake T, Munger K, Lambert PF, Arbeit JM. Dissection of human papillomavirus E6 and E7 function in transgenic mouse models of cervical carcinogenesis. *Cancer research.* 2003 Aug 15;63(16):4862-71.

[45] Demers GW, Espling E, Harry JB, Etscheid BG, Galloway DA. Abrogation of growth arrest signals by human papillomavirus type 16 E7 is mediated by sequences required for transformation. *Journal of virology.* 1996 Oct;70(10):6862-9.

[46] Zerfass K, Schulze A, Spitkovsky D, Friedman V, Henglein B, Jansen-Durr P. Sequential activation of cyclin E and cyclin A gene expression by human papillomavirus type 16 E7 through sequences necessary for transformation. *Journal of virology.* 1995 Oct;69(10):6389-99.

[47] Nguyen DX, Westbrook TF, McCance DJ. Human papillomavirus type 16 E7 maintains elevated levels of the cdc25A tyrosine phosphatase during deregulation of cell cycle arrest. *Journal of virology.* 2002 Jan;76(2):619-32.

[48] Villa LL, Vieira KB, Pei XF, Schlegel R. Differential effect of tumor necrosis factor on proliferation of primary human keratinocytes and cell lines containing human papillomavirus types 16 and 18. *Molecular carcinogenesis.* 1992;6(1):5-9.

[49] Basile JR, Zacny V, Munger K. The cytokines tumor necrosis factor-alpha (TNF-alpha) and TNF-related apoptosis-inducing ligand differentially modulate proliferation and apoptotic pathways in human keratinocytes expressing the human papillomavirus-16 E7 oncoprotein. *The Journal of biological chemistry.* 2001 Jun 22;276(25):22522-8.

[50] Pietenpol JA, Stein RW, Moran E, Yaciuk P, Schlegel R, Lyons RM, et al. TGF-beta 1 inhibition of c-myc transcription and growth in keratinocytes is abrogated by viral transforming proteins with pRB binding domains. *Cell.* 1990 Jun 1;61(5):777-85.

[51] Barbosa MS, Edmonds C, Fisher C, Schiller JT, Lowy DR, Vousden KH. The region of the HPV E7 oncoprotein homologous to adenovirus E1a and Sv40 large T antigen contains separate domains for Rb binding and casein kinase II phosphorylation. *The EMBO journal.* 1990 Jan;9(1):153-60.
[52] Massimi P, Banks L. Differential phosphorylation of the HPV-16 E7 oncoprotein during the cell cycle. *Virology.* 2000 Oct 25;276(2):388-94.
[53] McIntyre MC, Frattini MG, Grossman SR, Laimins LA. Human papillomavirus type 18 E7 protein requires intact Cys-X-X-Cys motifs for zinc binding, dimerization, and transformation but not for Rb binding. *Journal of virology.* 1993 Jun;67(6):3142-50.
[54] Ohlenschlager O, Seiboth T, Zengerling H, Briese L, Marchanka A, Ramachandran R, et al. Solution structure of the partially folded high-risk human papilloma virus 45 oncoprotein E7. *Oncogene.* 2006 Sep 28;25(44):5953-9.
[55] Watanabe S, Sato H, Furuno A, Yoshiike K. Changing the spacing between metal-binding motifs decreases stability and transforming activity of the human papillomavirus type 18 E7 oncoprotein. *Virology.* 1992 Oct;190(2):872-5.
[56] Chellappan S, Kraus VB, Kroger B, Munger K, Howley PM, Phelps WC, et al. Adenovirus E1A, simian virus 40 tumor antigen, and human papillomavirus E7 protein share the capacity to disrupt the interaction between transcription factor E2F and the retinoblastoma gene product. *Proceedings of the National Academy of Sciences of the United States of America.* 1992 May 15;89(10):4549-53.
[57] Clements A, Johnston K, Mazzarelli JM, Ricciardi RP, Marmorstein R. Oligomerization properties of the viral oncoproteins adenovirus E1A and human papillomavirus E7 and their complexes with the retinoblastoma protein. *Biochemistry.* 2000 Dec 26;39(51):16033-45.
[58] Jewers RJ, Hildebrandt P, Ludlow JW, Kell B, McCance DJ. Regions of human papillomavirus type 16 E7 oncoprotein required for immortalization of human keratinocytes. *Journal of virology.* 1992 Mar;66(3):1329-35.
[59] Pahel G, Aulabaugh A, Short SA, Barnes JA, Painter GR, Ray P, et al. Structural and functional characterization of the HPV16 E7 protein expressed in bacteria. *The Journal of biological chemistry.* 1993 Dec 5;268(34):26018-25.
[60] Alonso LG, Garcia-Alai MM, Nadra AD, Lapena AN, Almeida FL, Gualfetti P, et al. High-risk (HPV16) human papillomavirus E7 oncoprotein is highly stable and extended, with conformational transitions that could explain its multiple cellular binding partners. *Biochemistry.* 2002 Aug 20;41(33):10510-8.
[61] Uversky VN, Roman A, Oldfield CJ, Dunker AK. Protein intrinsic disorder and human papillomaviruses: increased amount of disorder in E6 and E7 oncoproteins from high risk HPVs. *Journal of proteome research.* 2006 Aug;5(8):1829-42.
[62] Smotkin D, Wettstein FO. The major human papillomavirus protein in cervical cancers is a cytoplasmic phosphoprotein. *J. Virol.* 1987 May;61(5):1686-9.
[63] Selvey LA, Dunn LA, Tindle RW, Park DS, Frazer IH. Human papillomavirus (HPV) type 18 E7 protein is a short-lived steroid-inducible phosphoprotein in HPV-transformed cell lines. *The Journal of general virology.* 1994 Jul;75 (Pt 7):1647-53.

[64] Reinstein E, Scheffner M, Oren M, Ciechanover A, Schwartz A. Degradation of the E7 human papillomavirus oncoprotein by the ubiquitin-proteasome system: targeting via ubiquitination of the N-terminal residue. *Oncogene.* 2000 Nov 30;19(51):5944-50.

[65] Oh KJ, Kalinina A, Wang J, Nakayama K, Nakayama KI, Bagchi S. The papillomavirus E7 oncoprotein is ubiquitinated by UbcH7 and Cullin 1- and Skp2-containing E3 ligase. *Journal of virology.* 2004 May;78(10):5338-46.

[66] Zatsepina O, Braspenning J, Robberson D, Hajibagheri MA, Blight KJ, Ely S, et al. The human papillomavirus type 16 E7 protein is associated with the nucleolus in mammalian and yeast cells. *Oncogene.* 1997 Mar 13;14(10):1137-45.

[67] Sato H, Watanabe S, Furuno A, Yoshiike K. Human papillomavirus type 16 E7 protein expressed in Escherichia coli and monkey COS-1 cells: immunofluorescence detection of the nuclear E7 protein. *Virology.* 1989 May;170(1):311-5.

[68] Smith-McCune K, Kalman D, Robbins C, Shivakumar S, Yuschenkoff L, Bishop JM. Intranuclear localization of human papillomavirus 16 E7 during transformation and preferential binding of E7 to the Rb family member p130. *Proceedings of the National Academy of Sciences of the United States of America.* 1999 Jun 8;96(12):6999-7004.

[69] Dyson N, Guida P, Munger K, Harlow E. Homologous sequences in adenovirus E1A and human papillomavirus E7 proteins mediate interaction with the same set of cellular proteins. *Journal of virology.* 1992 Dec;66(12):6893-902.

[70] Cobrinik D. Pocket proteins and cell cycle control. *Oncogene.* 2005 Apr 18;24(17):2796-809.

[71] Rowland BD, Bernards R. Re-evaluating cell-cycle regulation by E2Fs. *Cell.* 2006 Dec 1;127(5):871-4.

[72] Cam H, Dynlacht BD. Emerging roles for E2F: beyond the G1/S transition and DNA replication. *Cancer cell.* 2003 Apr;3(4):311-6.

[73] Harbour JW, Dean DC. Chromatin remodeling and Rb activity. *Current opinion in cell biology.* 2000 Dec;12(6):685-9.

[74] Godefroy N, Lemaire C, Mignotte B, Vayssiere JL. p53 and Retinoblastoma protein (pRb): a complex network of interactions. *Apoptosis.* 2006 May;11(5):659-61.

[75] Boyer SN, Wazer DE, Band V. E7 protein of human papilloma virus-16 induces degradation of retinoblastoma protein through the ubiquitin-proteasome pathway. *Cancer* research. 1996 Oct 15;56(20):4620-4.

[76] Huh K, Zhou X, Hayakawa H, Cho JY, Libermann TA, Jin J, et al. Human papillomavirus type 16 e7 oncoprotein associates with the cullin 2 ubiquitin ligase complex, which contributes to degradation of the retinoblastoma tumor suppressor. *Journal of virology.* 2007 Sep;81(18):9737-47.

[77] Gonzalez SL, Stremlau M, He X, Basile JR, Munger K. Degradation of the retinoblastoma tumor suppressor by the human papillomavirus type 16 E7 oncoprotein is important for functional inactivation and is separable from proteasomal degradation of E7. *Journal of virology.* 2001 Aug;75(16):7583-91.

[78] Zhang B, Chen W, Roman A. The E7 proteins of low- and high-risk human papillomaviruses share the ability to target the pRB family member p130 for degradation. *Proceedings of the National Academy of Sciences of the United States of America.* 2006 Jan 10;103(2):437-42.

[79] Funk JO, Waga S, Harry JB, Espling E, Stillman B, Galloway DA. Inhibition of CDK activity and PCNA-dependent DNA replication by p21 is blocked by interaction with the HPV-16 E7 oncoprotein. *Genes and development.* 1997 Aug 15;11(16):2090-100.

[80] Jones DL, Alani RM, Munger K. The human papillomavirus E7 oncoprotein can uncouple cellular differentiation and proliferation in human keratinocytes by abrogating p21Cip1-mediated inhibition of cdk2. *Genes and development.* 1997 Aug 15;11(16):2101-11.

[81] Zerfass-Thome K, Zwerschke W, Mannhardt B, Tindle R, Botz JW, Jansen-Durr P. Inactivation of the cdk inhibitor p27KIP1 by the human papillomavirus type 16 E7 oncoprotein. *Oncogene.* 1996 Dec 5;13(11):2323-30.

[82] Antinore MJ, Birrer MJ, Patel D, Nader L, McCance DJ. The human papillomavirus type 16 E7 gene product interacts with and trans-activates the AP1 family of transcription factors. *The EMBO journal.* 1996 Apr 15;15(8):1950-60.

[83] Massimi P, Pim D, Storey A, Banks L. HPV-16 E7 and adenovirus E1a complex formation with TATA box binding protein is enhanced by casein kinase II phosphorylation. *Oncogene.* 1996 Jun 6;12(11):2325-30.

[84] Brehm A, Nielsen SJ, Miska EA, McCance DJ, Reid JL, Bannister AJ, et al. The E7 oncoprotein associates with Mi2 and histone deacetylase activity to promote cell growth. *The EMBO journal.* 1999 May 4;18(9):2449-58.

[85] Zwerschke W, Mazurek S, Massimi P, Banks L, Eigenbrodt E, Jansen-Durr P. Modulation of type M2 pyruvate kinase activity by the human papillomavirus type 16 E7 oncoprotein. *Proceedings of the National Academy of Sciences of the United States of America.* 1999 Feb 16;96(4):1291-6.

[86] Zwerschke W, Mannhardt B, Massimi P, Nauenburg S, Pim D, Nickel W, et al. Allosteric activation of acid alpha-glucosidase by the human papillomavirus E7 protein. *The Journal of biological chemistry.* 2000 Mar 31;275(13):9534-41.

[87] Berezutskaya E, Bagchi S. The human papillomavirus E7 oncoprotein functionally interacts with the S4 subunit of the 26 S proteasome. *The Journal of biological chemistry.* 1997 Nov 28;272(48):30135-40.

[88] Park JS, Kim EJ, Kwon HJ, Hwang ES, Namkoong SE, Um SJ. Inactivation of interferon regulatory factor-1 tumor suppressor protein by HPV E7 oncoprotein. Implication for the E7-mediated immune evasion mechanism in cervical carcinogenesis. *The Journal of biological chemistry.* 2000 Mar 10;275(10):6764-9.

[89] Perea SE, Massimi P, Banks L. Human papillomavirus type 16 E7 impairs the activation of the interferon regulatory factor-1. *International journal of molecular medicine.* 2000 Jun;5(6):661-6.

[90] Mazzarelli JM, Atkins GB, Geisberg JV, Ricciardi RP. The viral oncoproteins Ad5 E1A, HPV16 E7 and SV40 TAg bind a common region of the TBP-associated factor-110. *Oncogene.* 1995 Nov 2;11(9):1859-64.

[91] Barnard P, McMillan NA. The human papillomavirus E7 oncoprotein abrogates signaling mediated by interferon-alpha. *Virology.* 1999 Jul 5;259(2):305-13.

[92] Rey O, Lee S, Baluda MA, Swee J, Ackerson B, Chiu R, et al. The E7 oncoprotein of human papillomavirus type 16 interacts with F-actin in vitro and in vivo. *Virology.* 2000 Mar 15;268(2):372-81.

[93] Luscher-Firzlaff JM, Westendorf JM, Zwicker J, Burkhardt H, Henriksson M, Muller R, et al. Interaction of the fork head domain transcription factor MPP2 with the human papilloma virus 16 E7 protein: enhancement of transformation and transactivation. *Oncogene.* 1999 Oct 7;18(41):5620-30.

[94] Schilling B, De-Medina T, Syken J, Vidal M, Munger K. A novel human DnaJ protein, hTid-1, a homolog of the Drosophila tumor suppressor protein Tid56, can interact with the human papillomavirus type 16 E7 oncoprotein. *Virology.* 1998 Jul 20;247(1):74-85.

[95] Scheffner M. Ubiquitin, E6-AP, and their role in p53 inactivation. *Pharmacology and therapeutics.* 1998 Jun;78(3):129-39.

[96] Mittnacht S. Control of pRB phosphorylation. *Current opinion in genetics and development.* 1998 Feb;8(1):21-7.

[97] Munger K, Werness BA, Dyson N, Phelps WC, Harlow E, Howley PM. Complex formation of human papillomavirus E7 proteins with the retinoblastoma tumor suppressor gene product. *The EMBO journal.* 1989 Dec 20;8(13):4099-105.

[98] Imai Y, Matsushima Y, Sugimura T, Terada M. Purification and characterization of human papillomavirus type 16 E7 protein with preferential binding capacity to the underphosphorylated form of retinoblastoma gene product. *Journal of virology.* 1991 Sep;65(9):4966-72.

[99] Martin LG, Demers GW, Galloway DA. Disruption of the G1/S transition in human papillomavirus type 16 E7-expressing human cells is associated with altered regulation of cyclin E. *Journal of virology.* 1998 Feb;72(2):975-85.

[100] Schulze A, Mannhardt B, Zerfass-Thome K, Zwerschke W, Jansen-Durr P. Anchorage-independent transcription of the cyclin A gene induced by the E7 oncoprotein of human papillomavirus type 16. *Journal of virology.* 1998 Mar;72(3):2323-34.

[101] Hwang SG, Lee D, Kim J, Seo T, Choe J. Human papillomavirus type 16 E7 binds to E2F1 and activates E2F1-driven transcription in a retinoblastoma protein-independent manner. *The Journal of biological chemistry.* 2002 Jan 25;277(4):2923-30.

[102] Lukas J, Parry D, Aagaard L, Mann DJ, Bartkova J, Strauss M, et al. Retinoblastoma-protein-dependent cell-cycle inhibition by the tumour suppressor p16. *Nature.* 1995 Jun 8;375(6531):503-6.

[103] Mann DJ, Jones NC. E2F-1 but not E2F-4 can overcome p16-induced G1 cell-cycle arrest. *Curr.Biol.* 1996 Apr 1;6(4):474-83.

[104] Giarre M, Caldeira S, Malanchi I, Ciccolini F, Leao MJ, Tommasino M. Induction of pRb degradation by the human papillomavirus type 16 E7 protein is essential to efficiently overcome p16INK4a-imposed G1 cell cycle Arrest. *J.Virol.* 2001 May;75(10):4705-12.

[105] Sano T, Oyama T, Kashiwabara K, Fukuda T, Nakajima T. Expression status of p16 protein is associated with human papillomavirus oncogenic potential in cervical and genital lesions. T*he American journal of pathology.* 1998 Dec;153(6):1741-8.

[106] Menges CW, Baglia LA, Lapoint R, McCance DJ. Human papillomavirus type 16 E7 up-regulates AKT activity through the retinoblastoma protein. *Cancer research.* 2006 Jun 1;66(11):5555-9.

[107] Charette ST, McCance DJ. The E7 protein from human papillomavirus type 16 enhances keratinocyte migration in an Akt-dependent manner. *Oncogene.* 2007 May 28.

[108] Khwaja A, Rodriguez-Viciana P, Wennstrom S, Warne PH, Downward J. Matrix adhesion and Ras transformation both activate a phosphoinositide 3-OH kinase and protein kinase B/Akt cellular survival pathway. *The EMBO journal.* 1997 May 15;16(10):2783-93.

[109] Rangarajan A, Syal R, Selvarajah S, Chakrabarti O, Sarin A, Krishna S. Activated Notch1 signaling cooperates with papillomavirus oncogenes in transformation and generates resistance to apoptosis on matrix withdrawal through PKB/Akt. *Virology.* 2001 Jul 20;286(1):23-30.

[110] DeCaprio JA, Ludlow JW, Figge J, Shew JY, Huang CM, Lee WH, et al. SV40 large tumor antigen forms a specific complex with the product of the retinoblastoma susceptibility gene. *Cell.* 1988 Jul 15;54(2):275-83.

[111] Chen S, Paucha E. Identification of a region of simian virus 40 large T antigen required for cell transformation. *Journal of virology.* 1990 Jul;64(7):3350-7.

[112] Hiebert SW, Chellappan SP, Horowitz JM, Nevins JR. The interaction of RB with E2F coincides with an inhibition of the transcriptional activity of E2F. *Genes and development.* 1992 Feb;6(2):177-85.

[113] Qin XQ, Chittenden T, Livingston DM, Kaelin WG, Jr. Identification of a growth suppression domain within the retinoblastoma gene product. *Genes and development.* 1992 Jun;6(6):953-64.

[114] Lee JO, Russo AA, Pavletich NP. Structure of the retinoblastoma tumour-suppressor pocket domain bound to a peptide from HPV E7. *Nature.* 1998 Feb 26;391(6670):859-65.

[115] Patrick DR, Oliff A, Heimbrook DC. Identification of a novel retinoblastoma gene product binding site on human papillomavirus type 16 E7 protein. *The Journal of biological chemistry.* 1994 Mar 4;269(9):6842-50.

[116] Xiao B, Spencer J, Clements A, Ali-Khan N, Mittnacht S, Broceno C, et al. Crystal structure of the retinoblastoma tumor suppressor protein bound to E2F and the molecular basis of its regulation. *Proceedings of the National Academy of Sciences of the United States of America.* 2003 Mar 4;100(5):2363-8.

[117] Liu X, Clements A, Zhao K, Marmorstein R. Structure of the human Papillomavirus E7 oncoprotein and its mechanism for inactivation of the retinoblastoma tumor suppressor. *The Journal of biological chemistry.* 2006 Jan 6;281(1):578-86.

[118] Jones DL, Munger K. Analysis of the p53-mediated G1 growth arrest pathway in cells expressing the human papillomavirus type 16 E7 oncoprotein. *Journal of virology.* 1997 Apr;71(4):2905-12.

[119] Ciccolini F, Di Pasquale G, Carlotti F, Crawford L, Tommasino M. Functional studies of E7 proteins from different HPV types. *Oncogene.* 1994 Sep;9(9):2633-8.

[120] Schmitt A, Harry JB, Rapp B, Wettstein FO, Iftner T. Comparison of the properties of the E6 and E7 genes of low- and high-risk cutaneous papillomaviruses reveals strongly transforming and high Rb-binding activity for the E7 protein of the low-risk human papillomavirus type 1. *Journal of virology.* 1994 Nov;68(11):7051-9.

[121] Davies R, Hicks R, Crook T, Morris J, Vousden K. Human papillomavirus type 16 E7 associates with a histone H1 kinase and with p107 through sequences necessary for transformation. *Journal of virology*. 1993 May;67(5):2521-8.

[122] Classon M, Dyson N. p107 and p130: versatile proteins with interesting pockets. *Experimental cell research*. 2001 Mar 10;264(1):135-47.

[123] Arroyo M, Bagchi S, Raychaudhuri P. Association of the human papillomavirus type 16 E7 protein with the S-phase-specific E2F-cyclin A complex. *Molecular and cellular biology*. 1993 Oct;13(10):6537-46.

[124] Lam EW, Morris JD, Davies R, Crook T, Watson RJ, Vousden KH. HPV16 E7 oncoprotein deregulates B-myb expression: correlation with targeting of p107/E2F complexes. *The EMBO journal*. 1994 Feb 15;13(4):871-8.

[125] Zerfass K, Levy LM, Cremonesi C, Ciccolini F, Jansen-Durr P, Crawford L, et al. Cell cycle-dependent disruption of E2F-p107 complexes by human papillomavirus type 16 E7. *The Journal of general virology*. 1995 Jul;76 (Pt 7):1815-20.

[126] Grana X, Garriga J, Mayol X. Role of the retinoblastoma protein family, pRB, p107 and p130 in the negative control of cell growth. *Oncogene*. 1998 Dec 24;17(25):3365-83.

[127] Berezutskaya E, Yu B, Morozov A, Raychaudhuri P, Bagchi S. Differential regulation of the pocket domains of the retinoblastoma family proteins by the HPV16 E7 oncoprotein. *Cell Growth Differ*. 1997 Dec;8(12):1277-86.

[128] Firzlaff JM, Luscher B, Eisenman RN. Negative charge at the casein kinase II phosphorylation site is important for transformation but not for Rb protein binding by the E7 protein of human papillomavirus type 16. *Proceedings of the National Academy of Sciences of the United States of America*. 1991 Jun 15;88(12):5187-91.

[129] Banks L, Edmonds C, Vousden KH. Ability of the HPV16 E7 protein to bind RB and induce DNA synthesis is not sufficient for efficient transforming activity in NIH3T3 cells. *Oncogene*. 1990 Sep;5(9):1383-9.

[130] Phelps WC, Munger K, Yee CL, Barnes JA, Howley PM. Structure-function analysis of the human papillomavirus type 16 E7 oncoprotein. *Journal of virology*. 1992 Apr;66(4):2418-27.

[131] Tommasino M, Adamczewski JP, Carlotti F, Barth CF, Manetti R, Contorni M, et al. HPV16 E7 protein associates with the protein kinase p33CDK2 and cyclin A. *Oncogene*. 1993 Jan;8(1):195-202.

[132] McIntyre MC, Ruesch MN, Laimins LA. Human papillomavirus E7 oncoproteins bind a single form of cyclin E in a complex with cdk2 and p107. *Virology*. 1996 Jan 1;215(1):73-82.

[133] He W, Staples D, Smith C, Fisher C. Direct activation of cyclin-dependent kinase 2 by human papillomavirus E7. *Journal of virology*. 2003 Oct;77(19):10566-74.

[134] Ruesch MN, Laimins LA. Initiation of DNA synthesis by human papillomavirus E7 oncoproteins is resistant to p21-mediated inhibition of cyclin E-cdk2 activity. *Journal of virology*. 1997 Jul;71(7):5570-8.

[135] Jian Y, Schmidt-Grimminger DC, Chien WM, Wu X, Broker TR, Chow LT. Post-transcriptional induction of p21cip1 protein by human papillomavirus E7 inhibits

unscheduled DNA synthesis reactivated in differentiated keratinocytes. *Oncogene.* 1998 Oct 22;17(16):2027-38.

[136] Noya F, Chien WM, Broker TR, Chow LT. p21cip1 Degradation in differentiated keratinocytes is abrogated by costabilization with cyclin E induced by human papillomavirus E7. *Journal of virology.* 2001 Jul;75(13):6121-34.

[137] Helt AM, Funk JO, Galloway DA. Inactivation of both the retinoblastoma tumor suppressor and p21 by the human papillomavirus type 16 E7 oncoprotein is necessary to inhibit cell cycle arrest in human epithelial cells. *Journal of virology.* 2002 Oct;76(20):10559-68.

[138] Besson A, Gurian-West M, Schmidt A, Hall A, Roberts JM. p27Kip1 modulates cell migration through the regulation of RhoA activation. *Genes and development.* 2004 Apr 15;18(8):862-76.

[139] Li R, Wheeler TM, Dai H, Sayeeduddin M, Scardino PT, Frolov A, et al. Biological correlates of p27 compartmental expression in prostate cancer. *The Journal of urology.* 2006 Feb;175(2):528-32.

[140] McAllister SS, Becker-Hapak M, Pintucci G, Pagano M, Dowdy SF. Novel p27(kip1) C-terminal scatter domain mediates Rac-dependent cell migration independent of cell cycle arrest functions. *Molecular and cellular biology.* 2003 Jan;23(1):216-28.

[141] Boutros R, Dozier C, Ducommun B. The when and wheres of CDC25 phosphatases. *Current opinion in cell biology.* 2006 Apr;18(2):185-91.

[142] Katich SC, Zerfass-Thome K, Hoffmann I. Regulation of the Cdc25A gene by the human papillomavirus Type 16 E7 oncogene. *Oncogene.* 2001 Feb 1;20(5):543-50.

[143] Brehm A, Miska EA, McCance DJ, Reid JL, Bannister AJ, Kouzarides T. Retinoblastoma protein recruits histone deacetylase to repress transcription. *Nature.* 1998 Feb 5;391(6667):597-601.

[144] Luo RX, Postigo AA, Dean DC. Rb interacts with histone deacetylase to repress transcription. *Cell.* 1998 Feb 20;92(4):463-73.

[145] Harbour JW, Dean DC. The Rb/E2F pathway: expanding roles and emerging paradigms. *Genes and development.* 2000 Oct 1;14(19):2393-409.

[146] Dick FA, Sailhamer E, Dyson NJ. Mutagenesis of the pRB pocket reveals that cell cycle arrest functions are separable from binding to viral oncoproteins. *Molecular and cellular biology.* 2000 May;20(10):3715-27.

[147] Li H, Ou X, Xiong J, Wang T. HPV16E7 mediates HADC chromatin repression and downregulation of MHC class I genes in HPV16 tumorigenic cells through interaction with an MHC class I promoter. *Biochemical and biophysical research communications.* 2006 Nov 3;349(4):1315-21.

[148] Huh KW, DeMasi J, Ogawa H, Nakatani Y, Howley PM, Munger K. Association of the human papillomavirus type 16 E7 oncoprotein with the 600-kDa retinoblastoma protein-associated factor, p600. *Proceedings of the National Academy of Sciences of the United States of America.* 2005 Aug 9;102(32):11492-7.

[149] Maldonado E, Cabrejos ME, Banks L, Allende JE. Human papillomavirus-16 E7 protein inhibits the DNA interaction of the TATA binding transcription factor. *Journal of cellular biochemistry.* 2002;85(4):663-9.

[150] Bernat A, Avvakumov N, Mymryk JS, Banks L. Interaction between the HPV E7 oncoprotein and the transcriptional coactivator p300. *Oncogene.* 2003 Sep 11;22(39):7871-81.

[151] Huang SM, McCance DJ. Down regulation of the interleukin-8 promoter by human papillomavirus type 16 E6 and E7 through effects on CREB binding protein/p300 and P/CAF. *Journal of virology.* 2002 Sep;76(17):8710-21.

[152] Turnell AS, Mymryk JS. Roles for the coactivators CBP and p300 and the APC/C E3 ubiquitin ligase in E1A-dependent cell transformation. *British journal of cancer.* 2006 Sep 4;95(5):555-60.

[153] Patel D, Huang SM, Baglia LA, McCance DJ. The E6 protein of human papillomavirus type 16 binds to and inhibits co-activation by CBP and p300. *The EMBO journal.* 1999 Sep 15;18(18):5061-72.

[154] Bernat A, Massimi P, Banks L. Complementation of a p300/CBP defective-binding mutant of adenovirus E1a by human papillomavirus E6 proteins. *The Journal of general virology.* 2002 Apr;83(Pt 4):829-33.

[155] Vousden KH, Vojtesek B, Fisher C, Lane D. HPV-16 E7 or adenovirus E1A can overcome the growth arrest of cells immortalized with a temperature-sensitive p53. *Oncogene.* 1993 Jun;8(6):1697-702.

[156] Demers GW, Foster SA, Halbert CL, Galloway DA. Growth arrest by induction of p53 in DNA damaged keratinocytes is bypassed by human papillomavirus 16 E7. Proceedings of the National Academy of Sciences of the United States of America. 1994 May 10;91(10):4382-6.

[157] Demers GW, Halbert CL, Galloway DA. Elevated wild-type p53 protein levels in human epithelial cell lines immortalized by the human papillomavirus type 16 E7 gene. *Virology.* 1994 Jan;198(1):169-74.

[158] Seavey SE, Holubar M, Saucedo LJ, Perry ME. The E7 oncoprotein of human papillomavirus type 16 stabilizes p53 through a mechanism independent of p19(ARF). *Journal of virology.* 1999 Sep;73(9):7590-8.

[159] Pan H, Griep AE. Altered cell cycle regulation in the lens of HPV-16 E6 or E7 transgenic mice: implications for tumor suppressor gene function in development. *Genes and development.* 1994 Jun 1;8(11):1285-99.

[160] Thomas M, Banks L. Inhibition of Bak-induced apoptosis by HPV-18 E6. *Oncogene.* 1998 Dec 10;17(23):2943-54.

[161] Massimi P, Banks L. Repression of p53 transcriptional activity by the HPV E7 proteins. *Virology.* 1997 Jan 6;227(1):255-9.

[162] Stoppler H, Stoppler MC, Johnson E, Simbulan-Rosenthal CM, Smulson ME, Iyer S, et al. The E7 protein of human papillomavirus type 16 sensitizes primary human keratinocytes to apoptosis. *Oncogene.* 1998 Sep 10;17(10):1207-14.

[163] Thyrell L, Sangfelt O, Zhivotovsky B, Pokrovskaja K, Wang Y, Einhorn S, et al. The HPV-16 E7 oncogene sensitizes malignant cells to IFN-alpha-induced apoptosis. *J Interferon Cytokine Res.* 2005 Feb;25(2):63-72.

[164] Yuan H, Fu F, Zhuo J, Wang W, Nishitani J, An DS, et al. Human papillomavirus type 16 E6 and E7 oncoproteins upregulate c-IAP2 gene expression and confer resistance to apoptosis. *Oncogene.* 2005 Jul 28;24(32):5069-78.

[165] Thompson DA, Zacny V, Belinsky GS, Classon M, Jones DL, Schlegel R, et al. The HPV E7 oncoprotein inhibits tumor necrosis factor alpha-mediated apoptosis in normal human fibroblasts. *Oncogene.* 2001 Jun 21;20(28):3629-40.
[166] Duensing S, Lee LY, Duensing A, Basile J, Piboonniyom S, Gonzalez S, et al. The human papillomavirus type 16 E6 and E7 oncoproteins cooperate to induce mitotic defects and genomic instability by uncoupling centrosome duplication from the cell division cycle. *Proceedings of the National Academy of Sciences of the United States of America.* 2000 Aug 29;97(18):10002-7.
[167] Hashida T, Yasumoto S. Induction of chromosome abnormalities in mouse and human epidermal keratinocytes by the human papillomavirus type 16 E7 oncogene. *The Journal of general virology.* 1991 Jul;72 (Pt 7):1569-77.
[168] Duensing S, Munger K. Human papillomavirus type 16 E7 oncoprotein can induce abnormal centrosome duplication through a mechanism independent of inactivation of retinoblastoma protein family members. *Journal of virology.* 2003 Nov;77(22):12331-5.
[169] Mantel C, Braun SE, Reid S, Henegariu O, Liu L, Hangoc G, et al. p21(cip-1/waf-1) deficiency causes deformed nuclear architecture, centriole overduplication, polyploidy, and relaxed microtubule damage checkpoints in human hematopoietic cells. *Blood.* 1999 Feb 15;93(4):1390-8.
[170] Lacey KR, Jackson PK, Stearns T. Cyclin-dependent kinase control of centrosome duplication. *Proceedings of the National Academy of Sciences of the United States of America.* 1999 Mar 16;96(6):2817-22.
[171] Duensing A, Liu Y, Perdreau SA, Kleylein-Sohn J, Nigg EA, Duensing S. Centriole overduplication through the concurrent formation of multiple daughter centrioles at single maternal templates. *Oncogene.* 2007 Apr 16.
[172] Nguyen CL, Munger K. The Human Papillomavirus Type 16 E7 Oncoprotein Associates with the Centrosomal Component {gamma}-Tubulin. *Journal of virology.* 2007 Oct 3.
[173] Duensing S, Munger K. The human papillomavirus type 16 E6 and E7 oncoproteins independently induce numerical and structural chromosome instability. *Cancer research.* 2002 Dec 1;62(23):7075-82.
[174] Koopman LA, Szuhai K, van Eendenburg JD, Bezrookove V, Kenter GG, Schuuring E, et al. Recurrent integration of human papillomaviruses 16, 45, and 67 near translocation breakpoints in new cervical cancer cell lines. *Cancer research.* 1999 Nov 1;59(21):5615-24.
[175] Song S, Gulliver GA, Lambert PF. Human papillomavirus type 16 E6 and E7 oncogenes abrogate radiation-induced DNA damage responses in vivo through p53-dependent and p53-independent pathways. *Proceedings of the National Academy of Sciences of the United States of America.* 1998 Mar 3;95(5):2290-5.
[176] Hasan UA, Caux C, Perrot I, Doffin AC, Menetrier-Caux C, Trinchieri G, et al. Cell proliferation and survival induced by Toll-like receptors is antagonized by type I IFNs. *Proceedings of the National Academy of Sciences of the United States of America.* 2007 May 8;104(19):8047-52.

[177] Tanaka N, Ishihara M, Lamphier MS, Nozawa H, Matsuyama T, Mak TW, et al. Cooperation of the tumour suppressors IRF-1 and p53 in response to DNA damage. *Nature.* 1996 Aug 29;382(6594):816-8.

[178] Georgopoulos NT, Proffitt JL, Blair GE. Transcriptional regulation of the major histocompatibility complex (MHC) class I heavy chain, TAP1 and LMP2 genes by the human papillomavirus (HPV) type 6b, 16 and 18 E7 oncoproteins. *Oncogene.* 2000 Oct 5;19(42):4930-5.

[179] Cromme FV, Meijer CJ, Snijders PJ, Uyterlinde A, Kenemans P, Helmerhorst T, et al. Analysis of MHC class I and II expression in relation to presence of HPV genotypes in premalignant and malignant cervical lesions. *British journal of cancer.* 1993 Jun;67(6):1372-80.

[180] Cromme FV, Airey J, Heemels MT, Ploegh HL, Keating PJ, Stern PL, et al. Loss of transporter protein, encoded by the TAP-1 gene, is highly correlated with loss of HLA expression in cervical carcinomas. *The Journal of experimental medicine.* 1994 Jan 1;179(1):335-40.

[181] Bottley G, Watherston OG, Hiew YL, Norrild B, Cook GP, Blair GE. High-risk human papillomavirus E7 expression reduces cell-surface MHC class I molecules and increases susceptibility to natural killer cells. *Oncogene.* 2007 Sep 10.

[182] Frazer IH, Thomas R, Zhou J, Leggatt GR, Dunn L, McMillan N, et al. Potential strategies utilised by papillomavirus to evade host immunity. *Immunological reviews.* 1999 Apr;168:131-42.

[183] zur Hausen H. Papillomaviruses in human cancers. *Proceedings of the Association of American Physicians.* 1999 Nov-Dec;111(6):581-7.

[184] Wright TC, Jr. Cervical cancer screening in the 21st century: is it time to retire the PAP smear? *Clinical obstetrics and gynecology.* 2007 Jun;50(2):313-23.

[185] Harper DM, Franco EL, Wheeler CM, Moscicki AB, Romanowski B, Roteli-Martins CM, et al. Sustained efficacy up to 4.5 years of a bivalent L1 virus-like particle vaccine against human papillomavirus types 16 and 18: follow-up from a randomised control trial. *Lancet.* 2006 Apr 15;367(9518):1247-55.

[186] Koutsky LA, Ault KA, Wheeler CM, Brown DR, Barr E, Alvarez FB, et al. A controlled trial of a human papillomavirus type 16 vaccine. *The New England journal of medicine.* 2002 Nov 21;347(21):1645-51.

[187] Moore DH. Chemotherapy for recurrent cervical carcinoma. *Current opinion in oncology.* 2006 Sep;18(5):516-9.

[188] Gammoh N, Grm HS, Massimi P, Banks L. Regulation of human papillomavirus type 16 E7 activity through direct protein interaction with the E2 transcriptional activator. *Journal of virology.* 2006 Feb;80(4):1787-97.

[189] Goodwin EC, Yang E, Lee CJ, Lee HW, DiMaio D, Hwang ES. Rapid induction of senescence in human cervical carcinoma cells. *Proceedings of the National Academy of Sciences of the United States of America.* 2000 Sep 26;97(20):10978-83.

[190] Lee CJ, Suh EJ, Kang HT, Im JS, Um SJ, Park JS, et al. Induction of senescence-like state and suppression of telomerase activity through inhibition of HPV E6/E7 gene expression in cells immortalized by HPV16 DNA. *Experimental cell research.* 2002 Jul 15;277(2):173-82.

[191] Desaintes C, Demeret C, Goyat S, Yaniv M, Thierry F. Expression of the papillomavirus E2 protein in HeLa cells leads to apoptosis. *The EMBO journal.* 1997 Feb 3;16(3):504-14.

[192] DeFilippis RA, Goodwin EC, Wu L, DiMaio D. Endogenous human papillomavirus E6 and E7 proteins differentially regulate proliferation, senescence, and apoptosis in HeLa cervical carcinoma cells. *Journal of virology.* 2003 Jan;77(2):1551-63.

[193] Wu S, Meng L, Wang S, Wang W, Xi L, Tian X, et al. Reversal of the malignant phenotype of cervical cancer CaSki cells through adeno-associated virus-mediated delivery of HPV16 E7 antisense RNA. *Clin. Cancer Res.* 2006 Apr 1;12(7 Pt 1):2032-7.

[194] Hamada K, Shirakawa T, Gotoh A, Roth JA, Follen M. Adenovirus-mediated transfer of human papillomavirus 16 E6/E7 antisense RNA and induction of apoptosis in cervical cancer. *Gynecologic oncology.* 2006 Dec;103(3):820-30.

[195] Kari I, Syrjanen S, Johansson B, Peri P, He B, Roizman B, et al. Antisense RNA directed to the human papillomavirus type 16 E7 mRNA from herpes simplex virus type 1 derived vectors is expressed in CaSki cells and downregulates E7 mRNA. *Virology journal.* 2007;4:47.

[196] Choo CK, Ling MT, Suen CK, Chan KW, Kwong YL. Retrovirus-mediated delivery of HPV16 E7 antisense RNA inhibited tumorigenicity of CaSki cells. *Gynecologic oncology.* 2000 Sep;78(3 Pt 1):293-301.

[197] Sima N, Wang S, Wang W, Kong D, Xu Q, Tian X, et al. Antisense targeting human papillomavirus type 16 E6 and E7 genes contributes to apoptosis and senescence in SiHa cervical carcinoma cells. *Gynecologic oncology.* 2007 Aug;106(2):299-304.

[198] Jiang M, Milner J. Selective silencing of viral gene expression in HPV-positive human cervical carcinoma cells treated with siRNA, a primer of RNA interference. *Oncogene.* 2002 Sep 5;21(39):6041-8.

[199] Hall AH, Alexander KA. RNA interference of human papillomavirus type 18 E6 and E7 induces senescence in HeLa cells. *Journal of virology.* 2003 May;77(10):6066-9.

[200] Fujii T, Saito M, Iwasaki E, Ochiya T, Takei Y, Hayashi S, et al. Intratumor injection of small interfering RNA-targeting human papillomavirus 18 E6 and E7 successfully inhibits the growth of cervical cancer. *International journal of oncology.* 2006 Sep;29(3):541-8.

[201] Tang S, Tao M, McCoy JP, Jr., Zheng ZM. Short-term induction and long-term suppression of HPV16 oncogene silencing by RNA interference in cervical cancer cells. *Oncogene.* 2006 Mar 30;25(14):2094-104.

[202] Gu W, Putral L, Hengst K, Minto K, Saunders NA, Leggatt G, et al. Inhibition of cervical cancer cell growth in vitro and in vivo with lentiviral-vector delivered short hairpin *RNA targeting human papillomavirus E6 and E7 oncogenes.* Cancer gene therapy. 2006 Nov;13(11):1023-32.

[203] Lu D, Chatterjee S, Brar D, Wong KK, Jr. Ribozyme-mediated in vitro cleavage of transcripts arising from thc major transforming genes of human papillomavirus type 16. *Cancer gene therapy.* 1994 Dec;1(4):267-77.

[204] Chen Z, Kamath P, Zhang S, Weil MM, Shillitoe EJ. Effectiveness of three ribozymes for cleavage of an RNA transcript from human papillomavirus type 18. *Cancer gene therapy.* 1995 Dec;2(4):263-71.

[205] Alvarez-Salas LM, Cullinan AE, Siwkowski A, Hampel A, DiPaolo JA. Inhibition of HPV-16 E6/E7 immortalization of normal keratinocytes by hairpin ribozymes. *Proceedings of the National Academy of Sciences of the United States of America.* 1998 Feb 3;95(3):1189-94.

[206] Zheng Y, Zhang J, Rao Z. Ribozyme targeting HPV16 E6E7 transcripts in cervical cancer cells suppresses cell growth and sensitizes cells to chemotherapy and radiotherapy. *Cancer biology and therapy.* 2004 Nov;3(11):1129-34; discussion 35-6.

[207] Kunke D, Grimm D, Denger S, Kreuzer J, Delius H, Komitowski D, et al. Preclinical study on gene therapy of cervical carcinoma using adeno-associated virus vectors. *Cancer gene therapy*. 2000 May;7(5):766-77.

[208] Nauenburg S, Zwerschke W, Jansen-Durr P. Induction of apoptosis in cervical carcinoma cells by peptide aptamers that bind to the HPV-16 E7 oncoprotein. *Faseb. J.* 2001 Mar;15(3):592-4.

[209] Accardi L, Dona MG, Di Bonito P, Giorgi C. Intracellular anti-E7 human antibodies in single-chain format inhibit proliferation of HPV16-positive cervical carcinoma cells. *International journal of cancer.* 2005 Sep 10;116(4):564-70.

[210] Griffin H, Elston R, Jackson D, Ansell K, Coleman M, Winter G, et al. Inhibition of papillomavirus protein function in cervical cancer cells by intrabody targeting. *Journal of molecular biology.* 2006 Jan 20;355(3):360-78.

[211] Karl T, Seibert N, Stohr M, Osswald H, Rosl F, Finzer P. Sulindac induces specific degradation of the HPV oncoprotein E7 and causes growth arrest and apoptosis in cervical carcinoma cells. *Cancer letters.* 2007 Jan 8;245(1-2):103-11.

[212] Andrei G, Snoeck R, Schols D, De Clercq E. Induction of apoptosis by cidofovir in human papillomavirus (HPV)-positive cells. *Oncology research.* 2000;12(9-10):397-408.

[213] Hostetler KY, Rought S, Aldern KA, Trahan J, Beadle JR, Corbeil J. Enhanced antiproliferative effects of alkoxyalkyl esters of cidofovir in human cervical cancer cells in vitro. *Molecular cancer therapeutics.* 2006 Jan;5(1):156-9.

[214] Divya CS, Pillai MR. Antitumor action of curcumin in human papillomavirus associated cells involves downregulation of viral oncogenes, prevention of NFkB and AP-1 translocation, and modulation of apoptosis. *Molecular carcinogenesis.* 2006 May;45(5):320-32.

[215] Yim EK, Lee MJ, Lee KH, Um SJ, Park JS. Antiproliferative and antiviral mechanisms of ursolic acid and dexamethasone in cervical carcinoma cell lines. *Int. J. Gynecol. Cancer.* 2006 Nov-Dec;16(6):2023-31.

[216] Beerheide W, Bernard HU, Tan YJ, Ganesan A, Rice WG, Ting AE. Potential drugs against cervical cancer: zinc-ejecting inhibitors of the human papillomavirus type 16 E6 oncoprotein. *Journal of the National Cancer Institute.* 1999 Jul 21;91(14):1211-20.

[217] Beerheide W, Sim MM, Tan YJ, Bernard HU, Ting AE. Inactivation of the human papillomavirus-16 E6 oncoprotein by organic disulfides. Bioorganic and medicinal chemistry. 2000 Nov;8(11):2549-60.

[218] Rogers JP, Beuscher AEt, Flajolet M, McAvoy T, Nairn AC, Olson AJ, et al. Discovery of protein phosphatase 2C inhibitors by virtual screening. *Journal of medicinal chemistry.* 2006 Mar 9;49(5):1658-67.

[219] Lu Y, Nikolovska-Coleska Z, Fang X, Gao W, Shangary S, Qiu S, et al. Discovery of a nanomolar inhibitor of the human murine double minute 2 (M*DM2)-p53 interaction through an integrated, virtual database screening strategy. Journal of medicinal chemistry.* 2006 Jun 29;49(13):3759-62.

[220] Nassar N, Cancelas J, Zheng J, Williams DA, Zheng Y. Structure-function based design of small molecule inhibitors targeting Rho family GTPases. *Current topics in medicinal chemistry.* 2006;6(11):1109-16.

[221] Trosset JY, Dalvit C, Knapp S, Fasolini M, Veronesi M, Mantegani S, et al. Inhibition of protein-protein interactions: the discovery of druglike beta-catenin inhibitors by combining virtual and biophysical screening. *Proteins.* 2006 Jul 1;64(1):60-7.

[222] Song H, Wang R, Wang S, Lin J. A low-molecular-weight compound discovered through virtual database screening inhibits Stat3 function in breast cancer cells. Proceedings of the National Academy of Sciences of the United States of America. 2005 Mar 29;102(13):4700-5.

[223] Cavasotto CN, Ortiz MA, Abagyan RA, Piedrafita FJ. In silico identification of novel EGFR inhibitors with antiproliferative activity against cancer cells. Bioorganic and medicinal chemistry letters. 2006 Apr 1;16(7):1969-74.

In: Oncogene Proteins: New Research
Editors: A. H. Malloy and E.C. Carson, pp.65-98 ISBN 978-1-60456-290-3

Chapter 2

Effect of Dietary Phytochemicals on Src, EGFR and HER2 in Cancer Cells

Elena P. Moiseeva[3]
Cancer Biomarkers and Prevention Group, Department of Cancer Studies,
University of Leicester, Leicester LE1 7RH, UK

Abstract

Cancer cells are often physiologically dependent on activation and/or increased expression of some oncogenes, which are essential for growth and maintenance of the malignant phenotype. EGFR and HER2 are among those "oncogenes of addiction" in many solid cancers, including breast cancer. These oncogenes often cooperate with another oncogene, Src, which is also overexpressed or constitutively activated in many solid cancers, such as breast cancer. Many studies show a connection between diet and cancer, with cancer incidence inversely correlated with consumption of fruit and vegetables. Moreover, occurrence of breast cancers with particular genotypes of the EGFR gene is related to a diet poor in vegetables. Till recently, the molecular basis of a connection between cancer and diet has been obscure, however, recent studies have shown that dietary compounds found in some vegetables affect oncogenes in cancer cells and, thereby, reduce tumor growth. Dietary phytochemicals, such as indole-3-carbinol, resveratrol, genistein, curcumin and epigallocatechin gallate, can reduce expression and/or activation of EGFR or HER2. This review summarizes known inhibitory effects of these phytochemicals on the activation and expression of EGFR, HER2 and Src, eventually resulting in downregulated signaling from these oncogenes and reduced cancer cell viability in breast, colon and other solid cancers.

3 Address for correspondence: Elena P. Moiseeva, Cancer Biomarkers and Prevention Group, Biocentre, University of Leicester, Leicester LE1 7RH, UK. em9@le.ac.uk, Tel: 44(0)116 223 1825, Fax: 44(0)116 223 1840.

Abbreviations:

BMP, bone morphogenetic protein-2;
CAS, Crk-associated substrate;
CDK, cyclin-dependent kinase;
COX II, cytochrome C oxidase subunit II;
DIM, diindolylmethane, EGCG, epigallocatechin-3-gallate;
EGF, epidermal growth factor,
EGFR, EGF receptor;
ER, estrogen receptor;
FAK, focal adhesion kinase;
ERK, extracellular signal-regulated kinase;
FOX, forkhead box transcription factor;
HER, human epidermal growth factor receptor;
I3C, indole-3-carbinol; NF-kappaB, nuclear transcription factor kappaB;
MAPK, mitogen-activated protein kinase;
MMP, matrix metalloproteinase;
PI3K, phosphatidylinositol 3-kinase;
STAT, signal transducer and activator of transcription;
TGF, tumor growth factor,
VEGF, vascular endothelial growth factor.

Introduction

Multiple studies show that that diet may be responsible for 20-60% of cancers and 35% of cancer-caused mortality [1]. In the previous decade, two international committees concluded that there is a convincing body of evidence that the consumption of foods rich in fruit and vegetables decreases the risk of cancers of the oropharynx, esophagus, stomach, colon, rectum and lung; and is also likely to decrease the risk of breast, ovary, endometrium, cervix, prostate, pancreas, liver, thyroid, kidney and bladder cancers [2, 3]. More recent EPIC (European Prospective Investigation into Cancer and nutrition) studies confirmed only associations between diet and lung and colorectal, but not breast or prostate, cancers [4].

There are several reasons for these discrepancies. First of all, diet patterns vary dramatically across Europe: the Mediterranean diet differs greatly from the diet in Northern Europe and the nutritional composition of fruit and vegetables consumed across Europe is variable. It is likely that only some specific foods may have a cancer preventive effect; some vegetables, e.g. potatoes, may have quite the opposite effect, particularly when they are consumed fried; hence combining all vegetables and fruit could remove this protective effect. Furthermore, vegetables of several botanical subgroups, e.g. tomatoes, cruciferous and allium vegetables, consistently show inverse correlation with the risk of several cancers [5, 6]. Secondly, the amount of consumed fruit and vegetables in the diet may have a threshold to achieve a preventive effect. For example, the protective effect of high vegetable intake (375 g) was found in a recent study on breast cancer in Germany [7]. Similarly, it was found that four servings of cabbage per week were required to decrease the risk of breast cancer [8]. Another important factor to consider is variability of cancers, e.g. steroid hormone-responsive breast and prostate cancers may respond to the presence of dietary phyto-estrogens (such as genistein or resveratrol) and phytochemicals, which modulate enzymes involved in steroid hormone metabolism (such as organosulphur compounds, indoles and terpenes) [6]. Moreover, specific genotypes may be associated with different risks to develop breast cancer depending on diet, e.g. the occurrence of breast cancers with specific EGFR genotypes is dependent on vegetable consumption [7]. Similarly, genetic polymorphisms influence the preventive effect of green tea on breast and lung cancer [9].

Specific phytochemicals present in foods are considered to be responsible for cancer prevention and delayed cancer development. Aggarwal and Shishodia recently reviewed a variety of preventive phytochemicals, such as curcumin (turmeric), resveratrol (red grapes, peanuts and berries), genistein (soybean), diallyl sulfide (allium), S-allyl cysteine (allium), allicin (garlic), lycopene (tomato), capsaicin (red chilli), diosgenin (fenugreek), 6-gingerol (ginger), ellagic acid (pomegranate), ursolic acid (apple, pears, prunes), silymarin (milk thistle), anethol (anise, camphor, and fennel), epigallocatechin-3-gallate and other catechins (green tea), eugenol (cloves), indole-3-carbinol (cruciferous vegetables), limonene (citrus fruits), beta carotene (carrots), and others [10]. Some of them, namely I3C and resveratrol (Figure 1), are present in the Western diet, whereas others, e.g. genistein, curcumin or epigallocatechin-3-gallate (Figure 1), are typically present in the Asian diet. Hopefully, products containing the latter will get a wider appreciation in the future. These dietary agents affect multiple molecular targets and signaling pathways [10]. The focus of this review is on the ability of these phytochemicals to modulate activities of Src and the EGFR family of oncogenes, which results in reduced cell growth and death of cancer cells.

EGFR, HER2, SRC are "Genes of Addiction" in Some Solid Cancers

Tumorigenesis is a multi-step process of progressive acquisition of genetic aberrations that initiate tumor development and maintain established tumors. A growing body of evidence from *in vitro* studies, animal models and clinical data indicate that only a few oncogenes may be sufficient for maintenance of the malignant phenotype of some tumors [11]. Cell growth and viability of such tumors are dependent on the expression of a mutated oncogene or overexpression of an oncogene. Cells become "addicted" to a particular oncogene according to the concept of "gene addiction", suggested by R. Weinstein [11]. Cell signaling in addicted cancer cells is distorted in such a way that downregulation of a signaling pathway driven by the oncogene of addiction results in apoptosis and cell cycle arrest in cancer cells, but not in normal cells. Epidermal growth factor receptor (EGFR), human epidermal growth factor receptor 2 (HER2, also called ErbB2, and Neu in the case of the mouse gene) and Src are such "oncogenes of addiction" in some cancers. They provide efficacious targets for clinical intervention using specific antibodies and inhibitors [12-14].

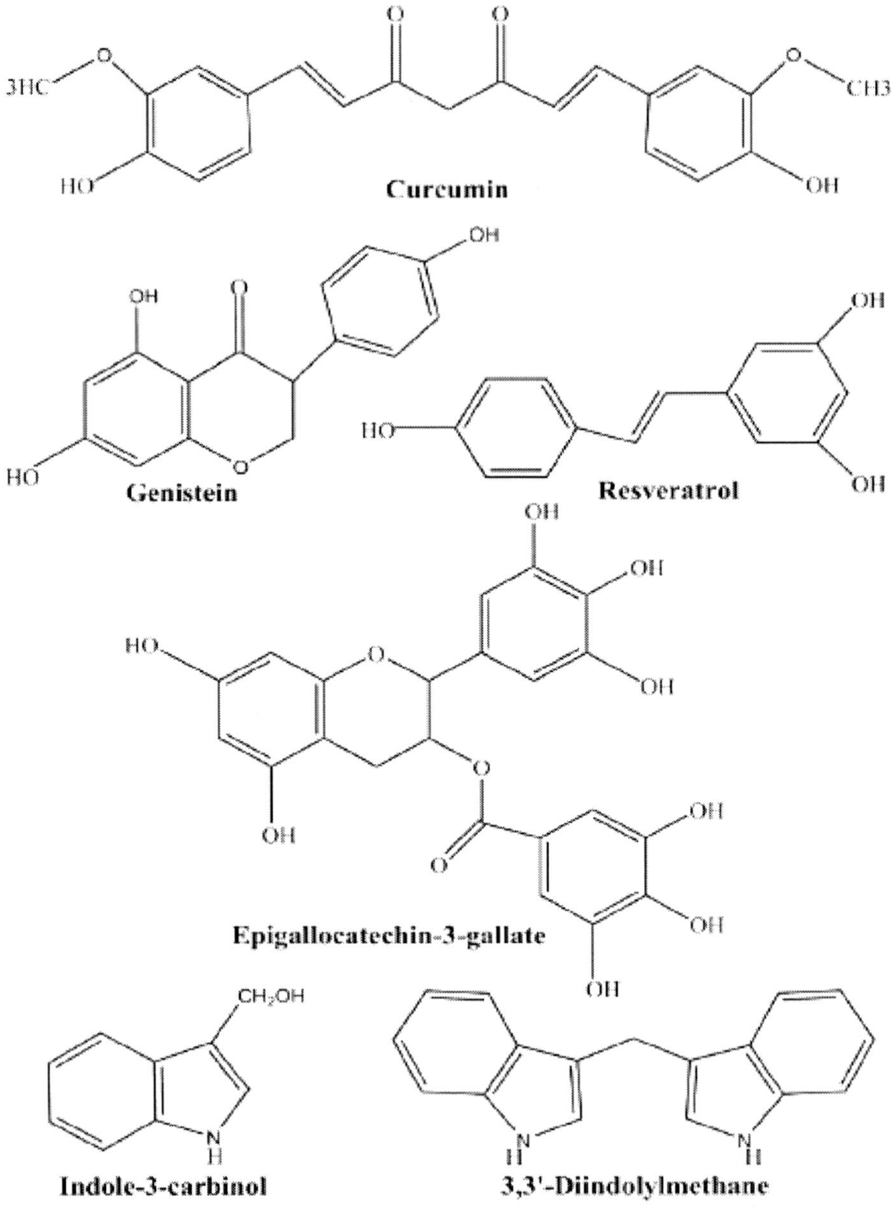

Figure 1. The chemical structures of phytochemicals.

EGFR and HER2 are members of the EGFR family of receptor tyrosine kinases; the structural, functional, clinical and other aspects of these proteins have been covered by a number of comprehensive reviews [14-17]. Cancer cells express various amounts of EGFR family members, which form homo- and hetero-dimers, with HER2 found mostly in heterodimers. EGFR, HER3 and HER4 are regulated by several ligands, whereas HER2 is not. The other characteristic feature of HER2 is that it is poised for an active conformation; therefore, it is likely to be the domineering partner in a heterodimer. Dimerization of the receptors leads to trans-phosphorylation of specific tyrosines, which bind to specific signaling proteins. The most recent data show that they have common and specific binding partners [18] Trafficking of the activated receptors seems to be regulated in a specific fashion for EGFR homodimers. Hence, a combination of specific ligand binding, preferred dimerization, common and unique signaling partners, and trafficking, makes an extremely complicated network of signaling pathways; as was recently shown by Oda *et al* [19]. However, this review will be focused on signaling of EGFR and HER2 and expression in cancer with particular attention to breast and colon tumors. Src will be reviewed mostly as a signaling partner and activator for EGFR and HER2.

EGFR and HER2 are Prognostic Factors in Cancer

EGFR and HER2 are overexpressed in many solid tumors [20], (see also data on EGFR in [16, 21]). Abnormal HER2 serum levels were found in breast, lung, prostate and ovarian carcinomas [22]. Expression of the EGFR family members in breast and colon cancer in summarized in Table 1. Expression levels of these oncogenes in breast and colon cancer cell lines are shown in Figure 2.

In breast cancer, EGFR expression is associated with tumor recurrence and poor prognosis [23-26]. Its overexpression is associated with the length and amplification of the CA repeat in the first intron of the EGFR gene][7, 27]. The CA repeat polymorphism is prognostic in breast and colon cancer [7, 28]. Similar to EGFR, overexpression of the full-length HER2p185 or its extracellular domain HER2p110 is associated with metastatic disease, shorter survival and bad prognosis in breast cancer [29-34]. In some studies only high expression of intracellular domain HER2p95 or phospho-HER2, but not full-length HER2p185, was independent predictors of poor disease-free survival [29, 33, 35-37]. HER2 overexpression is associated with the gene amplification [38] and prevalent cytoplasmic HER2 staining in breast cancer [32]. HER2 gene amplification occurs at the early stages of the disease, in a younger population of patients and is important for progression to invasive cancer, suggesting a role for HER2 at the initiating and progression stages of carcinogenesis [17]. Some studies showed that co-expression of HER2 and EGFR in a subpopulation of HER2-expressing patients resulted in activation of EGFR and the presence of phosphorylated EGFR in the tumor samples, which was associated with the worst prognosis [37, 39-43].

Analysis of gene expression profiles revealed five distinct breast cancer subtypes with the HER2 and basal-like subtypes associated with the poor prognosis [57]. The HER2 subtype was characterized as ER-alpha negative with increased expression of the HER2 amplicon [57], whereas the basal-like subtype was characterized as negative for ER-alpha and HER2, and associated with increased EGFR expression [58-60] and stem cell-like properties [61, 62]. Thus, data obtained by analyses of gene expression profiles and tissue immunostaining are not in complete agreement, but all implicate both EGFR and HER2 in unfavorable prognosis for breast cancer.

Table 1. Expression of the EGFR family members in breast and colon cancer.

Tissue	EGFR	HER2	serum HERp110 fragment	HER3/HER4
Breast	15-41% of total cases, higher in metastases (23, 24, 39-41, 44).	20-42% of total cases (20, 22, 33, 34, 36, 39-41, 45); 42-50% in metastatic disease (22, 31); 22% HER2p95 and 14.3% both (45).	15-19% of total cases (46-48); increased with stage to 35-40% in metastatic disease (32, 48).	26% and 82% of total cases for HER3 and HER4, respectively (39).
Colon	47-71% % of total cases (49-51).	6-83% of total cases (20, 22, 49, 50, 52-54); 12-88% in metastatic disease (22, 55, 56); cytoplasmic HER2 was 3 fold frequent than membranous (49)	Present (22)	89% of total cases for HER3, in 77% HER3 was co-expressed with HER2 (52); 50% of cases for HER3 in Duke's stage A-B, 15% of cases fir HER3 in Dukes' stage C-D, 18% in early stage for HER4 versus 28% for HER4 in late stage (50)

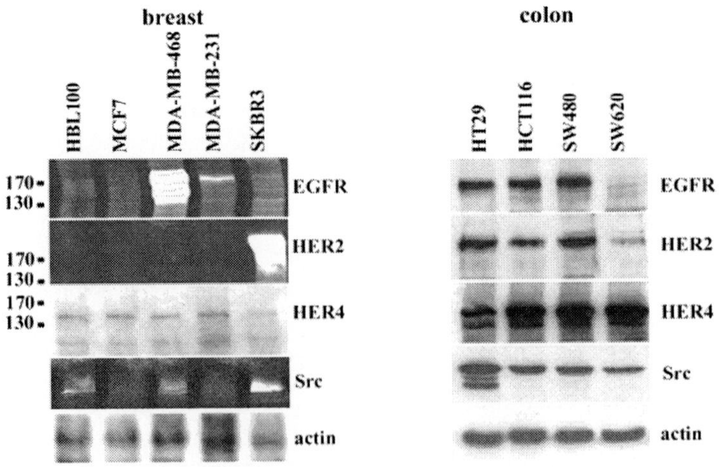

Figure 2. Expression levels of EGFR, HER2, HER4 and Src, determined by western blotting, in breast and colon cancer cell lines are shown. EGFR and HER2 expression is reproduced from Figure 2, published by Moiseeva et al [63].

In colon cancer, high-level gene amplification for HER2 and EGFR was only detected in a very small population of 1% and 0.6% of patients, respectively [54, 64, 65]. Other studies detected some HER2 chromosomal gain and membrane staining in a significant number of patients, both correlating with poor survival [55, 66]. In another study, HER2 seems to be expressed in the early stages of colorectal cancer [49]. HER2 overexpression constitutes an independent factor for poor prognosis in colon cancer [54, 66]. However, high membranous HER2 levels or absence of EGFR expression are associated with improved survival after partial liver resection for colorectal metastasis [67, 68]. Overexpression of EGFR and other members of the EGFR family alone was not significantly associated with a shortened survival, but patients with a simultaneous overexpression of HER2 and HER4 had a shorter survival in colon cancer [50].

Src is a Favorite Partner for EGFR and HER2

Src activation is found in a variety of cancers (see [69]). Src deregulation is found in several clusters of breast cancer, including those with bad prognosis [13]. Levels of activated Src-pY418 are significantly higher in invasive breast carcinomas and co-expressed with constitutively active STAT3 or HER2 [70, 71]. Neu-overexpressing mouse mammary tumors possessed 6-8 fold-increased c-Src kinase activity compared to the adjacent epithelium [72]. The increased c-Src tyrosine kinase activity was a result of elevation of its specific activity. In colon tumors, progression to metastatic disease correlates with increased c-Src kinase activity and protein levels [73, 74]. Curiously, whilst c-Src overexpression does not produce a malignant phenotype [75], its activity is associated with tumorigenic properties and enhances metastatic potential in cooperation with other tumorigenic pathways (reviewed in [69, 76]). Src contributes to the capacity of breast cancer to metastasize to bone in animal models [77]. Its activity correlates with the high invasive potential of colon cancer cells and directly regulates a set of genes *in vitro*, also characteristic to advanced colon tumors [78]. Src drives expression of several clusters of genes, some of them induce upregulation of expression of a second cascade of genes [79]. Importantly, a number of genes from both cascades are involved in the invasive potential of colon cancer cells [79]. Beta-catenin is one of the Src substrates in cancer cells [80].

In animal models, co-overexpression of c-Src and EGFR increased tumorigenesis in nude mice [81]. Src potentiates the survival, proliferation and tumorigenesis of EGFR family members, in part by associating with them [82, 83],(reviewed in [69]). It co-precipitates with EGFR and HER2 and phosphorylates their catalytic domains in colon, breast cancer and other cells [71, 74, 84-87]. Src recruitment by HER2, but not EGFR, results in loss of epithelial polarity [86]. Src, stimulated by stromal cell-derived factor, is involved in HER2 transactivation in breast cancer [88]. Moreover, recent data show that it enhances HER2/HER3 signaling and biological functions by positively modulating the association between them [89]. In turn, HER2 supports Src via an unusual mechanism: HER2 increases Src translation via activation of the Akt pathway and increases Src stability [83].

Pro-Survival EGFR/HER2 Signaling and Tumor Dependence on these Oncogenes

Both EGFR and HER2 are implicated in regulation of cell cycle and prevention of apoptosis (summarized in [16, 17]). Activation of the receptors induces tyrosine phosphorylation in a number of proteins, as well as autophosphorylation of the receptors. Tyrosine phosphorylation was increased in 198 proteins in HER2-transfected fibroblasts, and a HER2 specific inhibitor reduced phosphorylation in about 110 of them [90]. Several pathways activated by the receptors contribute to pro-survival signaling and tumorigenic activities of EGFR and HER2. The number of elucidated pathways has not yet been exhausted, as evidenced by the recent unexpected findings of interactions between EGFR and mitochondrial cytochrome c oxidase subunit II (COX II) [91], by a phosphoproteomic analysis of HER2-induced phosphorylation [90] or by EGFR-induced Src-mediated elimination of a checkpoint protein Fhit [92]. Activation of EGFR increases Src activity, which induces Fhit phosphorylation and its subsequent proteasomal degradation in breast cancer cells [92].

Three pathways are considered the most important in pro-survival signaling; however, they have not been completely delineated (Figure 3). All EGFR family members bind to and signal through Shc and Grb2; nonetheless, each of them has a specific set of phosphotyrosine sites for binding signaling proteins: only HER3 and HER4 have binding sites for PI3K, and only EGFR and HER4 bind to STAT5 [18]. Conversely, direct interactions between PI3K p85 and EGFR were described previously [93]. EGFR signaling to PI3K via Gab1, as well as EGFR or HER2 dimerization with either HER3 or HER4 to activate PI3K, contributes to downstream activation of Akt. Signaling via Shc and Grb2 maintains MAPK activity. The third pathway involves STAT1/3/5 and is also supported by Src [70, 91, 94]. HER2 signaling via STAT3 significantly contributes to tumorigenesis and loss of epithelial polarity [95]. Src-mediated specific phosphorylation of EGFR-Y845 and HER2 (via an unidentified tyrosine in the catalytic domain, likely to be Y877) [85, 86, 96, 97] enhance activation of the receptors and downstream signaling. EGFR-pY845 and HER2-pY877 are critical for tumorigenic and pro-survival signaling [97-99]. Two alternate signaling pathways mediated by phosphorylated Y845: one involves the transcription factor STAT5b [97], the other involves mitochondrial COX II [91].

The tumorigenic potential of EGFR and HER2, and tumor dependence on the oncogenes are shown in multiple models, both in cell culture and in animal models (HER2 dependence is summarized in [17]). Overexpression of EGFR, HER2 or v-Src can prevent cell death in breast epithelial cells by activating Akt and ERK pathways, consequently leading to degradation of the pro-apoptotic protein BIM [100, 101]. The molecular mechanisms of gene addiction have been revealed by Sharma *et al* [102], who showed that oncogene dependency arises from an imbalance of survival and apoptotic pathways. Inactivation of EGFR and Src in dependent cells results in rapid attenuation of pro-survival ERK, Akt and STAT3/5 signaling and a slow rise in apoptotic p38 MAPK signaling. Clinical success of therapies targeting EGFR and HER2 using specific inhibitors and antibodies supports the central role of these oncogenes of addiction in the maintenance of tumors [14, 16].

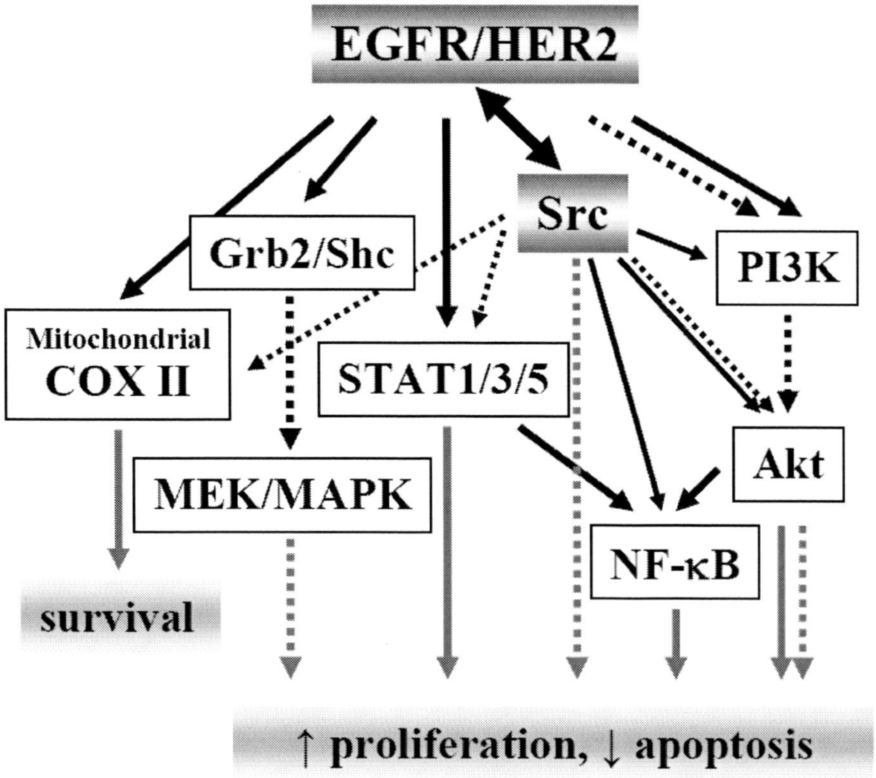

Figure 3. A diagram summarizes the EGFR- and HER2-mediated pro-survival signaling pathways. Black solid arrows indicate direct activation and/or modification of proteins. Black broken arrows indicate effect mediated by secondary messenger proteins. Gray solid arrows show effect via direct modification (Akt, Src) of proteins, involved in apoptosis, or transcriptional activation (STAT1/3/5, NF-kappaB) of genes, involved in regulation of cell proliferation and survival. Gray broken arrows show effect mediated by intermediate transcription factors.

EGFR, HER2, Src are Molecular Targets of Dietary Phytochemicals

Current development of anti-EGFR and anti–HER2 agents indicates that these oncogenes are valuable targets in cancer treatment. Remarkably and perhaps not surprisingly, they are also targets of dietary chemopreventive agents. Studies in vitro and in animal models have shown that dietary compounds found in some vegetables affect oncogenes in cancer cells and reduce tumor growth. Unfortunately, there is only one population-based study to date showing the connection between diet and oncogenes. The presence of two long alleles of the CA (≥19) repeat in the EGFR gene coincided with the increased breast cancer risk, which was also influenced by dietary factors, i.e. consumption of large amounts of vegetables was associated with a reduced cancer risk for individuals with these alleles [7].

This review will focus on the dietary phytochemicals, such as indole-3-carbinol, resveratrol, genistein, curcumin and epigallocatechin gallate, which can reduce expression and/or activation of EGFR or HER2.

Indole-3-Carbinol and 3,3'-Diindolylmethane

Indole-3-carbinol (I3C), derived from vegetables of the Cruciferae family, such as radish, broccoli, brussels sprouts and cabbage, is an FDA-approved nutritional supplement. Several epidemiological studies found a negative association between high consumption of cruciferous vegetables and lung, colorectal, breast and other cancers [8], (reviewed in [103]); it is effective in treatment of vulval and cervical intraepithelial neoplasias and recurrent papillomatosis, and is associated with very few side effects [104-106]. I3C can be converted into 3,3'-diindolylmethane (DIM), a major *in vivo* acid-catalyzed condensation product of I3C. Both I3C and DIM have been shown to inhibit development of tumors in a variety of tissues in animal models. Concentrations of 28/[24] and 170/(200) μM of I3C/(DIM) were detected in plasma and tissues in animal models [107].

Several molecular mechanisms have been suggested for cancer cell death induced by I3C and DIM. They involve (i) induction of enzymes metabolizing steroid hormones, (ii) downregulation of estrogen and androgen receptors, (iii) modulation of cell cycle-related proteins and (iv) downregulation of Akt⇒NF-kappaB signaling [108]. Recent studies show that I3C and DIM are effective against prostate and breast cancer, which are not dependent on steroid hormones, but rely on expression of such oncogenes as EGFR and HER2. In prostate cancer PC-3 cells, I3C (60-100 μM) induced EGFR downregulation, concomitant with reduced Bcl-xL expression and apoptosis [109]. Similarly, DIM (30-60 μM) inhibited EGFR expression and induced apoptosis in these cells, but not in the DIM-resistant non-tumorigenic prostate epithelial CRL-2221 cells [110]. DIM-induced EGFR downregulation was concurrent with the reduction in Akt-S473 phosphorylation, reduced NF-kappaB DNA binding and reduced Bcl-xL expression. Pre-treatment with DIM abrogated induction of PI3K activity by EGF [110]. Significant modulation of gene expression was observed in the I3C- and DIM-treated PC-3 cells: a total of 738 genes that showed a greater than two fold change after 24/48 h of 40 μM DIM treatment [111]. EGFR mRNA levels were reduced by 2.6 and 2.7 fold after 24 h exposure to 60 μM I3C and 40 μM DIM, respectively [111]. Several genes related to the EGFR signaling pathway were also reduced by I3C.

In breast cancer cells, EGFR downregulation was followed by or coincided with apoptosis via the mitochondrial pathway in ER-positive MCF7 cells, and ER-negative MDA-MB-468 and MDA-MB-231 cells, but not in HBL100 cells [63, 112]. Since MDA-MB-468 and MDA-MB-231 cells were EGFR-dependent [112], I3C-induced apoptosis was likely to be associated with EGFR downregulation. The MDA-MB-468 cells are the second highest EGFR expresser among cancer cell lines (Figure 2) and are extremely sensitive to I3C. In MCF7 cells, I3C-induced apoptosis may be related to the combined effects of I3C on ER-alpha and EGFR. Further investigations showed that I3C modulated tyrosine phosphorylation in all four cell lines studied: it decreased for EGFR and STAT-1 after 0.5-4.5 h treatment, but increased for several proteins, including Src, EGFR and p130CAS, after 5 h treatment [63]. Likewise, tyrosine kinase activity was significantly changed in the liver of rats receiving dietary I3C [113]. A detailed analysis of I3C-induced events showed sequential activation of Src, EGFR, STAT-1 and STAT-3, followed by EGFR degradation between 4.5 and 7 hours of treatment with 125 μM I3C. I3C also induced physical association of activated Src and EGFR [63]. The agent induced EGFR downregulation without affecting pAkt in four breast lines in these studies [63, 112]. In MCF7 and MDA-MB-231 cells, I3C modulated expression of cell cycle-related proteins, p21Cip1, p27Kip1, cyclin E, cyclin D1 and CDK6, with upregulation of p21Cip1 and cyclin E being dependent on Src in both cell lines and downregulation of cyclin D1 being dependent on EGFR in MCF7 cells (Figure 4) [63].

Modulation of either Src or EGFR significantly altered I3C-induced responses. Inhibition of EGFR by inhibitors PD153035 or ZD1839 (gefitinib) increased susceptibility to I3C-induced apoptosis of MCF7, MDA-MB-468 and MDA-MB-231 cells [63]. Inhibition of Src sensitized MDA-MB-468 and MDA-MB-231 cells to I3C, whereas overexpression of c-Src induced resistance to I3C in MDA-MB-468 and HBL100 cells [63]. These data indicated that Src and EGFR were critical in I3C-induced apoptosis in breast cancer cells [63].

In breast cancer MDA-MB-468 cells, I3C induced NF-kappaB p65/p50 DNA binding at 6.5 h dependent on Src activity, preceded by association of IKK-beta with the Src/EGFR complex and increased phosphorylation of IkappaB-alpha [114]. Nonetheless, no change in NF-kappaB DNA binding was detected in MDA-MB-231 and HBL100 cells, but STAT5 DNA binding was increased in contrast to MDA-MB-468 cells [114]. These data show that Src induces activation of various downstream targets in a cell-dependent manner.

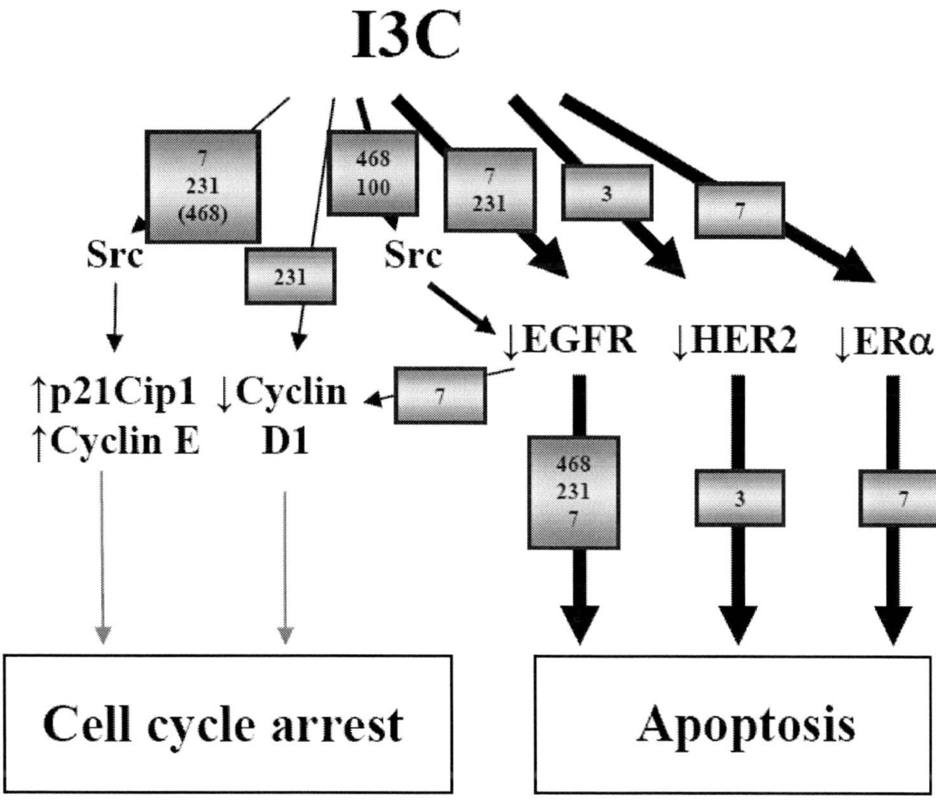

Figure 4. An updated diagram, adapted from Figure 7, published by Moiseeva et al [63], summarizes the molecular mechanisms of I3C-induced cell cycle arrest and apoptosis in breast cancer cells via reduced expression and/or activation of EGFR and HER2. The breast cancer cell lines are corresponded to the numbers in the gray boxes: HBL100, MCF7, MDA-MB-231, MDA-MB-468 and SKBR3.

They are also in agreement with STAT5 activation being dependent on Src in MDA-MB-231 cells [97]. Altogether, I3C influenced the NF-kappaB pathway in a cell-specific manner, which was not related to apoptosis in breast cancer cells.

Pilot data from our lab indicate that I3C (150 μM) is effective not only against EGFR-overexpressing breast cancer cells, but also against HER2-overexpressing SKBR3 cells and primary cultures from Tag mouse mammary carcinomas, which expressed both EGFR and HER2 [115]. I3C reduced HER2 autophosphorylation concomitantly with the induction of apoptosis in SKBR3 cells. In primary Tag tumor cultures, susceptibility to I3C was associated with sensitivity to Src inhibition, whereas apoptosis induced by Src inhibition correlated to that induced by EGFR inhibitor ZD1839 [115]. These data imply that susceptibility to I3C in primary Tag mouse tumors is related to Src, which is also likely to determine the response to EGFR inhibition.

Analogous to I3C, DIM (15 μM) decreased HER2 phosphorylation and total levels, concurrent with reduced ERK activation and increased apoptosis in breast cancer MDA-MB-435 with the forced expression of Neu/ErbB2/HER2 [116]. Similar changes were also induced by the chemotherapeutic tubulin-binding drug paclitaxel in these cells, with a synergistic response to combined treatment with DIM and paclitaxel [116].

Collectively, these data suggest that I3C-induced pathways depend on specific features of cancer cells. The cancer types, which rely on "EGFR/HER2 addiction" and/or Src deregulation, are likely to be susceptible to I3C.

Resveratrol

Resveratrol is a polyphenolic phytochemical found in grapes, red wine, peanuts, berries and other plant sources. It has multiple activities, such as cancer chemopreventive and anti-aging potentials and cardiovascular protection [117, 118]. It is also a phytoestrogen [119, 120] and an antioxidant (see [118]. Although resveratrol has attracted much attention as a potential anticancer agent *in vitro* and *in vivo* (reviewed in [118]), an epidemiological negative link between grapes consumption and cancer is established just in a few studies [121, 122]. The molecular mechanisms of anti-cancer activities of resveratrol are sketchy and poorly understood; nonetheless, a body of evidence below implicates Src, EGFR and HER2 in the mechanism of action of resveratrol in cancer and normal cells in concentrations comparable with physiologically achievable plasma levels in humans (about 2 μM) [107].

Resveratrol (25-50 μM) suppressed EGFR-dependent ERK1/2 activation pathways, stimulated by EGF, in prostate cancer PC2 cells [123]. At 1-2.5 μM it blocked Src kinase activation induced by VEGF in human umbilical endothelial cells [124]. Resveratrol attenuated VEGF-induced increase in reactive oxygen species (ROS), which caused Src activation. Similarly, resveratrol (10-20 μM) inhibited v-Src activity in v-Src-transformed mouse fibroblasts (NIH3T3/v-Src), downregulated constitutively-active STAT3 and consequently induced irreversible cell cycle arrest [125]. It also reduced phosphorylation of STAT3, its nuclear translocation and DNA-binding activity in variety of breast, prostate and pancreatic metastatic cancer cells, followed by cell cycle arrest, loss of viability and apoptosis [125]. By contrast, cells lacking aberrant STAT3 activity showed reversible growth arrest and minimal loss of viability, when treated with resveratrol [125]. In osteoblastic cells resveratrol (1-5 μM) increased expression of bone morphogenetic protein-2 (BMP-2) in cell culture and *in vivo,* and increased BMP-2-mediated expression of osteogenic differentiation biomarkers [126]. In this study, resveratrol induced activation and auto-phosphorylation of Src followed by Src-mediated ER-alpha nuclear translocation and DNA-binding, and increased transcription of the BMP-2 promoter, which required cooperation between ER-alpha and FOXA1 [126].

Several lines of evidence implicate resveratrol in modulation of EGFR signaling. Resveratrol treatment significantly decreased EGF expression, concomitant with the reduction in p21 and Bax levels, in endometrial cancer Ishiwaka cells [127]. Hence, resveratrol was likely to suppress proliferation of Ishikawa cells through down-regulation of EGF and EGFR signaling. In prostate cancer DU145 and LNCaP cells, pre-treatment with resveratrol at 10 μM for 4 hours modulated EGF-induced signaling [128]. Similarly, resveratrol inhibited EGF-induced PI3K/Akt signaling and Akt-mediated phosphorylation of members of the transcription factor FOXO family, followed by reduced DNA synthesis in human liver fibroblasts [129].

In normal vascular smooth muscle cells, resveratrol did not affect EGFR autophosphorylation or levels within 30 min of EGF stimulation, but did interfere with downstream EGFR signaling: it reduced tyrosine phosphorylation of an adapter protein Gab1, its association with the PI3K regulatory subunit p85 and their translocation to the plasma membrane, which consequently diminished phosphorylation and activation of Akt [130]. These effects were mediated by resveratrol-induced activation of tyrosine phosphatase Shp2 and were abolished in $Shp2^{-/-}$ cells; moreover, resveratrol activated purified Shp2 *in vitro* at 0.05-5 μM. [130].

In breast cancer MCF7 cells, resveratrol (2-10 μM) significantly suppressed heregulin-beta1-induced signaling via heterodimeric complexes of HER2 with either HER3 or HER4 [131]. It had negligible effects on phosphorylation of HER2, but inhibited ERK activation after 30 min exposure to heregulin, followed by inhibition of downstream MMP-9 activation after 24-hour treatment [131]. Resveratrol at 4.4-50 μM significantly reduced HER2 mRNA levels in HER2 overexpressing mouse mammary tumor N202/1 cells and human breast cancer SKBR3 cells in cell culture [132]. Similarly, resveratrol (3-30 μM) decreased transcription of the HER2 promoter in breast cancer MCF-7 cells in estrogen-free medium [133]. Conversely, it was increased in the presence of 10 nM 7-β-estradiol; this upregulation was associated with transient nuclear translocation of the transcription factor AP-2 and required two AP-2 binding sites in the HER2 promoter [133].

In animal models, resveratrol supplementation delayed development of spontaneous mammary tumors, diminished the mean number and size of mammary tumors and reduced the number of lung metastases by 3.3 fold in HER2/neu transgenic mice [132]. The anti-tumor effects of resveratrol were associated with decreased HER2/neu gene expression and increased apoptosis in mammary glands. It also decreased tumor growth in the mouse xenograft models with breast cancer MCF7 and MDA-MB-231 cells [126]. The tumor-suppressive effects of resveratrol in this model were the consequence of Akt inactivation-mediated FOXO3a nuclear accumulation and activation.

Curcumin

Curcumin is a major yellow pigment of turmeric, widely used in Asian cooking, but present in just a few western products, such as mustard. It is also used in negligible amounts as a food coloring agent. It has shown a positive effect in some cancers in phase I/II trials [134]. A body of evidence suggests that it is a potent inhibitor of the EGFR signaling.

Early studies by Kumar and co-authors indicated that curcumin inhibited intrinsic EGFR kinase activity and EGF-induced activation in a dose-dependent manner in human epidermoid carcinoma A431 cells and the mouse fibroblast NIH 3T3 cell line with forced expression of EGFR [135, 136]. Curcumin-induced EGFR downregulation resulted in inhibited growth of EGF-stimulated cells. Further studies confirmed that curcumin interferes with EGF-induced phosphorylation and activation of EGFR followed by reduced signaling via ERK and Akt in human breast cancer HBL100 and MDA-MB-468 cells [137]. Similarly, it inhibited EGF-induced autophosphorylation of EGFR and downstream PI3K/Akt signaling in human prostate cancer PC-3 cells [138]. Chen *et al* found that curcumin reduced EGFR expression at RNA and protein levels in colon cancer cell lines [139, 140]. Curcumin blocked EGF signaling by suppressing tyrosine phosphorylation and expression of EGFR, caused by activation of peroxisome proliferator-activated receptor-gamma [139]. The reduction of EGFR expression was mediated by curcumin-induced downregulation of the transcription factor Egr-1: curcumin significantly reduced the DNA-binding activity of Egr-1 to the curcumin response element in the EGFR promoter and reduced the trans-activation activity of Egr-1 by suppressing Egr-1 gene expression, which was caused by reduced phosphorylation of another transcription factor Elk-1, dependent on ERK activity [140]. Altogether these data are summarized in Figure 5.

Not only does curcumin inhibit cancer cell growth via suppression of the EGFR pathway, but it also interferes with the HER2 pathway by inhibiting HER2 activation and expression. Hong *et al* [141] found that curcumin inhibited autophosphorylation and activity of HER2 *in vitro* (IC50 ≅ 24 and 9 μM, respectively) and in cells in a dose-dependent manner, as well as reducing the basal levels of HER2 (IC50≅32 μM upon 12 h treatment) in breast cancer AU-565 cells. The mechanism of HER2 depletion was distinct from that of EGFR downregulation: curcumin reduced HER2 levels via HER2 dissociation from its complex with the molecular chaperone GRP94 followed by protein degradation. The agent preferentially reduced growth and colony formation of HER2-overexpressing breast cancer cells [141].

Curcumin also inhibited the kinase activity of v-Src in the mouse fibroblast NIH 3T3 cell line with the forced expression of the oncogene, which led to a decrease in tyrosine phosphorylation of Src downstream targets: Shc, cortactin, and FAK [142]. Consistent with the abrogation of Src activity was the reduction of Src-Tyr-416 auto-phosphorylation, followed by decreased Src-mediated Shc-Tyr-317 phosphorylation, decreased ERK activation and cell proliferation in v-Src transformed cells. Remarkably, curcumin not only exerted its negative effect on proliferation, but also on migration of the transformed cells by reducing Src-mediated FAK phosphorylation and activation.

Curcumin was also shown to inhibit tyrosine kinase activity of pp60c-Src and v-Src *in vitro* by 40% (100μM) and 30% (20 μM), respectively [142, 143].

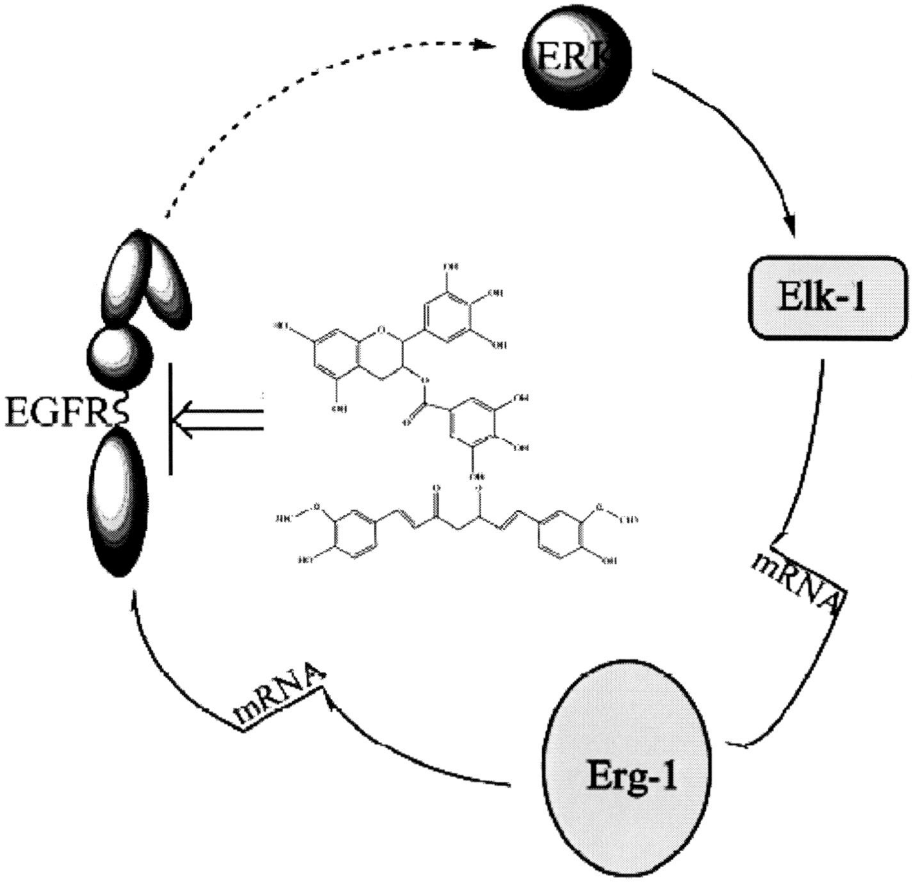

Figure 5. A diagram illustrates a suggested mechanism of action of curcumin and EGCG.

It inhibited v-Src with IC50≅50 μM and decreased FAK autophosphorylation *in vitro* with IC50 ≅34 μM, suggesting a direct effect of curcumin on this tyrosine kinase [142].

The concentrations used for *in vitro* inhibition of tyrosine kinases (Src, FAK, HER2) are generally higher than those required for inhibition of endogenous enzymes in cells and much higher than those which are achievable physiologically (<2 μM) [107]; therefore, it is unlikely that the effect of curcumin on these enzymes in cells is a result of direct inhibition.]

Epigallocatechin-3-Gallate

Constituents of green tea attract significant attention among health professionals in connection with anti-cancer activity and other health properties. Epigallocatechin gallate (EGCG) is the most abundant polyphenolic catechin, present in green tea, with powerful antioxidant and chemopreventive properties. The significant inverse association between risk of some cancers and green tea intake was demonstrated in some studies; the most recent analysis of about 150 epidemiological studies by Yang and co-authors [9] did not result in any clear conclusion, but a meta-analysis indicated a reduced risk for colorectal cancer; the preventive effect of green tea was associated with genetic polymorphisms for breast and lung cancer; some inconsistencies were attributed to the temperature of tea consumed, a risk factor for esophageal cancer. Encouragingly, an interventional study showed a clear preventive effect of green tea in human prostate cancer [144]. Many animal studies support the protective effect of tea and EGCG. Analysis of about 120 published studies, conducted by Yang and co-authors [9], revealed a protective effect of green and black tea constituents against cancer in various sites in 117 studies against 13 studies with no protective effect. Bioavailability of dietary EGCG is fairly low reaching nM concentrations in human plasma, but dietary supplementation produced concentrations up to 7 μM [107].

EGCG was a strong inhibitor of tyrosine kinase activity of EGF, PDGF and FGF receptors with IC50 of 1.1, 2.3 and 2.3 μM, respectively, but it did not affect the protein tyrosine kinase activities of pp60v-src, PKC, and PKA (IC50 > 22 μM) (145). EGCG (10 μM) directly inhibited various protein kinases of mitogen-activated pathways, such as ERK2 and p38δ, and also reduced activity of PDK1, which mediates Akt phosphorylation and activation, and is involved in the PI3K/Akt pathway [146]. The p38-regulated protein kinases PRAK and DYRK1A had the lowest IC50≤1μM EGCG [146]. Hence, downregulation of MAPK activity in cells could be attributed to a direct inhibiting effect of EGCG on particular enzymes, as well as downregulation of growth factor receptor signaling. Therefore, this review will concentrate only on the effect of EGCG mediated specifically by EGFR and HER2.

In epidermoid carcinoma A431 cells, EGCG (11 μM) reduced the EGFR autophosphorylation by EGF and reduced EGF binding to the receptor [145]. EGCG also significantly diminished DNA synthesis in a dose-dependent manner [145]. In mouse mammary tumor virus (MMTV)-HER22/neu-overexpressing cancer NF639 cells, EGCG inhibited HER2 auto-phosphorylation at 44-88 μM, reduced PI3K/Akt and NF-kappaB signaling, concomitant with reduced growth [147]. The importance of HER2 in the mechanism of EGCG action was also supported by the loss of tyrosine phosphorylation of HER2 in EGCG-resistant NF639 HER2/neu-driven mammary cancer cells [148].

Further investigation in other cancer cells confirmed that EGCG inhibited EGFR/HER2 activation and downstream signaling. In human head and neck squamous cell carcinoma YCU-H891 and breast cancer MDA-MB-231 cells, EGCG at 22 and 65 μM, respectively, inhibited basal levels and TGF-alpha -stimulated EGFR autophosphorylation and downstream signaling via STAT3 and ERK, with the former being responsible for VEGF production and secretion [149, 150]. Similarly, EGCG reduced the basal levels and TGF-alpha-stimulated HER2 autophosphorylation, downstream STAT3 signaling and growth in squamous carcinoma YCU-H891 and breast cancer BT-474 cells [151]. Moreover, EGCG at 0.2-2.2 μM markedly enhanced the growth-inhibitory effects of the chemotherapeutic drugs 5-fluorouracil and taxol in head and neck squamous carcinoma and breast cancer BT-474 cells [149, 151].

In cervical cancer cells, pre-treatment with 30-50 μM EGCG inhibited EGF-dependent auto-activation of EGFR, concurrent with EGFR-dependent activation of ERK1/2 and Akt, followed by reduced phosphorylation of downstream targets, such as BAD, and apoptosis [152]. In this study, EGCG directly inhibited ERK and Akt activity *in vitro* at 5-30 μM. EGCG did not affect activation of p38 signaling in these cells [152]. In colon cancer HT29 cells, EGCG at 44 μM inhibited both EGFR and HER2 autophosphorylation levels after 6 hours, which consequently reduced signaling via ERK and Akt, followed by reduced cell growth and apoptosis [153]. In colon cancer SW837 cells, the same EGCG treatment decreased phosphorylation of HER3 in addition to EGFR and HER2 [154].

Yang's group [155] found that prolonged treatment with EGCG (20 μM for 8-24 hours) resulted in a decreased level of phosphorylated EGFR (by 32-85%) and also decreased EGFR protein level by 80% in esophageal squamous carcinoma KYSE 150 cells. Similarly, EGCG treatment for 24 hours caused decreased HER2 expression in esophageal adenocarcinoma OE19 cells [155]. Since these EGCG effects were prevented or diminished by the addition of superoxide dismutase (SOD), or combined SOD with catalase, to the cell culture medium, the authors concluded that auto-oxidation of EGCG leads to H_2O_2 production, which in turn leads to EGFR inactivation, but the inhibition of cell growth is due to other mechanisms [155]. However, EGCG auto-oxidation, as an inhibitory mechanism of EGFR phosphorylation, was not confirmed by another group [156], which found that EGCG caused disruption of membrane organization and function, which resulted in reduced EGF binding to cells, diminished EGFR dimerization and autophosphorylation. EGCG-induced disruption of ordered domains in lipid membranes was proposed as the molecular mechanism involved in inhibition of several receptor tyrosine kinases, such as EGFR family members [156]. Furthermore, Fu and Chen demonstrated that the mechanism of reduced EGFR expression is caused by reduced transcription of the EGFR promoter, caused by decreased activity of Egr-1 according to the same mechanism as this group showed for curcumin-induced downregulation of EGFR (Figure 5) [157]. Collectively, these data indicate that EGCG downregulates activity of the members of the EGFR family and their signaling pathways.

Genistein

Genistein is an isoflavone phytoestrogen, which is present in the highest amounts in soy products, widely consumed in Asian diets. The main western product containing the highest amounts of this phytochemical is split peas, but it is also present in other products [158]. Genistein interacts with the estrogen receptors ER-alpha [159] and ER-beta [160] and stimulates the transcriptional activity of both ER subtypes at concentrations of 1-10 nM, however it has a much higher binding affinity to human ER-beta (8.5 nM) than to human ER-alpha (145 nM) [160]. A high level of consumption of soy products, containing genistein, is thought to be related to decreased occurrence of prostate cancer in Asian populations [161]. Consumption of soy isoflavones, including genistein, is also associated with decreased risk of breast, endometrial and ovarian cancer in some studies [162-165], but not others[166, 167]. A study of breast cancer in Germany found an inverse correlation between dietary intake of different phytoestrogens and breast cancer risk only in premenopausal women [168], confirmed for the total phytoestrogen intake [169, 170], but plasma genistein concentrations were not related to premenopausal breast cancer risk [170], which may be caused by low consumption of genistein-containing products and very low plasma concentrations of this phytochemical in European populations [171].

More recent data indicate that genistein may also affect the EGFR/HER2 pathways. The data, summarized below, imply that the effect of genistein may be mediated by either the ER-alpha or EGFR/HER2 pathways or via their cross-talk, which has particular importance for ER-alpha-positive breast cancers.

In Vitro and Cell Culture

An early study by a Japanese group showed that genistein is a specific inhibitor of tyrosine kinase activity of EGFR (IC50=22 μM) and p60v-src (IC50=26 μM) *in vitro*, but it did not affect serine- or threonine-specific protein kinases [172]. Autophosphorylation of EGFR was inhibited by genistein with IC50=2.6 μM. They demonstrated that the phytochemical at 150 μM inhibited the EGF-stimulated increase in phosphotyrosine level by 9 fold in epidermoid cancer A431 cells [172]. Genistein also inhibited EGF-stimulated mitogenesis with an IC50≈10 μM and other EGFR-mediated signaling dependent on tyrosine kinase activation in mouse 10T1/2 fibroblasts and rat liver T51B cells [173]. Inhibition of another tyrosine kinase, platelet derived growth factor receptor (PDGF), required a 4 times higher concentration of the phytochemical. Genistein inhibited EGFR tyrosine kinase activity of the A431 and breast ER-positive cancer ZR-75-1 cells independently of the expression level, whereas proliferation of these cells was inhibited in the range of IC50~0.1-2 μM after 6 days, correlated to the number of EGFR per cell [174].

Genistein (25 μM) alone or in combination with tamoxifen decreased mRNA levels of EGFR, concomitant with G1 cell cycle arrest and growth inhibition in ER-positive breast cancer MCF7 cells [175]. Genistein interfered with EGF-induced signaling in breast cancer SKBR3 and MCF-7 cells: it decreased Shc tyrosine phosphorylation, association of Shc with Grb-2 and decreased MAP kinase activity [176]. Consequently, genistein decreased proliferation of both cell lines in a dose-dependent manner [176]. Interestingly, genistein inhibited growth of both cell lines to similar extents despite their significantly different levels of HER2 or absence of ER-alpha in SKBR3 cells. However, in the bladder cancer cell lines, genistein at 15-37 μM inhibited proliferation more efficiently in the cell lines with higher expression levels of EGFR [177]. Moreover, invasiveness of the bladder cancer T24 cells, dependent on EGFR, was inhibited by genistein at a much higher concentration (74 μM) [177]. In contrast, the effect of genistein on growth of malignant primary brain tumor glioblastoma multiforme cell lines (IC20=10 μM in a monolayer) was much smaller that its effect on invasiveness: infiltration of human glioblastoma cell lines into fetal rat brain aggregates in co-culture was completely inhibited by genistein at 10 μM [178]. No change in EGFR-associated tyrosine phosphorylation was detected in that study.

Growth of pancreatic cancer Panc-1 cells was decreased by genistein in the range of 20-50 μM [179]. Interestingly, physiological concentrations of genistein (10 μM) were sufficient to change the transcription of 47 genes significantly, with the expression of 5 genes related to EGFR signaling, including EGFR, Egr-1, AKT2, CYP1B1, and NELL2, markedly reduced [179]. In a number of pancreatic cell lines (BxPC-3, CAPAN-2, and AsPC-1, but not HPAC and MIAPaCa cells) genistein significantly increased growth inhibition and apoptosis induced by an EGFR/HER2 inhibitor, erlotinib [180]. In pancreatic cancer BxPC-3 cells, the combined treatment resulted in significant downregulation of EGFR, phosphorylated Akt, NF-kappaB activation and survivin, suggesting genistein-enhanced inhibition of downstream EGFR targets, such as Akt and NF-kappaB, as the mechanisms for the potentiation of erlotinib-induced cell death by genistein [180]. Likewise, genistein induced apoptosis in breast cancer MDA-MB-231 cells concomitant with reduced NF-kappaB DNA binding and reduced Akt signaling; the latter was modulated by EGF indicating EGFR involvement [181].

In addition to the EGFR pathway, genistein also downregulates the HER2 pathway. It inhibited growth and induced apoptosis (2.5-10 μM) in breast cancer HER2-overexpressing 184-B5 cells, also expressing EGFR, by down-regulation of HER2 expression and phosphorylation (10 μM) [182]. Downregulation of HER2, NF-kappaB DNA binding and downstream NF-kappaB targets, MMP-2 and MMP-9, by genistein (25-50 μM), concurrent with reduced invasiveness at 50 μM genistein, was also found in head and neck cancer cells [183]. The combination of genistein with an anti-estrogen drug, tamoxifen, inhibited the growth of ER-positive HER2-overexpressing human breast cancer BT-474 cells in a synergistic manner, concurrent with the downregulation of EGFR, HER2, HER3, ER-alpha and survivin [184]. In HER2-expressing breast cancer cells, genistein at low doses enhanced a cytotoxic effect of a chemotherapy agent, adriamycin, by inducing necrotic-like cell death, but not apoptotic death, concomitant with marked inactivation of HER2 and Akt [185].

Low concentrations of genistein required for inhibition of EGFR activity *in vitro* and in cell culture, are comparable with physiologically achievable plasma concentrations (27 μM) in humans [186], suggesting that downregulation of the EGFR activity is a feasible mechanism of action of genistein *in vivo.*

In Animal Models

Lamartiniere's group showed that genistein has a significant effect on the EGFR pathway *in vivo*, which was related to chemoprotective effect of genistein.

Mammary glands of 21 day-old rats, injected with genistein, showed a significantly increased number of terminal end buds and ductal branching, with increased expression of total and tyrosine phosphorylated EGFR, compared with control animals [187]. This effect of genistein was counteracted by pre-treatment with a steroidal-anti-estrogen ICI182,780, and was likely to be mediated by estrogen receptors. Dietary genistein protected rats from carcinogen-induced mammary tumorigenesis when administered during whole life-time, but also showed some protection when administered to prepubertal rats [188] Lamartiniere and coauthors also showed that when injected prepubertally, genistein enhanced mammary gland differentiation: genistein treatment increased TGF-alpha and EGFR mRNA levels by 2.7 and 3 fold, respectively, in mammary terminal ductal structures at day 21 postpartum in rats, compared to control animals [189]. Mammary glands of genistein-treated rats were more differentiated: they had more lobules and fewer terminal end buds and terminal ducts [189]. This genistein-accelerated differentiation resulted in reduced EGFR-signaling in later life [189]. At day 50 postpartum, EGFR immunostaining intensity was decreased in terminal end buds, whereas EGF was increased in terminal end buds and terminal ducts at day 50. No changes in TGF-alpha, EGF and EGFR were observed in the mammary gland stroma and fat pad of genistein-treated rats. Hence, prepubertal genistein exposure leads to reduced EGF signaling, which render the mature mammary gland less proliferative and less susceptible to cancer, however, only life-time genistein supplementation provided the highest level of protection against mammary tumors [187, 188, 190]. Consistent with previous studies of this group, genistein, injected in rat uterus, increased EGFR levels, which peaked 16 h after treatment, and also increased cell proliferation, with the effect mediated by estrogen receptors and abolished by anti-estrogen pretreatment [191].

The involvement of EGFR in the chemopreventive action of genistein was confirmed in the mouse xenograft model in another study: tamoxifen and genistein synergistically delayed growth of MCF-7 tumors by inducing apoptosis and inhibiting proliferation of tumor cells, consistent with the effect of combined genistein and tamoxifen on decreased estrogen level and activity, and down-regulation of EGFR expression [175]. On the contrary, in a (MMTV)-Neu-overexpressing mouse model, dietary genistein increased mammary tumor latency, but did not reduce the number or size of tumors after they were formed [192].

Genistein is found to be an active agent against not only breast cancer, but also EGFR-dependent prostate cancer. When dietary genistein was given to Lobund-Wistar rats, which develop prostate metastatic adenocarcinomas with high incidence, expression of EGFR was reduced by 54% and its phosphorylation was reduced by 50%; moreover, expression of HER2 was also decreased by genistein, but to a lower extent [193]. These changes were associated with decreased incidence of prostate tumors [194]. Similarly, in transgenic TRAMP mice, dietary genistein reduced mRNA and protein expression levels of several key biomarkers of EGFR signaling, e.g. EGFR, Neu, EGF, TGF-alpha; these changes were associated with reduced cell proliferation and prostate adenomas [188, 195]. Thus, life-time dietary genistein protected against development and growth of prostate cancer in rats and mice.

Analysis of the genistein effect in animal models suggests that the chemopreventive effect of genistein on the development of mammary tumors is mediated by estrogen receptors, which modulate differentiation and the EGFR pathway in mammary glands. However, genistein delays growth of EGFR-expressing mammary tumors, e.g. MCF7 xenograft tumors, or EGFR/HER2-expressing prostate tumors, which may be caused by a direct action of genistein on the activities of EGFR and HER2, as well as their cross-talk with steroid receptors.

Conclusions

Comprehensive analyses of multiple molecular targets of the dietary phytochemicals by several researchers indicate a significant overlap of the molecular targets [196] by compounds of various chemical compositions. The phytochemicals, reviewed here, are shown to downregulate a set of similar cell cycle-promoting proteins and upregulate the same inhibitors of cell cycle in various cancer cells. They generally interfere with similar pathways in different cancers, such as signaling via MAPK, Akt and NF-kappaB. These similarities in action of various phytochemicals imply similar or the same upstream molecular targets, which are likely to be the "oncogenes of addiction", such as EGFR, HER2 and/or Src, as summarized in this review. The molecular mechanisms of modulation of the activities of the oncogenes by the compounds are likely to be different, but the downstream effects are bound to be similar, culminating in decreased cancer cell growth and increased cell death, tumor prevention or reduced growth and diminished metastatic spread across a variety of tissues and tumor types. Further investigations into the efficacy of these dietary agents, alone or in combination with existing treatment regimes, in cancer prevention and treatment are urgently required.

Acknowledgements

The author is grateful to colleagues in Cancer Biomarkers and Prevention Group, University of Leicester, and everyone else involved for critical advice and assistance in preparation of the manuscript.

References

[1] Doll, R. The Lessons of Life: Keynote Address to the Nutrition and Cancer Conference. *Cancer Res.* 1992 52, 2024s-9.

[2] WCRF-AICR. Food, nutrition and the prevention of cancer: a global perspective. Washington, DC: World Cancer Research Fund-American Institute for Cancer Research; 1997.

[3] COMA. Report of the Working Group on Diet and Cancer. Nutritional aspects of the development of cancer. . London: Stationery Office, Chief Medical Officer's Committee on Medical Aspects of Food and Nutrition Policy of the UK 1998.

[4] Gonzalez, CA. The European Prospective Investigation into Cancer and Nutrition (EPIC). *Public Health Nutr.* 2006 9, 124-6.

[5] McCullough, ML; Giovannucci, EL. Diet and cancer prevention. *Oncogene.* 2004 23, 6349-64.

[6] Greenwald, P; Clifford, CK; Milner, JA. Diet and cancer prevention. *Eur. J. Cancer.* 2001 37, 948-65.

[7] Brandt, B; Hermann, S; Straif, K; Tidow, N; Buerger, H; Chang-Claude, J. Modification of breast cancer risk in young women by a polymorphic sequence in the egfr gene. *Cancer Res.* 2004 64, 7-12.

[8] Pathak, DR; He, JP; Charzewska, J. Cabbage/sauerkraut intake at 12-13 years of age and adulthood and breast cancer risk in Polish migrant women: Results from the US component of the Polish women's health study. *Am. J. Epidemiol.* 2005 161, S17-S.

[9] Ju, J; Lu, G; Lambert, JD; Yang, CS. Inhibition of carcinogenesis by tea constituents. *Semin. Cancer Biol.* 2007.

[10] Aggarwal, BB; Shishodia, S. Molecular targets of dietary agents for prevention and therapy of cancer. *Biochem. Pharmacol.* 2006 71, 1397-421.

[11] Weinstein, IB. Cancer. Addiction to oncogenes--the Achilles heal of cancer. *Science.* 2002 297, 63-4.

[12] Gazdar, AF; Shigematsu, H; Herz, J; Minna, JD. Mutations and addiction to EGFR: the Achilles 'heal' of lung cancers? *Trends Mol. Med.*. 2004 10, 481-6.

[13] Bild, AH; Yao, G; Chang, JT; Wang, Q; Potti, A; Chasse, D, et al. Oncogenic pathway signatures in human cancers as a guide to targeted therapies. *Nature.* 2006 439, 353-7.

[14] Moasser, MM. Targeting the function of the HER2 oncogene in human cancer therapeutics. *Oncogene.* 2007 Epub ahead of print.

[15] Linggi, B; Carpenter, G. ErbB receptors: new insights on mechanisms and biology. *Trends Cell Biol.* 2006 16, 649-56.

[16] Sebastian, S; Settleman, J; Reshkin, SJ; Azzariti, A; Bellizzi, A; Paradiso, A. The complexity of targeting EGFR signalling in cancer: from expression to turnover. *Biochim. Biophys. Acta.* 2006 1766, 120-39.

[17] Moasser, MM. The oncogene HER2: its signaling and transforming functions and its role in human cancer pathogenesis. *Oncogene.* 2007 Epub ahead of print.

[18] Schulze, WX; Deng, L; Mann, M. Phosphotyrosine interactome of the ErbB-receptor kinase family. *Mol. Syst. Biol.* 2005 1, E1 - E13. .

[19] Oda, K; Matsuoka, Y; Funahashi, A; Kitano, H. A comprehensive pathway map of epidermal growth factor receptor signaling. *Molecular Systems Biology* 2005 1, E1 - E17.

[20] Koeppen, HK; Wright, BD; Burt, AD; Quirke, P; McNicol, AM; Dybdal, NO, et al. Overexpression of HER2/neu in solid tumours: an immunohistochemical survey. *Histopathology.* 2001 38, 96-104.

[21] Laskin, JJ; Sandler, AB. Epidermal growth factor receptor: a promising target in solid tumours. *Cancer Treat. Rev.* 2004 30, 1-17.

[22] Molina, R; Jo, J; Filella, X; Bruix, J; Castells, A; Hague, M, et al. Serum levels of C-erbB-2 (HER-2/neu) in patients with malignant and non-malignant diseases. *Tumour Biol.* 1997 18, 188-96.

[23] Sainsbury, JR; Farndon, JR; Needham, GK; Malcolm, AJ; Harris, AL. Epidermal-growth-factor receptor status as predictor of early recurrence of and death from breast cancer. *Lancet.* 1987 1, 1398-402.

[24] Nicholson, S; Sainsbury, JR; Halcrow, P; Chambers, P; Farndon, JR; Harris, AL. Expression of epidermal growth factor receptors associated with lack of response to endocrine therapy in recurrent breast cancer. *Lancet.* 1989 1, 182-5.

[25] Klijn, JG; Look, MP; Portengen, H; Alexieva-Figusch, J; van Putten, WL; Foekens, JA. The prognostic value of epidermal growth factor receptor (EGF-R) in primary breast cancer: results of a 10 year follow-up study. *Breast Cancer Res. Treat.* 1994 29, 73-83.

[26] Bucci, B; D'Agnano, I; Botti, C; Mottolese, M; Carico, E; Zupi, G, et al. EGF-R expression in ductal breast cancer: proliferation and prognostic implications. *Anticancer. Res.* 1997 17, 769-74.

[27] Tidow, N; Boecker, A; Schmidt, H; Agelopoulos, K; Boecker, W; Buerger, H, et al. Distinct amplification of an untranslated regulatory sequence in the egfr gene contributes to early steps in breast cancer development. *Cancer Res.* 2003 63, 1172-8.

[28] Zhang, W; Stoehlmacher, J; Park, DJ; Yang, D; Borchard, E; Gil, J, et al. Gene polymorphisms of epidermal growth factor receptor and its downstream effector,

interleukin-8, predict oxaliplatin efficacy in patients with advanced colorectal cancer. *Clin. Colorectal. Cancer.* 2005 5, 124-31.

[29] Wright, C; Angus, B; Nicholson, S; Sainsbury, JR; Cairns, J; Gullick, WJ, et al. Expression of c-erbB-2 oncoprotein: a prognostic indicator in human breast cancer. *Cancer Res.* 1989 49, 2087-90.

[30] Thor, AD; Schwartz, LH; Koerner, FC; Edgerton, SM; Skates, SJ; Yin, S, et al. Analysis of c-erbB-2 expression in breast carcinomas with clinical follow-up. *Cancer Res.* 1989 49, 7147-52.

[31] Vargas-Roig, LM; Gago, FE; Tello, O; Martin de Civetta, MT; Ciocca, DR. c-erbB-2 (HER-2/neu) protein and drug resistance in breast cancer patients treated with induction chemotherapy. *Int. J. Cancer.* 1999 84, 129-34.

[32] Hayes, DF; Yamauchi, H; Broadwater, G; Cirrincione, CT; Rodrigue, SP; Berry, DA, et al. Circulating HER-2/erbB-2/c-neu (HER-2) extracellular domain as a prognostic factor in patients with metastatic breast cancer: Cancer and Leukemia Group B Study 8662. *Clin. Cancer Res.* 2001 7, 2703-11.

[33] Thor, AD; Liu, S; Edgerton, S; Moore, D, 2nd; Kasowitz, KM; Benz, CC, et al. Activation (tyrosine phosphorylation) of ErbB-2 (HER-2/neu): a study of incidence and correlation with outcome in breast cancer. *J. Clin. Oncol.* 2000 18, 3230-9.

[34] Paik, S; Hazan, R; Fisher, ER; Sass, RE; Fisher, B; Redmond, C, et al. Pathologic findings from the National Surgical Adjuvant Breast and Bowel Project: prognostic significance of erbB-2 protein overexpression in primary breast cancer. *J. Clin. Oncol.* 1990 8, 103-12.

[35] Saez, R; Molina, MA; Ramsey, EE; Rojo, F; Keenan, EJ; Albanell, J, et al. p95HER-2 predicts worse outcome in patients with HER-2-positive breast cancer. *Clin. Cancer. Res.* 2006 12, 424-31.

[36] Gasparini, G; Gullick, WJ; Bevilacqua, P; Sainsbury, JR; Meli, S; Boracchi, P, et al. Human breast cancer: prognostic significance of the c-erbB-2 oncoprotein compared with epidermal growth factor receptor, DNA ploidy, and conventional pathologic features. *J. Clin. Oncol.* 1992 10, 686-95.

[37] Harris, AL; Nicholson, S; Sainsbury, JR; Farndon, J; Wright, C. Epidermal growth factor receptors in breast cancer: association with early relapse and death, poor response to hormones and interactions with neu. *J. Steroid. Biochem.* 1989 34, 123-31.

[38] Ro, JS; el-Naggar, A; Ro, JY; Blick, M; Frye, D; Fraschini, G, et al. c-erbB-2 amplification in node-negative human breast cancer. *Cancer Res.* 1989 49, 6941-4.

[39] Suo, Z; Risberg, B; Kalsson, MG; Willman, K; Tierens, A; Skovlund, E, et al. EGFR family expression in breast carcinomas. c-erbB-2 and c-erbB-4 receptors have different effects on survival. *J. Pathol.* 2002 196, 17-25.

[40] Thor, AD; Edgerton, SM; Liu, S; Moore, DH, 2nd; Kwiatkowski, DJ. Gelsolin as a negative prognostic factor and effector of motility in erbB-2-positive epidermal growth factor receptor-positive breast cancers. *Clin. Cancer Res.* 2001 7, 2415-24.

[41] Osaki, A; Toi, M; Yamada, H; Kawami, H; Kuroi, K; Toge, T. Prognostic significance of co-expression of c-erbB-2 oncoprotein and epidermal growth factor receptor in breast cancer patients. *Am. J. Surg.* 1992 164, 323-6.

[42] DiGiovanna, MP; Stern, DF; Edgerton, SM; Whalen, SG; Moore, D, 2nd; Thor, AD. Relationship of epidermal growth factor receptor expression to ErbB-2 signaling activity and prognosis in breast cancer patients. *J. Clin. Oncol.* 2005 23, 1152-60.

[43] Gschwantler-Kaulich, D; Hudelist, G; Koestler, WJ; Czerwenka, K; Mueller, R; Helmy, S, et al. EGFR activity in HER-2 over-expressing metastatic breast cancer: evidence for simultaneous phosphorylation of Her-2/neu and EGFR. *Oncol. Rep.* 2005 14, 305-11.

[44] Toi, M; Osaki, A; Yamada, H; Toge, T. Epidermal growth factor receptor expression as a prognostic indicator in breast cancer. *Eur. J. Cancer.* 1991 27, 977-80.

[45] Christianson, TA; Doherty, JK; Lin, YJ; Ramsey, EE; Holmes, R; Keenan, EJ, et al. NH2-terminally truncated HER-2/neu protein: relationship with shedding of the extracellular domain and with prognostic factors in breast cancer. *Cancer Res.* 1998 58, 5123-9.

[46] Pupa, SM; Menard, S; Morelli, D; Pozzi, B; De Palo, G; Colnaghi, MI. The extracellular domain of the c-erbB-2 oncoprotein is released from tumor cells by proteolytic cleavage. *Oncogene.* 1993 8, 2917-23.

[47] Volas, GH; Leitzel, K; Teramoto, Y; Grossberg, H; Demers, L; Lipton, A. Serial serum c-erbB-2 levels in patients with breast carcinoma. *Cancer.* 1996 78, 267-72.

[48] Chung, FL; Schwartz, J; Herzog, CR; Yang, YM. Tea and cancer prevention: studies in animals and humans. *J Nutr.* 2003 133, 3268S-74S.

[49] Kountourakis, P; Pavlakis, K; Psyrri, A; Rontogianni, D; Xiros, N; Patsouris, E, et al. Clinicopathologic significance of EGFR and Her-2/neu in colorectal adenocarcinomas. *Cancer J.* 2006 12, 229-36.

[50] Lee, JC; Wang, ST; Chow, NH; Yang, HB. Investigation of the prognostic value of coexpressed erbB family members for the survival of colorectal cancer patients after curative surgery. *Eur. J. Cancer.* 2002 38, 1065-71.

[51] Hong, T; Nakagawa, T; Pan, W; Kim, MY; Kraus, WL; Ikehara, T, et al. Isoflavones stimulate estrogen receptor-mediated core histone acetylation. *Biochem. Biophys. Res. Commun.* 2004 317, 259-64.

[52] Maurer, CA; Friess, H; Kretschmann, B; Zimmermann, A; Stauffer, A; Baer, HU, et al. Increased expression of erbB3 in colorectal cancer is associated with concomitant increase in the level of erbB2. *Hum. Pathol.* 1998 29, 771-7.

[53] Ward, RL; Hawkins, NJ; Coomber, D; Disis, ML. Antibody immunity to the HER-2/neu oncogenic protein in patients with colorectal cancer. *Hum. Immunol.* 1999 60, 510-5.

[54] Park, DI; Kang, MS; Oh, SJ; Kim, HJ; Cho, YK; Sohn, CI, et al. HER-2/neu overexpression is an independent prognostic factor in colorectal cancer. *Int. J. Colorectal. Dis.* 2007 22, 491-7.

[55] Half, E; Broaddus, R; Danenberg, KD; Danenberg, PV; Ayers, GD; Sinicrope, FA. HER-2 receptor expression, localization, and activation in colorectal cancer cell lines and human tumors. *Int. J. Cancer.* 2004 108, 540-8.

[56] Yang, JL; Yu, Y; Markovic, B; Russell, PJ; Crowe, PJ. Overexpression of c-erbB-2 mRNA and/or c-neu oncoprotein is a predictor for metastases from colorectal cancer. *Anticancer. Res.* 1997 17, 1023-6.

[57] Sorlie, T. Molecular portraits of breast cancer: tumour subtypes as distinct disease entities. *Eur. J. Cancer.* 2004 40, 2667-75.

[58] Nielsen, TO; Hsu, FD; Jensen, K; Cheang, M; Karaca, G; Hu, Z, et al. Immunohistochemical and clinical characterization of the basal-like subtype of invasive breast carcinoma. *Clin. Cancer Res.* 2004 10, 5367-74.

[59] Rakha, E; Putti, T; Abd El-Rehim, D; Paish, C; Green, A; Powe, D, et al. Morphological and immunophenotypic analysis of breast carcinomas with basal and myoepithelial differentiation. *J. Pathol.* 2006 208, 495-506.

[60] Livasy, CA; Karaca, G; Nanda, R; Tretiakova, MS; Olopade, OI; Moore, DT, et al. Phenotypic evaluation of the basal-like subtype of invasive breast carcinoma. *Mod. Pathol.* 2006 19, 264-71.

[61] Asselin-Labat, ML; Shackleton, M; Stingl, J; Vaillant, F; Forrest, NC; Eaves, CJ, et al. Steroid hormone receptor status of mouse mammary stem cells. *J. Natl. Cancer Inst.* 2006 98, 1011-4.

[62] Tsuda, H; Tani, Y; Weisenberger, J; Kitada, S; Hasegawa, T; Murata, T, et al. Frequent KIT and epidermal growth factor receptor overexpressions in undifferentiated-type breast carcinomas with 'stem-cell-like' features. *Cancer Sci.* 2005 96, 333-9.

[63] Moiseeva, EP; Heukers, R; Manson, MM. EGFR and Src are involved in indole-3-carbinol-induced death and cell cycle arrest of human breast cancer cells. *Carcinogenesis.* 2007 28, 435-45.

[64] Al-Kuraya, K; Novotny, H; Bavi, PP; Siraj, AK; Uddin, S; Ezzat, A, et al. HER2, TOP2A, CCND1, EGFR, And C-MYC oncogene amplification in colorectal cancer. *J. Clin. Pathol.* 2007 60 768-72.

[65] Cianciulli, A; Cosimelli, M; Marzano, R; Merola, R; Piperno, G; Sperduti, I, et al. Genetic and pathologic significance of 1p, 17p, and 18q aneusomy and the ERBB2 gene in colorectal cancer and related normal colonic mucosa. *Cancer Genet. Cytogenet.* 2004 151, 52-9.

[66] Knosel, T; Yu, Y; Stein, U; Schwabe, H; Schluns, K; Schlag, PM, et al. Overexpression of c-erbB-2 protein correlates with chromosomal gain at the c-erbB-2 locus and patient survival in advanced colorectal carcinomas. *Clin. Exp. Metastasis.* 2002 19, 401-7.

[67] De Jong, KP; Stellema, R; Karrenbeld, A; Koudstaal, J; Gouw, AS; Sluiter, WJ, et al. Clinical relevance of transforming growth factor alpha, epidermal growth factor receptor, p53, and Ki67 in colorectal liver metastases and corresponding primary tumors. *Hepatology.* 1998 28, 971-9.

[68] Vizoso, F; Villar, C; Rodriguez, JC; Raigoso, P; Garcia-Muniz, JL; Allende, MT, et al. C-erbB-2 oncoprotein content in colorectal cancer and in surrounding mucosa: relationship with clinicopathologic parameters and prognostic significance. *Int. J. Surg. Investig.* 2000 1, 483-93.

[69] Ishizawar, R; Parsons, SJ. c-Src and cooperating partners in human cancer. *Cancer Cell.* 2004 6, 209-14.

[70] Diaz, N; Minton, S; Cox, C; Bowman, T; Gritsko, T; Garcia, R, et al. Activation of stat3 in primary tumors from high-risk breast cancer patients is associated with elevated levels of activated SRC and survivin expression. *Clin. Cancer Res.* 2006 12, 20-8.

[71] Belsches-Jablonski, AP; Biscardi, JS; Peavy, DR; Tice, DA; Romney, DA; Parsons, SJ. Src family kinases and HER2 interactions in human breast cancer cell growth and survival. *Oncogene.* 2001 20, 1465-75.

[72] Muthuswamy, SK; Siegel, PM; Dankort, DL; Webster, MA; Muller, WJ. Mammary tumors expressing the neu proto-oncogene possess elevated c-Src tyrosine kinase activity. *Mol. Cell Biol.* 1994 14, 735-43.

[73] Talamonti, MS; Roh, MS; Curley, SA; Gallick, GE. Increase in activity and level of pp60c-src in progressive stages of human colorectal cancer. *J. Clin. Invest.* 1993 91, 53-60.

[74] Mao, W; Irby, R; Coppola, D; Fu, L; Wloch, M; Turner, J, et al. Activation of c-Src by receptor tyrosine kinases in human colon cancer cells with high metastatic potential. *Oncogene.* 1997 15, 3083-90.

[75] Webster, MA; Cardiff, RD; Muller, WJ. Induction of mammary epithelial hyperplasias and mammary tumors in transgenic mice expressing a murine mammary tumor virus/activated c-src fusion gene. *Proc. Natl. Acad. Sci. U S A.* 1995 92, 7849-53.

[76] Frame, MC. Src in cancer: deregulation and consequences for cell behaviour. *Biochim Biophys. Acta.* 2002 1602, 114-30.

[77] Myoui, A; Nishimura, R; Williams, PJ; Hiraga, T; Tamura, D; Michigami, T, et al. C-SRC tyrosine kinase activity is associated with tumor colonization in bone and lung in an animal model of human breast cancer metastasis. *Cancer Res.* 2003 63, 5028-33.

[78] Malek, RL; Irby, RB; Guo, QM; Lee, K; Wong, S; He, M, et al. Identification of Src transformation fingerprint in human colon cancer. *Oncogene.* 2002 21, 7256-65.

[79] Irby, RB; Malek, RL; Bloom, G; Tsai, J; Letwin, N; Frank, BC, et al. Iterative microarray and RNA interference-based interrogation of the SRC-induced invasive phenotype. *Cancer Res.* 2005 65, 1814-21.

[80] Coluccia, AM; Benati, D; Dekhil, H; De Filippo, A; Lan, C; Gambacorti-Passerini, C. SKI-606 decreases growth and motility of colorectal cancer cells by preventing pp60(c-Src)-dependent tyrosine phosphorylation of beta-catenin and its nuclear signaling. *Cancer Res.* 2006 66, 2279-86.

[81] Maa, MC; Leu, TH; McCarley, DJ; Schatzman, RC; Parsons, SJ. Potentiation of epidermal growth factor receptor-mediated oncogenesis by c-Src: implications for the etiology of multiple human cancers. *Proc. Natl .Acad. Sci. U S A.* 1995 92, 6981-5.

[82] Tice, DA; Biscardi, JS; Nickles, AL; Parsons, SJ. Mechanism of biological synergy between cellular Src and epidermal growth factor receptor. *Proc. Natl. Acad. Sci. U S A.* 1999 96, 1415-20.

[83] Tan, M; Li, P; Klos, KS; Lu, J; Lan, KH; Nagata, Y, et al. ErbB2 promotes Src synthesis and stability: novel mechanisms of Src activation that confer breast cancer metastasis. *Cancer Res.* 2005 65, 1858-67.

[84] Biscardi, JS; Belsches, AP; Parsons, SJ. Characterization of human epidermal growth factor receptor and c-Src interactions in human breast tumor cells. *Mol. Carcinog.* 1998 21, 261-72.

[85] Biscardi, JS; Maa, MC; Tice, DA; Cox, ME; Leu, TH; Parsons, SJ. c-Src-mediated phosphorylation of the epidermal growth factor receptor on Tyr845 and Tyr1101 is associated with modulation of receptor function. *J. Biol. Chem.* 1999 274, 8335-43.

[86] Kim, H; Chan, R; Dankort, DL; Zuo, D; Najoukas, M; Park, M, et al. The c-Src tyrosine kinase associates with the catalytic domain of ErbB-2: implications for ErbB-2 mediated signaling and transformation. *Oncogene.* 2005 24, 7599-607.

[87] Luttrell, DK; Lee, A; Lansing, TJ; Crosby, RM; Jung, KD; Willard, D, et al. Involvement of pp60c-src with two major signaling pathways in human breast cancer. *Proc. Natl. Acad. Sci. U S A.* 1994 91, 83-7.

[88] Cabioglu, N; Summy, J; Miller, C; Parikh, NU; Sahin, AA; Tuzlali, S, et al. CXCL-12/stromal cell-derived factor-1alpha transactivates HER2-neu in breast cancer cells by a novel pathway involving Src kinase activation. *Cancer Res.* 2005 65, 6493-7.

[89] Ishizawar, RC; Miyake, T; Parsons, SJ. c-Src modulates ErbB2 and ErbB3 heterocomplex formation and function. *Oncogene.* 2007 26, 3503-10.

[90] Bose, R; Molina, H; Patterson, AS; Bitok, JK; Periaswamy, B; Bader, JS, et al. Phosphoproteomic analysis of Her2/neu signaling and inhibition. *Proc. Natl. Acad. Sci. U S A.* 2006 103, 9773-8.

[91] Boerner, JL; Demory, ML; Silva, C; Parsons, SJ. Phosphorylation of Y845 on the epidermal growth factor receptor mediates binding to the mitochondrial protein cytochrome c oxidase subunit II. *Mol. Cell Biol.* 2004 24, 7059-71.

[92] Bianchi, F; Magnifico, A; Olgiati, C; Zanesi, N; Pekarsky, Y; Tagliabue, E, et al. FHIT-proteasome degradation caused by mitogenic stimulation of the EGF receptor family in cancer cells. *Proc. Natl. Acad. Sci. U S A.* 2006 103, 18981-6.

[93] Stover, DR; Becker, M; Liebetanz, J; Lydon, NB. Src phosphorylation of the epidermal growth factor receptor at novel sites mediates receptor interaction with Src and P85 alpha. *J.Biol. Chem.* 1995 270, 15591-7.

[94] Garcia, R; Bowman, TL; Niu, G; Yu, H; Minton, S; Muro-Cacho, CA, et al. Constitutive activation of Stat3 by the Src and JAK tyrosine kinases participates in growth regulation of human breast carcinoma cells. *Oncogene.* 2001 20, 2499-513.

[95] Guo, W; Pylayeva, Y; Pepe, A; Yoshioka, T; Muller, WJ; Inghirami, G, et al. Beta 4 integrin amplifies ErbB2 signaling to promote mammary tumorigenesis. *Cell.* 2006 126, 489-502.

[96] Sato, K; Sato, A; Aoto, M; Fukami, Y. c-Src phosphorylates epidermal growth factor receptor on tyrosine 845. *Biochem. Biophy.s Res. Commun.* 1995 215, 1078-87.

[97] Kloth, MT; Laughlin, KK; Biscardi, JS; Boerner, JL; Parsons, SJ; Silva, CM. STAT5b, a Mediator of Synergism between c-Src and the Epidermal Growth Factor Receptor. *J. Biol. Chem.* 2003 278, 1671-9.

[98] Zhang, HT; O'Rourke, DM; Zhao, H; Murali, R; Mikami, Y; Davis, JG, et al. Absence of autophosphorylation site Y882 in the p185neu oncogene product correlates with a reduction of transforming potential. *Oncogene.* 1998 16, 2835-42.

[99] Boerner, JL; Biscardi, JS; Silva, CM; Parsons, SJ. Transactivating agonists of the EGF receptor require Tyr 845 phosphorylation for induction of DNA synthesis. *Mol. Carcinog.* 2005 44, 262-73.

[100] Reginato, MJ; Mills, KR; Becker, EB; Lynch, DK; Bonni, A; Muthuswamy, SK, et al. Bim regulation of lumen formation in cultured mammary epithelial acini is targeted by oncogenes. *Mol. Cell Biol.* 2005 25, 4591-601.

[101] Reginato, MJ; Mills, KR; Paulus, JK; Lynch, DK; Sgroi, DC; Debnath, J, et al. Integrins and EGFR coordinately regulate the pro-apoptotic protein Bim to prevent anoikis. *Nat. Cell Biol.* 2003 5, 733-40.

[102] Sharma, SV; Gajowniczek, P; Way, IP; Lee, DY; Jiang, J; Yuza, Y, et al. A common signaling cascade may underlie "addiction" to the Src, BCR-ABL, and EGF receptor oncogenes. *Cancer Cell.* 2006 10, 425-35.

[103] Higdon, JV; Delage, B; Williams, DE; Dashwood, RH. Cruciferous vegetables and human cancer risk: epidemiologic evidence and mechanistic basis. *Pharmacol. Res.* 2007 55, 224-36.

[104] Naik, R; Nixon, S; Lopes, A; Godfrey, K; Hatem, MH; Monaghan, JM. A randomized phase II trial of indole-3-carbinol in the treatment of vulvar intraepithelial neoplasia. *Int. J. Gynecol. Cancer.* 2006 16, 786-90.

[105] Rosen, CA; Bryson, PC. Indole-3-carbinol for recurrent respiratory papillomatosis: long-term results. *J. Voice.* 2004 18, 248-53.

[106] Bell, MC; Crowley-Nowick, P; Bradlow, HL; Sepkovic, DW; Schmidt-Grimminger, D; Howell, P, et al. Placebo-controlled trial of indole-3-carbinol in the treatment of CIN. *Gynecol. Oncol.* 2000 78, 123-9.

[107] Howells, LM; Moiseeva, EP; Neal, CP; Foreman, BE; Andreadi, CK; Sun, YY, et al. Predicting the physiological relevance of in vitro cancer preventive activities of phytochemicals. *Acta Pharmacol. Sin.* 2007 28, 1274-304.

[108] Aggarwal, BB; Ichikawa, H. Molecular Targets and Anticancer Potential of Indole-3-Carbinol and Its Derivatives. *Cell Cycle.* 2005 4.

[109] Chinni, SR; Sarkar, FH. Akt inactivation is a key event in indole-3-carbinol-induced apoptosis in PC-3 cells. *Clin. Cancer Res.* 2002 8, 1228-36.

[110] Li, Y; Chinni, SR; Sarkar, FH. Selective growth regulatory and pro-apoptotic effects of DIM is mediated by AKT and NF-kappaB pathways in prostate cancer cells. *Front Biosci.* 2005 10, 236-43.

[111] Li, Y; Li, X; Sarkar, FH. Gene expression profiles of I3C- and DIM-treated PC3 human prostate cancer cells determined by cDNA microarray analysis. *J. Nutr.* 2003 133, 1011-9.

[112] Moiseeva, EP; Fox, LH; Howells, LM; Temple, LA; Manson, MM. Indole-3-carbinol-induced death in cancer cells involves EGFR downregulation and is exacerbated in a 3D environment. *Apoptosis.* 2006 11, 799-812.

[113] Manson, MM; Hudson, EA; Ball, HW; Barrett, MC; Clark, HL; Judah, DJ, et al. Chemoprevention of aflatoxin B1-induced carcinogenesis by indole-3-carbinol in rat liver - predicting the outcome using early biomarkers. *Carcinogenesis.* 1998 19, 1829-36.

[114] Moiseeva, EP; Heukers, R. Indole-3-carbinol-induced modulation of NF-kappaB signalling is breast cancer cell-specific and does not correlate with cell death. *Breast Cancer Res Treat.* 2007 Epub ahead of print.

[115] Moiseeva, EP; Verschoyle, RD; Manson, MM. Susceptibility to indole-3-carbinol is related to Src in primary Tag mouse mammary tumours and EGFR/Her2 overexpression in human breast cancer cell lines. B37. NCRI conference; 2007; Birmigham, UK; 2007.

[116] McGuire, KP; Ngoubilly, N; Neavyn, M; Lanza-Jacoby, S. 3,3'-Diindolylmethane and Paclitaxel Act Synergistically to Promote Apoptosis in HER2/Neu Human Breast Cancer Cells. *J Surg Res.* 2006 132, 208-13.

[117] Pervaiz, S. Resveratrol: from grapevines to mammalian biology. *Faseb. J.* 2003 17, 1975-85.

[118] Shankar, S; Singh, G; Srivastava, RK. Chemoprevention by resveratrol: molecular mechanisms and therapeutic potential. *Front Biosci.* 2007 12, 4839-54.

[119] Gehm, BD; McAndrews, JM; Chien, PY; Jameson, JL. Resveratrol, a polyphenolic compound found in grapes and wine, is an agonist for the estrogen receptor. *Proc. Natl. Acad. Sci. U S A.* 1997 94, 14138-43.

[120] Bowers, JL; Tyulmenkov, VV; Jernigan, SC; Klinge, CM. Resveratrol acts as a mixed agonist/antagonist for estrogen receptors alpha and beta. *Endocrinology.* 2000 141, 3657-67.

[121] Levi, F; Pasche, C; Lucchini, F; Ghidoni, R; Ferraroni, M; La Vecchia, C. Resveratrol and breast cancer risk. *Eur. J. Cancer Prev.* 2005 14, 139-42.

[122] Do, MH; Lee, SS; Jung, PJ; Lee, MH. Intake of fruits, vegetables, and soy foods in relation to breast cancer risk in Korean women: a case-control study. *Nutr Cancer.* 2007 57, 20-7.

[123] Stewart, JR; O'Brian, CA. Resveratrol antagonizes EGFR-dependent Erk1/2 activation in human androgen-independent prostate cancer cells with associated isozyme-selective PKC alpha inhibition. *Invest New Drugs.* 2004 22, 107-17.

[124] Lin, MT; Yen, ML; Lin, CY; Kuo, ML. Inhibition of vascular endothelial growth factor-induced angiogenesis by resveratrol through interruption of Src-dependent vascular endothelial cadherin tyrosine phosphorylation. *Mol. Pharmacol.* 2003 64, 1029-36.

[125] Kotha, A; Sekharam, M; Cilenti, L; Siddiquee, K; Khaled, A; Zervos, AS, et al. Resveratrol inhibits Src and Stat3 signaling and induces the apoptosis of malignant cells containing activated Stat3 protein. *Mol. Cancer Ther.* 2006 5, 621-9.

[126] Su, JL; Yang, CY; Zhao, M; Kuo, ML; Yen, ML. Forkhead proteins are critical for bone morphogenetic protein-2 regulation and anti-tumor activity of resveratrol. *J. Biol. Chem.* 2007 282, 19385-98.

[127] Kaneuchi, M; Sasaki, M; Tanaka, Y; Yamamoto, R; Sakuragi, N; Dahiya, R. Resveratrol suppresses growth of Ishikawa cells through down-regulation of EGF. *Int. J. Oncol.* 2003 23, 1167-72.

[128] Shih, A; Zhang, S; Cao, HJ; Boswell, S; Wu, YH; Tang, HY, et al. Inhibitory effect of epidermal growth factor on resveratrol-induced apoptosis in prostate cancer cells is mediated by protein kinase C-alpha. *Mol. Cancer Ther.* 2004 3, 1355-64.

[129] Godichaud, S; Si-Tayeb, K; Auge, N; Desmouliere, A; Balabaud, C; Payrastre, B, et al. The grape-derived polyphenol resveratrol differentially affects epidermal and platelet-derived growth factor signaling in human liver myofibroblasts. *Int. J. Biochem. Cell Biol.* 2006 38, 629-37.

[130] Haider, UG; Roos, TU; Kontaridis, MI; Neel, BG; Sorescu, D; Griendling, KK, et al. Resveratrol inhibits angiotensin II- and epidermal growth factor-mediated Akt activation: role of Gab1 and Shp2. *Mol Pharmacol.* 2005 68, 41-8.

[131] Tang, FY; Chiang, EP; Sun, YC. Resveratrol inhibits heregulin-beta1-mediated matrix metalloproteinase-9 expression and cell invasion in human breast cancer cells. *J. Nutr. Biochem.* 2007 Epub ahead of print.

[132] Provinciali, M; Re, F; Donnini, A; Orlando, F; Bartozzi, B; Di Stasio, G, et al. Effect of resveratrol on the development of spontaneous mammary tumors in HER-2/neu transgenic mice. *Int. J. Cancer.* 2005 115, 36-45.

[133] Choi, HK; Yang, JW; Kang, KW. Bifunctional effect of resveratrol on the expression of ErbB2 in human breast cancer cell. *Cancer Lett.* 2006 242, 198-206.

[134] Thomasset, SC; Berry, DP; Garcea, G; Marczylo, T; Steward, WP; Gescher, AJ. Dietary polyphenolic phytochemicals--promising cancer chemopreventive agents in humans? A review of their clinical properties. *Int. J. Cancer.* 2007 120, 451-8.

[135] Korutla, L; Cheung, JY; Mendelsohn, J; Kumar, R. Inhibition of ligand-induced activation of epidermal growth factor receptor tyrosine phosphorylation by curcumin. *Carcinogenesis.* 1995 16, 1741-5.

[136] Korutla, L; Kumar, R. Inhibitory effect of curcumin on epidermal growth factor receptor kinase activity in A431 cells. *Biochim Biophys Acta.* 1994 1224, 597-600.

[137] Squires, MS; Hudson, EA; Howells, L; Sale, S; Houghton, CE; Jones, JL, et al. Relevance of mitogen activated protein kinase (MAPK) and phosphotidylinositol-3-kinase/protein kinase B (PI3K/PKB) pathways to induction of apoptosis by curcumin in breast cells. *Biochem. Pharmacol.* 2003 65, 361-76.

[138] Kim, JH; Xu, C; Keum, YS; Reddy, B; Conney, A; Kong, AN. Inhibition of EGFR signaling in human prostate cancer PC-3 cells by combination treatment with beta-phenylethyl isothiocyanate and curcumin. *Carcinogenesis.* 2006 27, 475-82.

[139] Chen, A; Xu, J. Activation of PPAR{gamma} by curcumin inhibits Moser cell growth and mediates suppression of gene expression of cyclin D1 and EGFR. *Am. J. Physiol. Gastrointest. Liver Physiol.* 2005 288, G447-56.

[140] Chen, A; Xu, J; Johnson, AC. Curcumin inhibits human colon cancer cell growth by suppressing gene expression of epidermal growth factor receptor through reducing the activity of the transcription factor Egr-1. *Oncogene.* 2006 25, 278-87.

[141] Hong, RL; Spohn, WH; Hung, MC. Curcumin inhibits tyrosine kinase activity of p185neu and also depletes p185neu. *Clin. Cancer Res.* 1999 5, 1884-91.

[142] Leu, TH; Su, SL; Chuang, YC; Maa, MC. Direct inhibitory effect of curcumin on Src and focal adhesion kinase activity. *Biochem Pharmacol.* 2003 66, 2323-31.

[143] Reddy, S; Aggarwal, BB. Curcumin is a non-competitive and selective inhibitor of phosphorylase kinase. *FEBS Lett.* 1994 341, 19-22.

[144] Bettuzzi, S; Brausi, M; Rizzi, F; Castagnetti, G; Peracchia, G; Corti, A. Chemoprevention of human prostate cancer by oral administration of green tea catechins in volunteers with high-grade prostate intraepithelial neoplasia: a preliminary report from a one-year proof-of-principle study. *Cancer Res.* 2006 66, 1234-40.

[145] Liang, YC; Lin-shiau, SY; Chen, CF; Lin, JK. Suppression of extracellular signals and cell proliferation through EGF receptor binding by (-)-epigallocatechin gallate in human A431 epidermoid carcinoma cells. *J. Cell. Biochem.* 1997 67, 55-65.

[146] Bain, J; McLauchlan, H; Elliott, M; Cohen, P. The specificities of protein kinase inhibitors: an update. *Biochem. J.* 2003 371, 199-204.

[147] Pianetti, S; Guo, S; Kavanagh, KT; Sonenshein, GE. Green tea polyphenol epigallocatechin-3 gallate inhibits Her-2/neu signaling, proliferation, and transformed phenotype of breast cancer cells. *Cancer Res.* 2002 62, 652-5.

[148] Guo, S; Lu, J; Subramanian, A; Sonenshein, GE. Microarray-assisted pathway analysis identifies mitogen-activated protein kinase signaling as a mediator of resistance to the green tea polyphenol epigallocatechin 3-gallate in her-2/neu-overexpressing breast cancer cells. *Cancer Res.* 2006 66, 5322-9.

[149] Masuda, M; Suzui, M; Weinstein, IB. Effects of epigallocatechin-3-gallate on growth, epidermal growth factor receptor signaling pathways, gene expression, and chemosensitivity in human head and neck squamous cell carcinoma cell lines. *Clin Cancer Res.* 2001 7, 4220-9.

[150] Masuda, M; Suzui, M; Lim, JT; Deguchi, A; Soh, JW; Weinstein, IB. Epigallocatechin-3-gallate decreases VEGF production in head and neck and breast carcinoma cells by inhibiting EGFR-related pathways of signal transduction. *J. Exp. Ther. Oncol.* 2002 2, 350-9.

[151] Masuda, M; Suzui, M; Lim, JT; Weinstein, IB. Epigallocatechin-3-gallate inhibits activation of HER-2/neu and downstream signaling pathways in human head and neck and breast carcinoma cells. *Clin. Cancer Res.* 2003 9, 3486-91.

[152] Sah, JF; Balasubramanian, S; Eckert, RL; Rorke, EA. Epigallocatechin-3-gallate inhibits epidermal growth factor receptor signaling pathway. Evidence for direct inhibition of ERK1/2 and AKT kinases. *J. Biol. Chem.* 2004 279, 12755-62.

[153] Shimizu, M; Deguchi, A; Lim, JT; Moriwaki, H; Kopelovich, L; Weinstein, IB. (-)-Epigallocatechin gallate and polyphenon E inhibit growth and activation of the epidermal growth factor receptor and human epidermal growth factor receptor-2 signaling pathways in human colon cancer cells. *Clin. Cancer Res.* 2005 11, 2735-46.

[154] Shimizu, M; Deguchi, A; Joe, AK; McKoy, JF; Moriwaki, H; Weinstein, IB. EGCG inhibits activation of HER3 and expression of cyclooxygenase-2 in human colon cancer cells. *J. Exp. Ther. Oncol.* 2005 5, 69-78.

[155] Hou, Z; Sang, S; You, H; Lee, MJ; Hong, J; Chin, KV, et al. Mechanism of action of (-)-epigallocatechin-3-gallate: auto-oxidation-dependent inactivation of epidermal growth factor receptor and direct effects on growth inhibition in human esophageal cancer KYSE 150 cells. *Cancer Res.* 2005 65, 8049-56.

[156] Adachi, S; Nagao, T; Ingolfsson, HI; Maxfield, FR; Andersen, OS; Kopelovich, L, et al. The inhibitory effect of (-)-epigallocatechin gallate on activation of the epidermal growth factor receptor is associated with altered lipid order in HT29 colon cancer cells. *Cancer Res.* 2007 67, 6493-501.

[157] Fu, Y; Chen, A. The phyto-chemical (-)-epigallocatechin gallate suppresses gene expression of epidermal growth factor receptor in rat hepatic stellate cells in vitro by reducing the activity of Egr-1. *Biochem Pharmacol.* 2006 72, 227-38.

[158] Thompson, LU; Boucher, BA; Liu, Z; Cotterchio, M; Kreiger, N. Phytoestrogen content of foods consumed in Canada, including isoflavones, lignans, and coumestan. *Nutr. Cancer.* 2006 54, 184-201.

[159] Martin, PM; Horwitz, KB; Ryan, DS; McGuire, WL. Phytoestrogen interaction with estrogen receptors in human breast cancer cells. *Endocrinology.* 1978 103, 1860-7.

[160] Kuiper, GG; Lemmen, JG; Carlsson, B; Corton, JC; Safe, SH; van der Saag, PT, et al. Interaction of estrogenic chemicals and phytoestrogens with estrogen receptor beta. *Endocrinology.* 1998 139, 4252-63.

[161] Hebert, JR; Hurley, TG; Olendzki, BC; Teas, J; Ma, Y; Hampl, JS. Nutritional and socioeconomic factors in relation to prostate cancer mortality: a cross-national study. *J. Natl. Cancer Inst.* 1998 90, 1637-47.

[162] Lee, HP; Gourley, L; Duffy, SW; Esteve, J; Lee, J; Day, NE. Dietary effects on breast-cancer risk in Singapore. *Lancet.* 1991 337, 1197-200.

[163] Dai, Q; Shu, XO; Jin, F; Potter, JD; Kushi, LH; Teas, J, et al. Population-based case-control study of soyfood intake and breast cancer risk in Shanghai. *Br. J. Cancer.* 2001 85, 372-8.

[164] Zhang, M; Xie, X; Lee, AH; Binns, CW. Soy and isoflavone intake are associated with reduced risk of ovarian cancer in southeast china. *Nutr. Cancer.* 2004 49, 125-30.

[165] Goodman, MT; Wilkens, LR; Hankin, JH; Lyu, LC; Wu, AH; Kolonel, LN. Association of soy and fiber consumption with the risk of endometrial cancer. *Am. J. Epidemiol.* 1997 146, 294-306.

[166] Keinan-Boker, L; van Der Schouw, YT; Grobbee, DE; Peeters, PH. Dietary phytoestrogens and breast cancer risk. *Am. J. Clin. Nutr.* 2004 79, 282-8.

[167] Grace, PB; Taylor, JI; Low, YL; Luben, RN; Mulligan, AA; Botting, NP, et al. Phytoestrogen concentrations in serum and spot urine as biomarkers for dietary phytoestrogen intake and their relation to breast cancer risk in European prospective investigation of cancer and nutrition-norfolk. *Cancer Epidemiol Biomarkers Prev.* 2004 13, 698-708.

[168] Linseisen, J; Piller, R; Hermann, S; Chang-Claude, J. Dietary phytoestrogen intake and premenopausal breast cancer risk in a German case-control study. *Int. J. Cancer.* 2004 110, 284-90.

[169] Ingram, D; Sanders, K; Kolybaba, M; Lopez, D. Case-control study of phyto-oestrogens and breast cancer. *Lancet.* 1997 350, 990-4.

[170] Piller, R; Chang-Claude, J; Linseisen, J. Plasma enterolactone and genistein and the risk of premenopausal breast cancer. *Eur J Cancer Prev.* 2006 15, 225-32.

[171] Peeters, PH; Slimani, N; van der Schouw, YT; Grace, PB; Navarro, C; Tjonneland, A, et al. Variations in Plasma Phytoestrogen Concentrations in European Adults. *J. Nutr.* 2007 137, 1294-300.

[172] Akiyama, T; Ishida, J; Nakagawa, S; Ogawara, H; Watanabe, S; Itoh, N, et al. Genistein, a specific inhibitor of tyrosine-specific protein kinases. *J. Biol. Chem.* 1987 262, 5592-5.

[173] Dean, NM; Kanemitsu, M; Boynton, AL. Effects of the tyrosine-kinase inhibitor genistein on DNA synthesis and phospholipid-derived second messenger generation in mouse 10T1/2 fibroblasts and rat liver T51B cells *Biochem Biophys. Res Commun.* 1989 165, 795-801.

[174] El-Zarruk, AA; van den Berg, HW. The anti-proliferative effects of tyrosine kinase inhibitors towards tamoxifen-sensitive and tamoxifen-resistant human breast cancer cell lines in relation to the expression of epidermal growth factor receptors (EGF-R) and the inhibition of EGF-R tyrosine kinase. *Cancer Lett.* 1999 142, 185-93.

[175] Mai, Z; Blackburn, GL; Zhou, JR. Soy phytochemicals synergistically enhance the preventive effect of tamoxifen on the growth of estrogen-dependent human breast carcinoma in mice. *Carcinogenesis.* 2007 28, 1217-23.

[176] Clark, JW; Santos-Moore, A; Stevenson, LE; Frackelton, AR, Jr. Effects of tyrosine kinase inhibitors on the proliferation of human breast cancer cell lines and proteins important in the ras signaling pathway. *Int. J. Cancer.* 1996 65, 186-91.

[177] Theodorescu, D; Laderoute, KR; Calaoagan, JM; Guilding, KM. Inhibition of human bladder cancer cell motility by genistein is dependent on epidermal growth factor receptor but not p21ras gene expression. *Int. J. Cancer.* 1998 78, 775-82.

[178] Penar, PL; Khoshyomn, S; Bhushan, A; Tritton, TR. Inhibition of epidermal growth factor receptor-associated tyrosine kinase blocks glioblastoma invasion of the brain. *Neurosurgery.* 1997 40, 141-51.

[179] Bai, J; Sata, N; Nagai, H; Wada, T; Yoshida, K; Mano, H, et al. Genistein-induced changes in gene expression in Panc 1 cells at physiological concentrations of genistein. *Pancreas.* 2004 29, 93-8.

[180] El-Rayes, BF; Ali, S; Ali, IF; Philip, PA; Abbruzzese, J; Sarkar, FH. Potentiation of the effect of erlotinib by genistein in pancreatic cancer: the role of Akt and nuclear factor-kappaB. *Cancer Res.* 2006 66, 10553-9.

[181] Gong, L; Li, Y; Nedeljkovic-Kurepa, A; Sarkar, FH. Inactivation of NF-kappaB by genistein is mediated via Akt signaling pathway in breast cancer cells. *Oncogene.* 2003 22, 4702-9.

[182] Katdare, M; Osborne, M; Telang, NT. Soy isoflavone genistein modulates cell cycle progression and induces apoptosis in HER-2/neu oncogene expressing human breast epithelial cells. *Int. J. Oncol.* 2002 21, 809-15.

[183] Alhasan, SA; Aranha, O; Sarkar, FH. Genistein elicits pleiotropic molecular effects on head and neck cancer cells. *Clin. Cancer Res.* 2001 7, 4174-81.

[184] Mai, Z; Blackburn, GL; Zhou, JR. Genistein sensitizes inhibitory effect of tamoxifen on the growth of estrogen receptor-positive and HER2-overexpressing human breast cancer cells. *Mol. Carcinog.* 2007.

[185] Satoh, H; Nishikawa, K; Suzuki, K; Asano, R; Virgona, N; Ichikawa, T, et al. Genistein, a soy isoflavone, enhances necrotic-like cell death in a breast cancer cell treated with a chemotherapeutic agent. *Res Commun. Mol. Pathol. Pharmacol.* 2003 113-114, 149-58.

[186] Manach, C; Williamson, G; Morand, C; Scalbert, A; Remesy, C. Bioavailability and bioefficacy of polyphenols in humans. I. Review of 97 bioavailability studies. *Am. J. Clin. Nutr.* 2005 81, 230S-42S.

[187] Cotroneo, MS; Wang, J; Fritz, WA; Eltoum, IE; Lamartiniere, CA. Genistein action in the prepubertal mammary gland in a chemoprevention model. *Carcinogenesis.* 2002 23, 1467-74.

[188] Lamartiniere, CA; Cotroneo, MS; Fritz, WA; Wang, J; Mentor-Marcel, R; Elgavish, A. Genistein chemoprevention: timing and mechanisms of action in murine mammary and prostate. *J. Nutr.* 2002 132, 552S-8S.

[189] Brown, NM; Wang, J; Cotroneo, MS; Zhao, YX; Lamartiniere, CA. Prepubertal genistein treatment modulates TGF-alpha, EGF and EGF-receptor mRNAs and proteins in the rat mammary gland. *Mol. Cell. Endocrinol.* 1998 144, 149-65.

[190] Rowell, C; Carpenter, DM; Lamartiniere, CA. Chemoprevention of breast cancer, proteomic discovery of genistein action in the rat mammary gland. *J. Nutr.* 2005 135, 2953S-9S.

[191] Cotroneo, MS; Fritz, WA; Lamartiniere, CA. Dynamic profiling of estrogen receptor and epidermal growth factor signaling in the uteri of genistein- and estrogen-treated rats. *Food Chem. Toxicol.* 2005 43, 637-45.

[192] Jin, Z; MacDonald, RS. Soy isoflavones increase latency of spontaneous mammary tumors in mice. *J. Nutr.* 2002 132, 3186-90.

[193] Dalu, A; Haskell, JF; Coward, L; Lamartiniere, CA. Genistein, a component of soy, inhibits the expression of the EGF and ErbB2/Neu receptors in the rat dorsolateral prostate. *Prostate.* 1998 37, 36-43.

[194] Wang, J; Eltoum, IE; Lamartiniere, CA. Dietary genistein suppresses chemically induced prostate cancer in Lobund-Wistar rats. *Cancer Lett.* 2002 186, 11-8.

[195] Wang, J; Eltoum, IE; Lamartiniere, CA. Genistein alters growth factor signaling in transgenic prostate model (TRAMP). *Mol Cell Endocrinol.* 2004 219, 171-80.

[196] Manson, MM. Inhibition of survival signalling by dietary polyphenols and indole-3-carbinol. *Eur. J. Cancer.* 2005 41, 1842-53.

In: Oncogene Proteins: New Research
Editors: A. H. Malloy and E. C. Carson, pp. 99-128
ISBN 978-1-60456-290-3

Chapter 3

Review: Cyclin D1 as a Therapeutic Target in Cancer

John P. Alao
Department of Cell and Molecular Biology, Gothenburg University, Gothenburg, Sweden

Abstract:

Cyclin dependent kinases (CDKs) and their activating cyclins are required for the ordered regulation of cell cycle progression throughout the cell cycle. Cyclin D1 forms active complexes with CDK4 or CDK6 that are essential for G1 to S phase progression. Together, these complexes phosphorylate the retinoblastoma (Rb) protein and facilitate E2F mediated gene expression. The E2F transcription factor regulates the expression of several genes required for progression into S phase. Cyclin D1 protein levels are elevated in a wide variety of cancers, including those of the breast, lung, oesophagus and prostate. The defective regulation of cyclin D1 degradation has been linked the elevated levels of cyclin D1 in many cancers. The ubiquitin dependent degradation pathway plays an important role in regulating cyclin D1 stability. Phosphorylation of threonine residue 286 and/ or 288 further enhances the rate of cyclin D1 degradation via this pathway. Additionally, cyclin D1 degradation can also occur independently of ubiquitin. Recent studies have identified the F-box proteins FBX4 and FBXW8 as E3-ligases that mediate cyclin D1 degradation. The loss of C-terminal regulatory sequences or expression of proteins that regulate cyclin D1 stability (and their cofactors) has been linked to deregulated cyclin D1 degradation. Numerous clinical and experimental pharmacological agents have been shown to induce cyclin D1 degradation. Generally, these agents induce the degradation of cyclin D1 via the ubiquitin dependent pathway. The functional characterization of FBX4 and FBXW8 is likely to increase current knowledge on the molecular pharmacology of these agents. Together, these findings suggest that cyclin D1 presents an important therapeutic target for cancer therapy. Currently, cyclin D1 ablative agents are in development and the utility of this treatment strategy has been demonstrated in clinical trials. In this chapter, recent advances in our understanding of the normal and pathological roles of cyclin D1 in cellular proliferation are discussed. The mechanisms that underlie the elevated levels of cyclin D1 in cancer cells are also explored. Finally, the development and current status of cyclin D1 as a therapeutic target in cancer therapy is discussed.

Introduction

Cyclin D1 is a small approximately 34 kD protein encoded by the *CCND1* gene and is important for G1 to S phase progression during the cell cycle. Cyclin D1 forms active complexes with the cyclin dependent kinases (CDKs) CDK4 and CDK6. Cyclin D1- CDK4/ 6 complexes phosphorylate several residues on the retinoblastoma (Rb) tumour suppressor protein resulting in its hyperphosphorylation. Rb hyperphosphorylation in turn results in the release of Rb-sequestered E2F transcription factors and the subsequent expression of genes required for progression through S phase (Baldin et al., 1993; Lukas et al., 1994; Weinberg, 1995). Cyclin D1 levels accumulate during G1 and then decline rapidly at the G1-S phase boundary (Baldin et al., 1993). The microinjection of antibodies directed against cyclin D1 prevented S phase entry thus demonstrating its importance during G1 cell cycle progression. Indeed, cyclin D1 was found to be essential for the entry of quiescent cells into the cell cycle (Baldin et al., 1993). Conversely, cyclin D1 degradation is required for the proper replication of DNA and S phase progression (Fukami-Kobayashi and Mitsui, 1999; Guo et al., 2005a; Guo et al., 2005b; Pagano et al., 1994). Cyclin D1 acts as a cellular sensor of the mitogenic signalling and couples environmental signals to the cell cycle regulation machinery. As a consequence, the loss of mitogenic stimulation (e.g. through serum withdrawal) attenuates *CCND1* expression and is accompanied by the rapid decline of cellular cyclin D1 levels (Aktas et al., 1997; Herber et al., 1994; Matsushime et al., 1991; Muller et al., 1994). Recent studies suggest that the mitogen dependent rise of cyclin D1 levels during G2 is essential for continuous proliferation (Guo et al., 2005a; Guo et al., 2005b; Hitomi and Stacey, 1999a; Hitomi and Stacey, 1999b; Hitomi and Stacey, 2001). Cell division is thus highly dependent on sustained mitogenic signalling from the extracellular environment. Several signalling pathways have been shown to regulate cyclin D1 expression. These include the ras regulated mitogen activated protein kinase (MAPK), the protein kinase B (PKB)/ Akt and the Wnt/ β-catenin pathways (Albanese et al., 1995; Roovers and Assoian, 2000; Tetsu and McCormick, 1999; Vivanco and Sawyers, 2002). In estrogen responsive tissues, estrogen receptor alpha (ERα) stimulation also leads to the induction of the *CCND1* promoter (Sabbah et al., 1999). These pathways are often deregulated in cancer where they positively influence the level of cyclin D1 expression.

In addition to regulating cell cycle progression, cyclin D1 has also been shown to regulate the activity of several transcription factors. In general, this regulation of transcription factor activity occurs independently of CDK4 activity. The role of cyclin D1 in regulating the activity of transcription factors has been extensively reviewed in (Coqueret, 2002; Knudsen, 2006). Subsequent section in this chapter will briefly consider the interaction of cyclin D1 with transcription factors in the context of cancer development and progression.

The activity of cyclin D1- CDK4/6 complexes is negatively regulated by several small proteins collectively known as cyclin dependent kinase inhibitors (CDIs) (Sherr and Roberts, 1999). The Cip/ Kip proteins $p21^{Cip1}$ and $p27^{Kip1}$ inhibit the activity of cyclin D1- CDK4/6 complexes less efficiently than they do other cyclin- CDK complexes. Interestingly, these CDIs are required for the formation of active cyclin D1- CDK4/6 complexes. One function of cyclin D1- CDK4/6 complexes is therefore to sequester $p21^{Cip1}$ and $p27^{Kip1}$ away from cyclin E- CDK2 complexes (Blain et al., 1997; el-Deiry et al., 1993; LaBaer et al., 1997; Soos et al., 1996; Toyoshima and Hunter, 1994; Xiong et al., 1993). The Cip/ Kip proteins additionally play a role in regulating the stability and nuclear localization of cyclin D1 (Cheng et al., 1999). Despite their central role in regulating the activity of cyclin D1- CDK4/6 complexes, the lack of $p21^{Cip1}$ and $p27^{Kip1}$ expression appears to be relatively well tolerated in mice (Cheng et al., 1999). Another CDI that is able to inhibit the activity of cyclin D1- CDK4/6 complexes is the $p27^{Kip1}$ related $p57^{Kip2}$ protein (Lee et al., 1995). The genomic region encoding $p57^{Kip2}$ is frequently lost in cancer suggesting that this CDI might function as a tumour suppressor gene (Matsuoka et al., 1995). The INK4 family of CDIs comprises the $p15^{INK4B}$, $p16^{INK4A}$, $p18^{INK4C}$ and $p19^{INK4D}$ proteins (Sherr and Roberts, 1999). These proteins inhibit the function of cyclin D1- CDK4/6 complexes by binding to CDK4/ 6 and displacing cyclin D1 from the complex (Guan et al., 1994; Hannon and Beach, 1994; Hirai et al., 1995; Serrano et al., 1993). The expression or function of $p16^{INK4A}$ is often lost in cancer cell lines (Castellano et al., 1997; Kamb et al., 1994a; Kamb et al., 1994b; Nobori et al., 1994). Furthermore, germline mutations within the INK4 locus have been shown to predispose carriers to multiple melanoma cancers (Borg et al., 2000; Hayward, 2000; Hussussian et al., 1994; Ranade et al., 1995). Together, these CDIs play a crucial role in regulating the activity of cyclin D1- CDK4/6 complexes. The perturbation of their activity can thus result in enhanced cyclin D1 activity and a loss of cell cycle control.

Posttranslational mechanisms play a central role in regulating the activity of cyclin D1. The ubiquitin dependent 26S proteasomal degradation pathway plays a major role in regulating cyclin D1 stability. Furthermore, the phosphorylation of specific threonine residues and protein- protein interactions also serve to regulate the subcellular localization and/ or stability of this cyclin. The various pathways that converge to regulate cyclin D1 stability are discussed in the following section.

Regulation of Cyclin D1 Degradation

Cyclin D1 is an unstable protein with a half life of about 20 min (Diehl et al., 1997). Following its accumulation during G1, cyclin D1 levels decline dramatically at the S phase boundary (Baldin et al., 1993). This degradation may be necessary to prevent cyclin D1 from interfering with DNA replication by binding to proliferating cell nuclear antigen (PCNA) and CDK2 (Fukami-Kobayashi and Mitsui, 1999; Pagano et al., 1994). Since the renewed accumulation of cyclin D1 during G2 acts as a signal for continued cycling, degradation may also provide a route for cell cycle exit under unfavourable conditions (Hitomi and Stacey, 1999a; Hitomi and Stacey, 1999b; Hitomi and Stacey, 2001). Cyclin D1 degradation occurs predominantly via the ubiquitin-dependent degradation pathway (Diehl et al., 1997). Phosphorylation of threonine residue 286 (T286) enhances the rate of cyclin D1 degradation by facilitating its CRM1 dependent nuclear export (Alt et al., 2000; Diehl et al., 1997). Mutation of T286 to alanine results in a constitutively nuclear protein with an extended half life (Diehl et al., 1997). Although CDK4 binding enhances cyclin D1 phosphorylation, this CDK is not required for T286 phosphorylation (Diehl et al., 1997). Glycogen synthase kinase 3β (GSK3β) has been shown to phosphorylate cyclin D1 on T286 both *in vitro* and *in vivo* (Diehl et al., 1997). Inhibition of GSK3β with lithium chloride (LiCl) or SB216763 stabilizes cyclin D1 and results in its nuclear accumulation (Alao et al., 2006a; Alt et al., 2000; Diehl et al., 1997). GSK3β accumulates in the nucleus at the S phase boundary when cyclin D1 relocates to the cytoplasm. GSK3β is thus believed to be important for regulating cyclin D1 subcellular localization and stability (Alt et al., 2000; Diehl et al., 1997). None the less, the precise role of GSK3β in regulating cyclin D1 stability during the cell cycle remains unclear. Studies by Guo *et al.*, (Guo et al., 2005a; Guo et al., 2005b) have confirmed that phosphorylation of T286 is required for the degradation of cyclin D1 at the G1-S phase boundary. These authors did not however, observe changes in GSK3β activity during S phase and inhibition of GSK3β did not abolish T286 phosphorylation (Yang et al., 2006). Similarly, cell cycle dependent changes in GSK3β localization were not observed in the MCF-7 breast cancer line (Alao et al., 2006a). GSK3β is clearly able to influence the stability and localization of cyclin D1 but its precise function in terms of cell cycle regulation remain unclear. It is possible that different experimental systems produce different results. Further studies should help to shed further light on the precise role of GSK3β in regulating cyclin D1 stability during the cell cycle.

Two additional kinases, $p38^{SAPK2a}$ and Mirk/ Dyrk1b, have also been shown to mediate cyclin D1 degradation by phosphorylating T286 and T288 respectively (Casanovas et al., 2000; Zou et al., 2004). $p38^{SAPK2a}$ is activated in response to diverse environmental stresses including ultraviolet radiation (UV), sodium arsenite, histone deacetylase (HDAC) inhibition, oxidative stress and osmotic stress (Casanovas et al., 2000; Chen et al., 1996; Mikhailov et al., 2004; Price et al., 1996). $p38^{SAPK2a}$ enhances the rate of cyclin D1 ubiquitin dependent degradation in response to osmotic stress, oxidative stress and sodium arsenite, by phosphorylating cyclin D1 on T286 (Casanovas et al., 2000). The specific inhibition of 26S proteasomes or $p38^{SAPK2a}$ abolished cyclin D1 degradation under these conditions. Furthermore, mutation of T286 to alanine abolished the *in vitro* ubiquitylation of $p38^{SAPK2a}$ phosphorylated cyclin D1 (Casanovas et al., 2000). Cellular exposure to environmental stresses creates an unfavourable environment for protein synthesis and function, DNA replication and cell division, which can result in genomic instability or loss of viability. In fact cellular exposure to osmotic stress conditions induces DNA strand breaks and a transient cell cycle arrest (Dmitrieva and Burg, 2005; Dmitrieva et al., 2005; Zhang et al., 2005). The $p38^{SAPK2a}$ mediated phosphorylation and subsequent degradation of cyclin D1 may thus facilitate cell cycle exit under environmental stress conditions. Mirk/ Dyrk1b is an arginine directed kinase that is active during the G0 and early G1 phases of the cell cycle (Friedman, 2007; Zou et al., 2004). Mirk/ Dyrk1b enhances the rate of cyclin D1 ubiquitin dependent degradation by phosphorylating T288. Mirk/ Dyrk1b mediated phosphorylation of residue T288 has been demonstrated both *in vivo* and *in vitro* (Zou et al., 2004). Small interfering RNA (siRNA) mediated knockdown of Mirk/ Dyrk1b resulted in the accumulation of cyclin D1 protein without any apparent effect on its mRNA levels. Mutation of cyclin D1 residue T288 to alanine abolished Mirk/ Dyrk1b mediated cyclin D1 degradation, while the inhibition of GSK3β or mutation of T286 did not (Zou et al., 2004). The precise cellular function of Mirk/ Dyrk1b mediated cyclin D1 degradation remains unclear. The periodic degradation of cyclin D1 occurs at the G1- S phase transition, when Mirk/ Dyrk1b is thought to be inactive. One possibility is that Mirk/ Dyrk1b enforces G0 arrest by preventing cyclin D1 accumulation (Friedman, 2007). Alternatively, Mirk/ Dyrk1b may cooperate with GSK3β to regulate cyclin D1 stability, since T268/ T288 double mutants have a shorter half life than the respective single mutants (Germain et al., 2000; Zou et al., 2004).

Cyclin D1 degradation can also occur independently of T286 or T288 phosphorylation (Germain et al., 2000). Mutation of T286 and/ or T288 to alanine attenuates but does not abolish cyclin D1 degradation. *In vitro* assays clearly demonstrate that cyclin D1 mutants lacking threonine residues at 286 and/ or 288 can still be ubiquitylated (Germain et al., 2000). Similarly, cyclin D1-KE mutants that are unable to bind CDK4 are also ubiquitylated *in vitro*. Together these findings indicate that non Cdk4 bound ("free") cyclin D1 is targeted for ubiquitylation independently of T286 phosphorylation. This pathway may thus serve to regulate the levels of free cyclin D1 within the cell (Germain et al., 2000). Cellular exposure to genotoxic agents such as ionizing radiation (IR) or cisplatin has also been shown to induce the rapid degradation of cyclin D1. Genotoxin mediated cyclin D1 degradation requires a conserved RXXL degradation box in the N-terminal region of the cyclin and is mediated by the anaphase promoting complex (APC) (Agami and Bernards, 2000). Mutation of T286 to alanine did not prevent APC mediated cyclin D1 degradation. Furthermore, the abolition of APC mediated cyclin D1 degradation did not result in the accumulation of cellular cyclin D1 levels (Agami and Bernards, 2000). This degradation pathway is thus clearly distinct from the pathway that regulates cyclin D1 stability during the cell cycle. Recent studies indicate that $p38^{SAPK2a}$ is important for mediating cell cycle arrest in cancer cells exposed to DNA damaging agents by negatively regulating cdc25 activity (Reinhardt et al., 2007). The APC mediated cyclin D1 degradation pathway may thus cooperate with $p38^{SAPK2a}$ to facilitate cell cycle arrest following DNA damage. Viral infection of human cancer cell lines has also been shown to result in the ubiquitin dependent degradation of cyclin D1. When HeLa cells were infected with the coxsackievirus group B3 (CVB3), a decline in cyclin D1 levels was detectable within one hour and was maximal at three hours (Luo et al., 2003). Interestingly, the decline in cyclin D1 protein levels was accompanied by an increase in its mRNA levels. Specific inhibition of 26S proteasomes abolished the CVB3 induced cyclin D1 degradation. Inhibition of GSK3β activity with lithium chloride did not however, prevent CVB3 induced cyclin D1 degradation. Together, these findings demonstrate that CVB3 infection can induce the ubiquitin dependent degradation of cyclin D1 independently of T286 phosphorylation (Luo et al., 2003). The arrest of cells in G1 or at the G1-S phase boundary has been shown to enhance the production of infectious CVB3 (Feuer et al., 2002). CVB3 induced cyclin D1 degradation may thus facilitate viral production by inducing G1 cell cycle arrest.

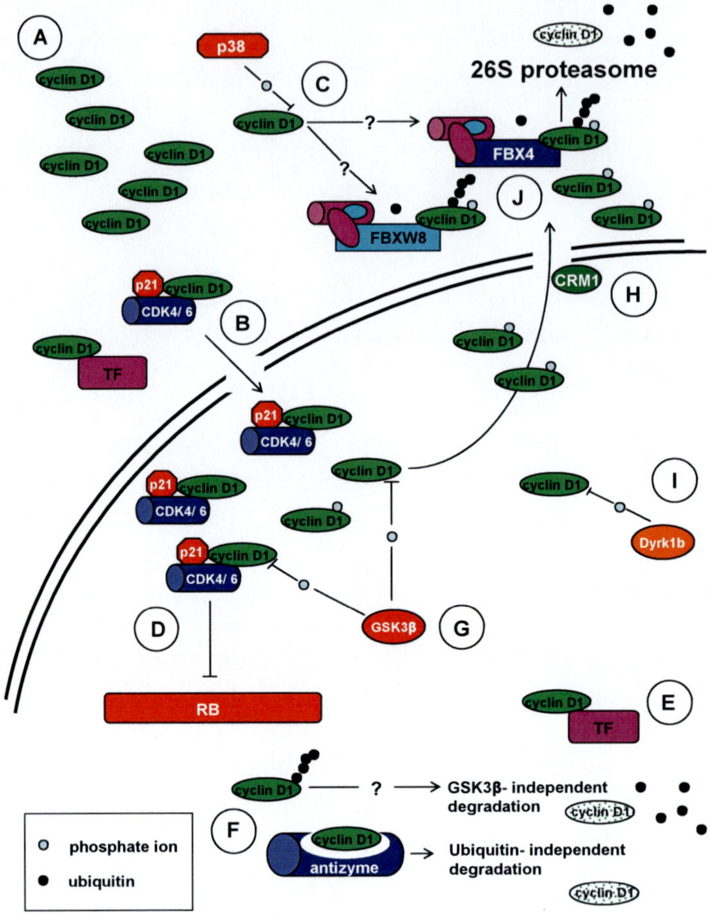

Figure 1: Regulation of cyclin D1 degradation. A. Cyclin D1 does not contain a nuclear localization signal (NLS) and its sequestration may result in accumulation within the cytoplasm. B. Cytoplasmic cyclin D1 is transported into the nucleus in association with its binding partners e.g. CDK4 and possibly various transcription factors (TF). C. p38^{SAPK2a} has been shown to phosphorylate cyclin D1 on threonine residue 286 (T286) and induce its proteasomal degradation. It is unclear if the F-box proteins FBX4 and FBXW8 are involved in mediating p38^{SAPK2a} induced cyclin D1 degradation in the cytoplasm. D, E. Within the nucleus, active cyclin dependent kinase 4 (CDK4) or CDK6- cyclin D1 complexes phosphorylate the retinoblastoma protein (RB). Cyclin D1 can also influence the activity of various transcription factors independently of CDK4/ 6. F. Free cyclin D1 is degraded through the ubiquitin dependent 26S proteasomal degradation pathway independently of glycogen synthase kinase 3β (GSK3β). Antizyme can also mediate cyclin D1 degradation via the 26S proteasome independently of ubiquitin. G, H. GSK3β phosphorylates cyclin D1 on T286 which facilitates its nuclear export by the exportin CRM1. GSK3β influences cyclin D1 stability since the phosphorylated form of the cyclin is subsequently degraded within the cytoplasm. I. Phosphorylation of T288 is mediated by the mirk/ dyrk

Interestingly, cyclin D1 degradation has also been shown to occur via the 26S proteasome independently of ubiquitin. Antizyme is a negative regulator of ornithine decarboxylase (ODC) and has recently been shown to target cyclin D1 for degradation via the 26S proteasome (Newman et al., 2004). Antizyme covalently binds cyclin D1 and induces its degradation in cell free systems independently of ubiquitin. Increasing the cellular levels of antizyme by overexpression or exposure to spermidine resulted in a reduction of cyclin D1 protein levels. Conversely, knockdown of antizyme expression by siRNA resulted in the increase of cellular cyclin D1 protein levels. Mutation of T286 to alanine did not prevent antizyme mediated cyclin D1 degradation. Antizyme thus appears to target cyclin D1 for degradation by 26S proteasomes independently of both phosphorylation and ubiquitylation (Newman et al., 2004). Increased levels of antizyme have been shown to negatively regulate cell cycle progression. Furthermore, the cellular levels of antizyme are reduced in late stage tumours. Antizyme may thus function together with the ubiquitin dependent cyclin D1 degradation pathway to regulate cell cycle progression ((Newman et al., 2004) and references therein).

The ubiquitin dependent and independent degradation of cyclin D1 clearly plays an important role in regulating the cellular levels of cyclin D1. It should be noted, however, that a number of studies suggest cyclin D1 degradation can occur independently of the 26S proteasome pathway (Alao et al., 2006b; Choi et al., 1997; Kawamata et al., 2007). Early studies by Choi *et al.*, (Choi et al., 1997) suggested that cellular cyclin D1 levels were regulated by calpain proteases. This family of calcium dependent cysteine proteases regulates numerous cellular processes either ubiquitously or in a tissue specific manner (Demarchi et al., 2007; Demarchi and Schneider, 2007). Following serum withdrawal, the calpain inhibitors leucyl-leucyl-norleucinal (LLnL, calpain inhibitor I) and leucyl-leucyl-methioninal (LLM, calpain inhibitor II) but not lactacystin (a 26S proteasome specific inhibitor) prevented the loss of cyclin D1 expression in NIH 3T3 cells. In contrast to LLnL, lactacystin also failed to prevent actinomycin D and lovastatin induced cyclin D1 degradation in PC-3-M prostate cancer cells (Choi et al., 1997). LLnL is now known to be a potent inhibitor of 26S proteasomes. This finding does not however, explain why LLnL and LLM but not lactacystin prevented the loss of cyclin D1 following serum starvation. Furthermore, calpain could induce cyclin D1 degradation *in vitro* and this activity could be inhibited by chelating agents or prior boiling of the calpain (Choi et al., 1997). Similarly, LLnL and MG132 effectively abolished cyclin D1 degradation in human breast cancer cell lines treated with the HDAC inhibitor trichostatin A (TSA), while lactacystin only partially abolished this activity (Alao et al., 2006b). Peptide aldehyde inhibitors of 26S proteasome activity like LLnL and MG132 can also inhibit other classes of proteases including calpains (Kisselev and Goldberg, 2001; Lee and Goldberg, 1998). Although further investigations are clearly desirable, these studies none the less suggest a possible role for calpains in regulating cyclin D1 stability. Further insights into the regulation of cyclin D1 stability have come from studies using cyclin D1 variants in which some or all lysine residues were substituted for arginine (Feng et al., 2007). The substitution of all lysine residues for arginine greatly enhanced the stability of cyclin D1 but did not completely abolish its degradation following the inhibition of protein synthesis with cycloheximide. The effect of lysine substitutions on the ability of cyclin D1 to bind antizyme has not been reported. It is thus unclear, if the delayed degradation of these cyclin

D1 variants occurs independently of ubiquitin via the 26S proteasome pathway or independently of both ubiquitin and proteasomes. Further evidence for the 26S proteasome independent degradation of cyclin D1 have come from studies using mantle cell lymphoma (MCL) cell lines. These studies suggest that cyclin D1 degradation occurs independently of the ubiquitin-26S pathway in MCL cells (Kawamata et al., 2007). Treatment of MCL cells with the HDAC inhibitor suberoylanilide hydroxamic acid (SAHA, vorinostat) induces a rapid decline in cellular cyclin D1 levels. Specific inhibition of 26S proteasomal activity with PS341 (Bortezmib, Velcade®) did not prevent the loss of cyclin D1. Intriguingly, treatment of MCL cells with PS341 alone also resulted in a decline of cyclin D1 levels. SAHA has been shown to inhibit cyclin D1 protein synthesis by inhibiting the Akt/ TOR/ eIF4E-BP/ eIF4E pathway without affecting cyclin D1 mRNA transcription (Kawamata et al., 2007). This study clearly shows that cyclin D1 degradation can occur independently of 26S proteasomes at least in MCL cells. The nature of this pathway and its role in regulating cyclin D1 stability in other cell lines remains to be determined.

Cyclin D1b, a Cyclin D1 Splice Variant with Enhanced Transforming Potential

Studies by Betticher *et al.*, (Betticher et al., 1995) identified cyclin D1b as a splice variant of cyclin D1 lacking C-terminal amino acid residues present in the full length protein (now referred to as cyclin D1a). This variant is likely to share many of the cellular activities of cyclin D1a. *In vitro* and *in vivo* studies suggest however, that cyclin D1b is more effective at inducing cellular transformation than cyclin D1a. Cyclin D1b expression has also been detected in several cancers (Burd et al., 2006; Hosokawa et al., 1997; Hosokawa et al., 1999; Krieger et al., 2006; Lu et al., 2003). For this reason, cyclin D1b is briefly discussed in this section. Reference to this variant will also be made where appropriate in subsequent sections. Alternative splicing results in exon 5 being replaced by amino acids encoded by intron 4 and results in the loss of C-terminal regulatory sequences. The C-terminal PEST sequence and threonine residues 286 and 288 have been shown to be important for the regulation of cyclin D1 stability and localization (reviewed in (Alao, 2007)). Cyclin D1b localization is thus confined mainly to the nuclear compartment throughout the cell cycle of most cancer cell lines (Lu et al., 2003; Solomon et al., 2003). Studies by Krieger *et al.*, (Krieger et al., 2006) indicate however, that cyclin D1b can also be detected in the cytoplasmic fraction of mantle cell lymphoma (MCL) cell lines and cells sampled from patients. Curiously, cyclin D1b is not effective at mediating Rb phosphorylation in vivo and its stability is comparable to that of cyclin D1a in some cell lines (Solomon et al., 2003). These finding also provide further evidence to show that cyclin D1 degradation can occur efficiently even in the absence of T286 phosphorylation. The effect of cyclin D1b on cellular transformation is thus likely to be independent of cell cycle regulation (Solomon et al., 2003). The constitutive nuclear localization may be important for cellular transformation. Alternatively, the modified C-terminus of cyclin D1b may confer a differential ability to bind and/ or activate transcription factors (Solomon et al., 2003). These studies also suggest that cyclin D1b may be as important a therapeutic target as cyclin D1a. Readers are encouraged to consult excellent reviews on cyclin D1b for more insight into the cellular activity of this splice variant (Gladden and Diehl, 2005; Knudsen, 2006; Knudsen et al., 2006).

F-Box Protein Mediated Cyclin D1 Degradation

F-box proteins are evolutionarily conserved components of the Skp/ Cullin/ F-box (SCF) complexes that are involved in targeting proteins for ubiquitin dependent degradation via 26s proteasomes (Cardozo and Pagano, 2004; Deshaies, 1999; Ho et al., 2006). To date, three distinct F-box proteins have been associated with the regulation of cyclin D1 stability (Ganiatsas et al., 2001; Lin et al., 2006; Okabe et al., 2006; Yu et al., 1998). Initial studies suggested a role for p45/ Skp2 in the regulation of cyclin D1 stability (Yu et al., 1998). Yu *et al.*, (Yu et al., 1998) demonstrated that the inhibition of Cul1, Skp1 or p45/ Skp2 expression with antisense oligodeoxynucleotides resulted in the accumulation of cyclin D1 and p21 in human keratinocyte, colon carcinoma, colorectal carcinoma and colon adenocarcinoma cell lines. Furthermore, Cul1-1 was shown to associate with both cyclin D1 and p21 in lysates from a human keratinocyte cell line (Yu et al., 1998). Studies on the regulation of cyclin D1 stability in the SK-UT-1B uterine carcinoma cell lines also suggested a role for p45/ Skp2 (Ganiatsas et al., 2001). SK-UT-1B cells are defective in cyclin D1 degradation, resulting in the cellular accumulation of this cyclin D1. The defective regulation of cyclin D1 in these cells was linked to the expression of a SK-UT-1B splice variant (Skp2-CTV) that is unable to migrate into the nucleus. Lysates from SK-UT-1B cells were unable to ubiquitylate p27 (another p45/ Skp2 substrate) *in vitro* and stable transfection of p45/ Skp2 resulted in a decrease of cellular cyclin D1 levels (Ganiatsas et al., 2001). A later study also suggested a role for p45/ Skp2 in mediating HDAC inhibitor induced cyclin D1 degradation in human breast cancer cells (Alao et al., 2004). These studies did not however, demonstrate the ability of p45/ Skp2 to mediate cyclin D1 or p21 ubiquitylation *in vitro*.

Two recent studies have identified two novel F-box proteins that are able to mediate cyclin D1 ubiquitylation both *in vitro* and *in vivo* (Lin et al., 2006; Okabe et al., 2006). These studies suggest that FBX4 and FBXW8 but not p45/ Skp2 are *bona fide* cyclin D1 ubiquitin ligases (E3's). FBX4 lacks a C-terminal protein- protein interaction motif and requires αB crystallin in order to target T286 phosphorylated cyclin D1 for ubiquitin dependent 26 S proteasomal degradation (Lin et al., 2006). In NIH3T3 cells, cyclin D1 phosphorylation of is required for its interaction with αB crystallin. These studies also showed that αB crystallin facilitates the binding of the Skp1/ Cul1/ FBX4 core SCF complex to phosphorylated cyclin D1. The inhibition of αB crystallin or FBX4 expression with small hairpin RNAs (shRNA) extended the half life of cyclin D1. Expression of a dominant negative allele of FBX4 resulted in the accumulation of wild type but not T286A mutant cyclin D1. In addition, overexpression of FBX4 induced the degradation of wild type but not T286A cyclin D1 in a αB crystallin dependent manner. *In vitro* assays demonstrated that the Skp1/ Cul1/ FBX4/ αB crystallin complex ubiquitylates cyclin D1 in a T286 phosphorylation dependent manner (Lin et al., 2006). Both FBX4 and αB crystallin are cytoplasmic proteins. This observation suggests that the Skp1/ Cul1/ FBX4/ αB crystallin complex targets cyclin D1 for ubiquitin dependent degradation in the cytoplasm, following its phosphorylation and consequent nuclear export. A second study has identified FBXW8 has an important regulator of cyclin D1 stability in human cancer cell lines (Okabe et al., 2006). FBXW8 ubiquitylates cyclin D1 *in vitro* and induces its degradation *in vivo* and these activities require phosphorylation of T286 (Okabe et al., 2006). Interestingly, this study also identified the extracellular regulated kinase 2 (ERK2), a mitogen activated protein kinase (MAPK), as a regulator of cyclin D1 T286 phosphorylation in human cancer cell lines. Accordingly, inhibition of ERK2 activity with U0126 prevented the accumulation of T286 phosphorylated cyclin D1 extended its half life. Furthermore, ERK2 was required for FBXW8 mediated cyclin D1 ubiquitylation *in vitro* (ref). Like the Skp1/ Cul1/ FBX4/ αB crystallin complex ($SCF^{FBX4\text{-}\ \alpha B\ crystallin}$), the Skp1/ Cul1/ FBXW8 complex is also localized to the cytoplasm. The siRNA mediated knockdown of FBXW8 resulted in the accumulation of T286 phosphorylated and non phosphorylated cyclin D1 predominantly within the cytoplasm (Okabe et al., 2006). Together, these studies demonstrate the importance of T286 phosphorylation and F-box proteins in regulating cellular cyclin D1 levels. The respective roles of these SCF complexes in the regulation of cyclin D1 stability remains to be fully characterized. Cyclin D1 degradation is clearly impaired in MCF-7 breast cancer cells (Lin et al., 2006; Russell et al., 1999), but its levels rapidly decline when protein synthesis is inhibited (Alao et al., 2006a; Alao et al., 2006b). This is despite the fact that αB crystallin expression is low or absent in this cell line (Lin et al., 2006). Knockdown of FBX4 or αB crystallin complex but not FBXW8 increases the rate of cell division, despite the elevation of cyclin D1 levels in cells lacking any of these proteins. In fact, knockdown of FBXW8 was observed to inhibit the proliferation of HCT116 human colon cancer cell lines (Lin et al., 2006; Okabe et al., 2006). Further studies are thus needed to elucidate in more detail, how the $SCF^{FBX4\text{-}\ \alpha B\ crystallin}$ and SCF^{FBXW8} mediated degradation of cyclin D1 is integrated.

The precise role of p45/ Skp2 in regulating cyclin D1 stability remains uncertain. One possibility is that p45/ Skp2 mediated degradation of cyclin D1 is cell and/ or context specific (Alao et al., 2004; Hiramatsu et al., 2006; Okabe et al., 2006; Schneider et al., 2007). Alternatively, p45/ Skp2 may indirectly target cyclin D1 bound to p21 or p27 for ubiquitin dependent degradation (Cheng et al., 1999; Ganiatsas et al., 2001; Lin et al., 2006; Yu et al., 1998). Further studies will be required to determine more clearly, the precise role of this F-box protein in regulating cyclin D1 stability in normal and cancer cells.

Roles of Cyclin D1 in Cancer Cell Proliferation

Several studies have provided evidence to support the notion, that cyclin D1 plays an important role in cancer development and progression. Cyclin D1 was initially identified as *PRAD1* in parathyroid adenomatosis, where it is overexpressed as a result of a result of chromosomal rearrangements (Arnold et al., 1989; Motokura and Arnold, 1993; Motokura et al., 1991; Rosenberg et al., 1993; Rosenberg et al., 1991). Cyclin D1 is highly expressed in parathyroid adenomas in which chromosomal translocations place the *CCND1* gene under the control of regulatory sequences that govern expression of the parathyroid hormone gene (PTH) (Arnold et al., 1989). Later studies identified *PRAD1* as *B-cell lymphoma 1* (*BCL1*), a candidate oncogene in B-cell lymphoma (BCL) (Rosenberg et al., 1991). Cyclin D1 is overexpressed in BCL as a consequence of chromosome rearrangement at the 11q13 locus. In their study, Rosenberg *et al.*, (Rosenberg et al., 1991) found high level expression of cyclin D1 mRNA in seven out of seven centrocytic lymphomas. Bodrug *et al.*, (Bodrug et al., 1994) studied the effect of constitutive cyclin D1 expression in lymphomagenesis, using transgenic mice in which cyclin D1 expression was under control of an immunoglobulin enhancer. In this study, cyclin D1 was shown to cooperate with Myc to induce the lymphoma development. Furthermore, the majority of spontaneous tumours in *myc* transgenic mice overexpressed cyclin D1. Similar results were obtained in a separate study by Lovec *et al.*, (Lovec et al., 1994). These studies indicate that cyclin D1 is a weak oncogene that can however, cooperate with other cell cycle regulators to induce cancer development.

Cyclin D1 has also been shown to play an important role in the development of breast cancer. Elevated cyclin D1 levels have been detected in approximately 43- 67% of breast cancers sampled (Elsheikh et al., 2007; Reis-Filho et al., 2006). Amplification of the *CCND1* gene however, accounts for only a subset of breast cancers with elevated cyclin D1 levels, indicating that the deregulation of mechanisms governing cyclin D1 stability is also important (Elsheikh et al., 2007; Reis-Filho et al., 2006). Ablation of cyclin D1 has been shown to protect against the development of breast cancers driven by the *neu* and *ras* oncogenes (Yu et al., 2001). These animals however, remain sensitive to breast cancers driven by oncogenic pathways independently of the *neu* and *ras* oncogenes. The findings of Yu *et al.*, never the less, demonstrate that cyclin D1 couples the Neu and ras signalling pathways to the cell cycle machinery (Yu et al., 2001). Cyclin D1 overexpression also correlates positively with estrogen receptor alpha (ERα) positivity (Elsheikh et al., 2007; Reis-Filho et al., 2006). In breast cancer cell lines, ERα often drives cyclin D1 expression and inhibition of the receptor substantially decreases the levels of cyclin D1 mRNA and protein (Ali and Coombes, 2002; Foster et al., 2001; Sabbah et al., 1999). The siRNA mediated knockdown of ERα similarly decreases the cellular levels of cyclin D1 in MCF7 breast cancer cells (Alao et al., 2004). In addition, the acquisition of signalling pathways that allow cyclin D1 expression independently of estrogen has been linked to the development of resistance to antiestrogens (Ahnstrom et al., 2005; Ali and Coombes, 2002). In fact, overexpression of cyclin D1 has been shown to confer resistance to antiestrogens in a breast cancer cell line (Hui et al., 2002). Together, these studies strongly suggest that targeting cyclin D1 may be an extremely useful therapeutic approach in breast cancer.

Cyclin D1 levels have also been shown to be elevated in numerous other cancer types, including those of the lung, colon, liver, oesophagus, head and neck (reviewed in (Tashiro et al., 2007)). Amplification of the *CCND1* gene however, accounts for only a subset of cancers with elevated cyclin D1 levels. The abolition or attenuation of cyclin D1 degradation must therefore, also contribute to the increased expression of this cyclin observed in cancer. Cyclin D1 degradation has in fact been shown to be defective in several cancer cell lines (Ganiatsas et al., 2001; Lin et al., 2006; Russell et al., 1999). A recent study demonstrated a role for the $SCF^{FBX4\text{-}\alpha B\ crystallin}$ complex in regulating cyclin D1 degradation. Interestingly, αB crystallin and FBX4 expression is often lost or downregulated in cancer and its reexpression in MCF-7 breast cancer cells reduces the half life of cyclin D1 in this cell line (Dohner et al., 1999; Lin et al., 2006). Knockdown of FBX4 or αB crystallin was also observed to accelerate the cell cycle progression of NIH 3T3 cells (Lin et al., 2006). This study demonstrates that the loss of this SCF can result in the accumulation of cyclin D1 levels. FBXW8 has similarly been shown to regulate the stability of cyclin D1 in cancer cell lines (Okabe et al., 2006). In contrast to the $SCF^{FBX4\text{-}\alpha B}$ complex however, inhibition of FBXW8 negatively regulates cell cycle progression (Lin et al., 2006; Okabe et al., 2006). The precise role of FBXW8 in cancer cell proliferation and its interaction with FBX4 thus needs to be further clarified.

The precise mechanisms that facilitate the oncogenic potential of cyclin D1 remain to be precisely determined. Its effect on Rb phosphorylation and thus cell division, have been shown to confer a proliferative advantage (Alt et al., 2000; Bartkova et al., 1994; Bates and Peters, 1995). The cyclin D1b splice variant which has little effect on Rb phosphorylation has however, also been shown to induce cellular transformation (Burd et al., 2006; Knudsen, 2006; Krieger et al., 2006; Lu et al., 2003; Solomon et al., 2003). Furthermore, cyclin D1 T268A mutants, which are constitutively nuclear, are more tumourigenic than the wild type protein (Alt et al., 2000; Gladden and Diehl, 2005; Lin et al., 2007) both *in vitro* and *in vivo*. It has thus been suggested, that the nuclear localization of cyclin D1 rather than its overexpression *per se*, is important in cancer development (Gladden and Diehl, 2005). Determining the exact role of cyclin D1 localization in the regulation of its activity has been problematic. Phosphorylation of T286 by GSK3β plays an important role in regulating the localization and stability of cyclin D1 (Alt et al., 2000; Diehl et al., 1998; Diehl et al., 1997; Lu et al., 2003).These studies clearly demonstrated that the inhibition of GSK3β or CRM1 results in the nuclear accumulation of cyclin D1. Mutations affecting the T286 phosphorylation site have been detected in tumour samples and inhibition of cyclin D1 nuclear export clearly enhances tumour formation (Benzeno et al., 2006; Lin et al., 2007). More recent studies have confirmed the importance of T286 phosphorylation in the degradation of cyclin D1 but have questioned the role of GSK3β in this activity (Guo et al., 2005b; Yang et al., 2006). In these studies, cell cycle dependent changes in the activity of GSK3β or its upstream regulators were not detected. Furthermore, cell cycle dependent changes in GSK3β subcellular localization were not observed in the MCF-7 breast cancer cell line (Alao et al., 2006a). It was also observed, that the stable expression of exogenous cyclin D1 was poorly tolerated by NIH 3T3 cells (Guo et al., 2005b). Early studies noted that cyclin D1 degradation at S phase was necessary for DNA replication (Baldin et al., 1993; Fukami-Kobayashi and Mitsui, 1999). High levels of cyclin D1 are believed to inhibit DNA replication by binding to proliferating cell nuclear antigen (PCNA) and negatively affecting its activity (Fukami-Kobayashi and Mitsui, 1999; Pagano et al., 1994). Taken together, these studies suggest that additional cellular adaptations are required for cells to cope with increased levels of cyclin D1. These adaptations allow cells to take advantage of the pro-proliferative effects of cyclin D1 while avoiding its potentially negative effect on DNA replication. *In vivo* studies have in fact demonstrated that overexpression of the T286A mutant is associated with increased levels of genomic instability in cancer cells (Lin et al., 2007). Furthermore, SK-UT-1B cells proliferate rapidly in culture despite the fact that they do not degrade cyclin D1 at S phase (Alao et al., 2006a). Cyclin D1 appears to be sequestered in the cytoplasm of several cancer cell lines and has been observed to exhibit a predominantly cytoplasmic localization in a high percentage of tumour samples (Alao et al., 2006a; Comstock et al., 2007). The observation that cytoplasmic cyclin D1 localization is often observed in samples from diverse tumour types suggests that localization is important for regulating its activity ((Comstock et al., 2007) and references therein). Cyclin D1 localization has been shown to regulate its activity in cardiomyocytes and postmitotic neurons (Sumrejkanchanakij et al., 2003; Tamamori-Adachi et al., 2003). Changes in the localization of cyclin D1 have also been shown to influence the proliferation of neoplastic hepatocytes (Yamamoto et al., 2006). Interestingly, the cyclin D1b splice variant, which lacks C-terminal amino acid residues including T286, has also been shown to be readily detectable in the cytoplasm of MCL (mantle cell lymphoma) cells (Krieger et al., 2006). In prostate cancer, the cytoplasmic localization of cyclin D1 is associated with a lower tumour grade and reduced proliferative rate (Comstock et al., 2007). Future studies are expected to shed more light on the impact of a predominantly cytoplasmic localization on its oncogenic potential.

It is likely, that the deregulation of additional pathways governing cyclin D1 stability also contributes to the elevation of this cyclin in cancer cells. FBX4 and FBXW8 specifically target T286 phosphorylated cyclin D1 for ubiquitin-dependent degradation. Cyclin D1 degradation can however, occur independently of both T286 phosphorylation and ubiquitin (Germain et al., 2000; Newman et al., 2004). Furthermore, the stability of cyclin D1b, which lacks amino acid residue T286, is not substantially different from that of full length cyclin D1 in some cell lines (Solomon et al., 2003). It has also been observed, that cyclin D1 degradation occurs relatively efficiently even when αB crystallin expression is low or absent (Alao et al., 2006a; Lin et al., 2006). These observations suggest the existence of additional pathways that regulate cyclin D1 stability independently of T286 phosphorylation. In fact, specific inhibition of 26S proteasomal activity failed to prevent cyclin D1 degradation in multiple myeloma cells treated with the HDAC inhibitor SAHA (Kawamata et al., 2007). The identity of these pathways and their role in cancer development remains to be determined. The role of antizyme mediated cyclin D1 degradation in cancer cells also remains to be clearly defined. The relative importance of these pathways in the regulation of cyclin D1 stability in both normal and cancer cells will probably be cell type specific. Their characterization is thus essential for the creation of a detailed and comprehensive model to describe how the stability of cyclin D1 is regulated.

It should be remembered, that cyclin D1 can also influence the activity of several transcription factors (reviewed in (Comstock et al., 2007; Knudsen, 2006)). Importantly, cyclin D1 has been shown to positively activate ERα independently of CDK4 activity (Zwijsen et al., 1997). Furthermore, the induction of cyclin D1 transcriptional programmes has been shown to occur independently of CDK4 activity (Lamb et al., 2003). This study also showed that the repressive effect of cyclin D1 on C/EBPβ activity may be important in cancer development. As discussed earlier in this chapter, the cyclin D1b variant is a poor mediator of Rb phosphorylation and has little impact on cell cycle progression (Solomon et al., 2003). This variant is however, more efficient at oncogenic transformation than the full length cyclin (Lu et al., 2003; Solomon et al., 2003). Similarly, the cyclin D1 T286A mutant is more efficient at inducing cellular transformation than the wild type protein (ref). The cell cycle independent activity of cyclin D1 is thus clearly important for its effects on cellular physiology. As such, the overexpression of cyclin D1 observed in many cancers can be expected to influence transcription factor activity in both a general and cell specific manner. Similarly, the effect of cyclin D1 on a transcription factor activity might also be expected to impact on cancer development.

Cyclin D1 as a Therapeutic Target

Numerous clinical and preclinical therapeutic agents have been shown to induce the reduction of cellular cyclin D1 levels. Given the importance of cyclin D1 in cancer cell proliferation, these studies suggest that the specific targeting of cyclin D1 may provide a useful strategy for treating cancer. Therapeutic agents that target cyclin D1 activity fall into three main classes. Class I consists of those agents that indirectly affect cyclin D1 activity by inhibiting the signalling pathways that regulate cyclin D1 expression. These include inhibitors of tyrosine receptor kinases, the extracellular regulated kinases (ERK1 and ERK2) and their upstream activator MEK1, Akt, mammalian target of rapamycin (mTOR) and ERα. (Musgrove, 2006). The inhibition of these proteins attenuates *CCND1* expression and induces the decline of cyclin D1 protein levels. Since a description of the function and regulation of the pathways regulated by these proteins is beyond the scope of this review, their inhibitors will not be discussed further. Readers are encouraged to consult the various review articles that discuss the activity of these proteins and their inhibitors. Class II compounds target the kinase activity of cyclin D1- CDK4/6 complexes such as PD0332991 (Fry et al., 2004; Lin et al., 2007; Shapiro, 2006). Class III includes those compounds that induce cyclin D1 degradation via the 26S proteasome with or without a requirement for cyclin D1 phosphorylation. Some of these compounds also inhibit cyclin D1 expression and their activity thus partially overlaps with that of the Class I compounds. In general, these compounds are referred to as cyclin D1 ablative agents (Huang et al., 2006). In this section, the effect of class III compounds on cyclin D1 stability and their potential as therapeutic ablative agents is discussed.

Retinoids such as All-*trans* retinoic acid (RA) activate nuclear retinoic acid receptors and have been shown to inhibit cancer cell proliferation by inducing G1 cell cycle arrest (Boyle et al., 1999; Langenfeld et al., 1997; Spinella et al., 1999). RA induces the ubiquitin dependent degradation of cyclin D1 in human bronchial carcinoma cell lines. RA induced cyclin D1 degradation was blocked by inhibitors of GSK3β or mutation of T286 to alanine, indicating a role for T286 phosphorylation (Dragnev et al., 2004; Ma et al., 2005). Retinoids have also been shown to inhibit the expression of cyclin D1 at the transcription level (Freemantle et al., 2007; Huang et al., 2007; Yu et al., 2007). Similarly to RA, retinoid X receptor agonists such as bexarotene have also been shown to induce cyclin D1 degradation in lung cancer cells both *in vitro* and *in vivo* (Dragnev et al., 2007; Dragnev et al., 2005). Bexarotene induced cyclin D1 degradation was observed in cell lines sensitive to this rexinoid but not in those resistant to the growth inhibitory effects of this agent (Dragnev et al., 2005; Rigas and Dragnev, 2005). The cotreatment of lung cancer cell lines with bexarotene and the EGFR inhibitor erlotinib enhanced the degradation of cyclin D1 *in vitro*. Clinically, combining bexarotene with erlotinib prolonged the survival of non-small-cell cancer (NSCLC) patients when compared with patients treated with either agent alone (Dragnev et al., 2005; Freemantle et al., 2007). Furthermore, EGFR activating mutations were not identified in patients who responded to combination therapy (Dragnev et al., 2005). Combining bexarotene with EGFR tyrosine kinase inhibitors (TKIs) may thus overcome the need for activating EGFR mutations to achieve clinical responses in patients treated with TKI inhibitors of this kinase (Dragnev et al., 2005). Encouragingly, these studies show that cyclin

D1 ablation is clinically achievable and thus a potentially useful therapeutic strategy. Recent studies have shown that RA does not induce cyclin D1 degradation in a human neuroblastoma cell line (De los Santos et al., 2007b). In fact, the ability of RA to induce cyclin D1 degradation varies widely between human bronchial epithelial cells and embryonal carcinoma cells (Langenfeld et al., 1997; Spinella et al., 1999). The efficacy of RA mediated cyclin D1 degradation is thus cell line dependent. Further studies will be necessary to characterize the mechanisms that underlie retinoid induced cyclin D1 degradation. This in turn will help to identify cancer subtypes in which retinoid induced cyclin D1 degradation is must likely to be of clinical benefit.

Activators of peroxisome proliferator activated receptor γ (PPARγ) have also been shown to induce cyclin D1 degradation in human cancer cell lines. The peroxisome proliferator activated receptor (PPAR) superfamily regulates numerous cellular processes, including proliferation and differentiation (Kota et al., 2005; Qi et al., 2000; Wang et al., 2006). Activation of PPARγ by prostaglandin J2 (PGJ2) or synthetic ligands induces the cytoplasmic localization and degradation of cyclin D1 in human breast cancer cell lines (Huang et al., 2006; Huang et al., 2005; Qin et al., 2003). PPARγ agonists also inhibit the expression of cyclin D1 mRNA in human cancer cell lines (Wang et al., 2001). Interestingly, the induction cyclin D1 degradation by synthetic PPARγ agonists occurs independently of their effect on receptor activation (Huang et al., 2006; Huang et al., 2005). Troglitazone (a PPARγ agonist belonging to the thiazolidinedione class) induced cyclin D1 degradation via the 26S proteasome pathway independently of GSK3β activity (Huang et al., 2006; Huang et al., 2005). Derivatives of thiazolidinediones such as troglitazone that lack PPARγ agonist activity have also been shown to induce cyclin D1 degradation. These derivatives induce cyclin D1 degradation with a potency that is similar or even higher than that of their parent compounds (Huang et al., 2006). Given their lack of PPARγ agonist activity, these derivatives can be expected to be less problematic in terms of lower side effects. They are currently being used as a platform for the development of cyclin D1 ablative agents.

Histone deacetylase inhibitors (HDACIs) are promising anticancer agents that are currently in clinical trials (Dokmanovic and Marks, 2005; Glaser, 2007; Johnstone, 2002; Marks and Jiang, 2005; Mehnert and Kelly, 2007; Vigushin and Coombes, 2002). Trichostatin A (TSA) has been shown to repress cyclin D1 expression and induce cyclin D1 degradation in several human cancer cell lines (Alao et al., 2006a; Alao et al., 2004; Alao et al., 2006b). HDACI induced cyclin D1 degradation was abolished by cotreatment with the proteasome inhibitor MG132 (Alao et al., 2004; Alao et al., 2006b). Inhibition of GSK3β activity or CRM1 dependent nuclear export only partially suppressed the effect of TSA on cyclin D1 stability. These studies suggest that HDACIs target cyclin D1 for degradation via 26S proteasomes in manner that is partially independent of T286 phosphorylation (Alao et al., 2006b). TSA was observed to induce the accumulation of the p45/ Skp2 F-box protein previously linked to cyclin D1 degradation. Furthermore, siRNA mediated knockdown of p45/ Skp2 delayed the effect of TSA on cyclin D1 stability (Alao et al., 2006a; Alao et al., 2004). Recent studies suggest that p45/ Skp2 is not a direct mediator of cyclin D1 ubiquitylation and that the effects of HDACIs on p45/ Skp2 may be cell line specific (Lin et al., 2006; Okabe et al., 2006). Other studies have previously demonstrated however, that knocking down p45/ Skp2 results in the elevation of $p21^{Cip1}$, $p27^{Kip1}$ and cyclin D1 levels (Yu et al., 1998). p45/ Skp2 may thus induce cyclin D1 indirectly, possibly via its interaction with $p27^{Kip1}$. The treatment of MCL cells with the suberoylanilide hydroxamic acid (SAHA) induces a decrease in cellular cyclin D1 levels (Kawamata et al., 2007). Interestingly, inhibition of 26S proteasomes did not abolish the effect of SAHA on cyclin D1 protein levels. SAHA did not affect cyclin D1 mRNA expression or the half life of the cyclin D1 protein. In MCL cell lines, SAHA induced a decline in the levels of phosphorylated Akt and its downstream targets mTOR and the eukaryotic translation initiation factor 4E binding protein (eIF4E-BP). SAHA thus induces a decline in the cyclin D1 level of MCL cells by inhibiting its translation (Kawamata et al., 2007). Interestingly, HDACIs have been shown to modulate the expression of the retinoic acid beta receptor (RARβ) in neuroblastoma cell lines (De los Santos et al., 2007b). HDACIs have also been shown cooperate with RA to inhibit the proliferation of a neuroblastoma cell line (De los Santos et al., 2007a). Similar mechanisms may thus underlie the effect of HDACIs, retinoids and rexinoids on cyclin D1 stability. Alternatively, combining these agents may prove to be more effective at inducing cyclin D1 degradation. HDACIs thus target cyclin D1 expression by different means including the inhibition of mRNA synthesis and protein translation. Additionally, these agents appear to specifically target cyclin D1 for proteasomal degradation. The mechanisms that underlie the ability of HDACIs to induce cyclin D1 degradation have not however, been determined. None the less, these studies suggest that HDACIs may also prove useful as cyclin D1 ablative agents.

As noted in this earlier chapter, mTOR is a component of the PI3K/ Akt regulated signalling pathway. mTOR couples mitogenic stimuli to the regulation of protein synthesis and cell cycle progression. The activity of mTOR is elevated in many cancers as a result of deregulated Akt activity. Furthermore, deregulated mTOR activity has been frequently associated with cancer development (Hay and Sonenberg, 2004; Janus et al., 2005; Mita et al., 2003a; Petroulakis et al., 2006). Rapamycin is a potential anticancer agent that inhibits cancer cell proliferation and induces a G1 cell cycle arrest (Wiederrecht et al., 1995). Rapamycin inhibits mTOR activity by binding to the FK506 binding protein 12 (FKBP12), which in turn forms an inactive complex with mTOR. The inhibition of mTOR activity prevents 4E-BP1 phosphorylation, the release of the eukaryotic initiation factor-4 (eIF-4F) and the formation of active eIF-4 complexes. These complexes regulate the translation of several cell cycle regulators including cyclin D1 (Gingras et al., 2004; Hay and Sonenberg, 2004; Hidalgo and Rowinsky, 2000; Wiederrecht et al., 1995). Rapamycin thus induces a decline in cellular cyclin D1 levels by inhibiting the translation of its mRNA by eIF-4 complexes. Rapamycin induces a decline in cyclin D1 mRNA levels in human glioma and prostate cancer cell lines (Gera et al., 2004). Rapamycin has also been shown to induce the ubiquitin dependent degradation of cyclin D1 in human breast cancer cell lines. Inhibition of GSK3β activity or mutation of T286 to alanine prevented rapamycin induced cyclin D1 degradation (Dong et al., 2005).Rapamycin thus targets cyclin D1 expression by decreasing mRNA levels, inhibiting the translation of its mRNA and targeting the protein for degradation via the 26S proteasome in a T286 phosphorylation dependent manner. The frequency of mTOR deregulation in malignancies makes it an attractive target for chemotherapy (Carraway and Hidalgo, 2004; Janus et al., 2005; Mita et al., 2003a; Mita et al., 2003b; Noh et al., 2004; Petroulakis et al., 2006; Wiederrecht et al., 1995). These studies suggest that rapamycin and other mTOR inhibitors may also be developed to target cyclin D1, especially when used together with other chemotherapeutic agents in combined therapies.

In addition to being a target for chemotherapy, cyclin D1 has also been proposed as a chemopreventative target in cancer (Dragnev et al., 2007; Freemantle et al., 2007). Interestingly, several compounds contained in dietary substances have been shown to induce cyclin D1 degradation in cancer cell lines. Resveratrol is a phytochemical found in many nutritional products including red wine (Baur and Sinclair, 2006; Fulda and Debatin, 2006; Jang et al., 1997; Joe et al., 2002). Resveratrol has been shown to induce cyclin D1 degradation via the 26S proteasome in human colon cancer cell line but not in a promyelocytic leukaemia cell line (Joe et al., 2002). Similarly, diferuloylmethane (curcumin), a major component of the spice turmeric, has also been shown to repress cyclin D1 mRNA expression and target the cyclin for proteasomal degradation (Mukhopadhyay et al., 2002; Sharma et al., 2005; Thangapazham et al., 2006). Resveratrol and diferuloylmethane thus inhibit cancer cell proliferation by targeting cyclin D1 at the mRNA and/ or protein level. It is currently unclear, if the degradation of cyclin D1 induced by these agents requires prior phosphorylation of threonine residues (e.g. T286). 15, 16-dihydrotanshinone I (DHTS), is a component of the traditional Chinese medicinal plant *Salvia miltiorrhiza* (Tanshen) and has been shown to inhibit the proliferation of cancer cells and *in vitro* and *in vivo* (Lee et al., 2006; Tsai et al., 2007). The precise mechanisms whereby DHTS induces cyclin D1 degradation in cancer cell remain unclear. The rapid decline of cyclin D1 protein levels following treatment with DHTS suggests however, that these mechanisms may share similarity with other agents that target cyclin D1 expression (Tsai et al., 2007). Aspirin, a commonly used analgesic agent has also been shown to induce cyclin D1 degradation in human cancer cell lines (Collier, 1969a; Collier, 1969b; Thoms et al., 2007). Aspirin induced cyclin D1 degradation requires p38SAPK2a mediated T286 phosphorylation and occurs via the ubiquitin- 26S proteasome pathway (Thoms et al., 2007). Other naturally occurring substances known to induce cyclin D1 degradation include differentiation inducing factor 1 (DIF1) (Mori et al., 2005; Morris et al., 1987; Takahashi-Yanaga et al., 2006; Takahashi-Yanaga et al., 2003), the metabolites 1-Methyl-4-phenylpyridinium ion (MPP^+) (Bai et al., 2004; Langston and Irwin, 1986), hypothemycin (Sonoda et al., 1999) and O-methyl deoxybouvardin (RA-VII) (Wakita et al., 2001). Interestingly, effective DIF1 induced cyclin D1 requires the prior phosphorylation of T286 and T288 (Takahashi-Yanaga et al., 2006). A requirement for T286 and/ or T288 phosphorylation has not yet been demonstrated for MPP+, hypothemycin or RA-VII induced cyclin D1 degradation. The specific inhibition of 26S proteasomes however, effectively blocked their effect on cyclin D1 stability thus indicating a role for this pathway (Bai et al., 2004; Sonoda et al., 1999; Wakita et al., 2001). Other compounds known to induce cyclin D1 degradation include GL331 (Lin et al., 2001), lovastain (Choi et al., 1997) and oncostatin M (Kim et al., 2007).

It is thus clear that several structurally diverse clinical, preclinical and dietary compounds have been shown to induce cyclin D1 degradation both *in vitro* and *in vivo* (as reviewed in (Alao, 2007) and this chapter). These studies suggest that targeting cyclin D1 in cancer is both feasible and useful. Targeting cyclin D1 may also prove useful in the context of chemoprevention (Dragnev et al., 2007; Freemantle et al., 2007; Petty et al., 2003). Further studies are required however, to determine more clearly, how these compounds target cyclin

D1 for degradation. The role of the FBX4 and FBXW8 in mediating drug induced cyclin D1 degradation has not been reported. Similarly, a possible role for $p38^{SAPK2a}$ in mediating the effects of at least some of these compounds on cyclin D1 stability has not yet been reported. Furthermore, the effect of these agents on cyclin D1b stability has also not been reported. In this regard, it will be necessary to identify compounds that can target cyclin D1 independently of T286 and/ or T288 phosphorylation. It is important to note, that APC mediated cyclin D1 degradation is important for checkpoint activation following exposure to genotoxic agents (Agami and Bernards, 2000). Inhibiting cyclin D1 degradation under these circumstances can thus enhance the cytotoxicity of genotoxic agents. This raises the possibility that cyclin D1 ablative agents might suppress the cytotoxicity of genotoxic agents. Agents that target the kinase activity of cyclin D1- CDK4/ 6 complexes are also in development (Shapiro, 2006). A number of studies suggest however, that the ability of cyclin D1 to promote tumour development is independent of its cell cycle activity (Coqueret, 2002; Knudsen, 2006; Lamb et al., 2003; Solomon et al., 2003). These studies suggest that agents targeting cyclin D1 expression and/ or stability may prove to be superior to CDK complex inhibitors. Questions of this nature demonstrate the desirability of further studies in this wide area of research.

Conclusion

This review highlights the importance of cyclin D1 in regulating the cell cycle progression of normal and cancer cell lines. Numerous clinical, experimental and naturally occurring compounds have been shown to induce cyclin D1 degradation. Together, these studies strongly suggest that targeting cyclin D1 expression and/ or stability will prove useful as a chemotherapeutic or chemopreventative strategy. Our knowledge of the mechanisms that regulate cyclin D1 expression and stability continues to expand at a rapid pace. In fact, several papers on cyclin D1 are published on a weekly basis. The identification of *bona fide* E3 ligases that regulate cyclin D1 degradation has been a major leap forward. Further studies are now required, to clearly define the mechanisms that underlie the ability of specific compounds to attenuate or abolish cyclin D1 expression. These studies will allow for the synthesis and development of novel compounds that more specifically target cyclin D1 expression, activity and stability. They will also allow for the optimization of treatment protocols in particular patients or patient subgroups. The study of drug induced cyclin D1 degradation will thus continue to be an important area of research in the near future and beyond.

References

Agami, R. and Bernards, R. (2000) Distinct initiation and maintenance mechanisms cooperate to induce G1 cell cycle arrest in response to DNA damage. *Cell*, 102, 55-66.

Ahnstrom, M., Nordenskjold, B., Rutqvist, L.E., Skoog, L. and Stal, O. (2005) Role of cyclin D1 in ErbB2-positive breast cancer and tamoxifen resistance. *Breast Cancer Res. Treat*, 91, 145-151.

Aktas, H., Cai, H. and Cooper, G.M. (1997) Ras links growth factor signaling to the cell cycle machinery via regulation of cyclin D1 and the Cdk inhibitor p27KIP1. *Mol. Cell Biol.*, 17, 3850-3857.

Alao, J.P. (2007) The regulation of cyclin D1 degradation: roles in cancer development and the potential for therapeutic invention. *Mol. Cancer*, 6, 24.

Alao, J.P., Gamble, S.C., Stavropoulou, A.V., Pomeranz, K.M., Lam, E.W., Coombes, R.C. and Vigushin, D.M. (2006a) The cyclin D1 proto-oncogene is sequestered in the cytoplasm of mammalian cancer cell lines. *Mol. Cancer*, 5, 7.

Alao, J.P., Lam, E.W., Ali, S., Buluwela, L., Bordogna, W., Lockey, P., Varshochi, R., Stavropoulou, A.V., Coombes, R.C. and Vigushin, D.M. (2004) Histone deacetylase inhibitor trichostatin A represses estrogen receptor alpha-dependent transcription and promotes proteasomal degradation of cyclin D1 in human breast carcinoma cell lines. *Clin. Cancer Res.*, 10, 8094-8104.

Alao, J.P., Stavropoulou, A.V., Lam, E.W., Coombes, R.C. and Vigushin, D.M. (2006b) Histone deacetylase inhibitor, trichostatin A induces ubiquitin-dependent cyclin D1 degradation in MCF-7 breast cancer cells. *Mol. Cancer*, 5, 8.

Albanese, C., Johnson, J., Watanabe, G., Eklund, N., Vu, D., Arnold, A. and Pestell, R.G. (1995) Transforming p21ras mutants and c-Ets-2 activate the cyclin D1 promoter through distinguishable regions. *J. Biol .Chem.*, 270, 23589-23597.

Ali, S. and Coombes, R.C. (2002) Endocrine-responsive breast cancer and strategies for combating resistance. *Nat. Rev. Cancer*, 2, 101-112.

Alt, J.R., Cleveland, J.L., Hannink, M. and Diehl, J.A. (2000) Phosphorylation-dependent regulation of cyclin D1 nuclear export and cyclin D1-dependent cellular transformation. *Genes Dev.*, 14, 3102-3114.

Arnold, A., Kim, H.G., Gaz, R.D., Eddy, R.L., Fukushima, Y., Byers, M.G., Shows, T.B. and Kronenberg, H.M. (1989) Molecular cloning and chromosomal mapping of DNA rearranged with the parathyroid hormone gene in a parathyroid adenoma. *J. Clin. Invest.*, 83, 2034-2040.

Bai, J., Nakamura, H., Ueda, S., Kwon, Y.W., Tanaka, T., Ban, S. and Yodoi, J. (2004) Proteasome-dependent degradation of cyclin D1 in 1-methyl-4-phenylpyridinium ion (MPP+)-induced cell cycle arrest. *J. Biol. Chem.*, 279, 38710-38714.

Baldin, V., Lukas, J., Marcote, M.J., Pagano, M. and Draetta, G. (1993) Cyclin D1 is a nuclear protein required for cell cycle progression in G1. *Genes Dev.*, 7, 812-821.

Bartkova, J., Lukas, J., Muller, H., Lutzhoft, D., Strauss, M. and Bartek, J. (1994) Cyclin D1 protein expression and function in human breast cancer. *Int. J. Cancer*, 57, 353-361.

Bates, S. and Peters, G. (1995) Cyclin D1 as a cellular proto-oncogene. *Semin. Cancer. Biol.*, 6, 73-82.

Baur, J.A. and Sinclair, D.A. (2006) Therapeutic potential of resveratrol: the in vivo evidence. *Nat. Rev. Drug Discov.*, 5, 493-506.

Benzeno, S., Lu, F., Guo, M., Barbash, O., Zhang, F., Herman, J.G., Klein, P.S., Rustgi, A. and Diehl, J.A. (2006) Identification of mutations that disrupt phosphorylation-dependent nuclear export of cyclin D1. *Oncogene*, 25, 6291-6303.

Betticher, D.C., Thatcher, N., Altermatt, H.J., Hoban, P., Ryder, W.D. and Heighway, J. (1995) Alternate splicing produces a novel cyclin D1 transcript. *Oncogene*, 11, 1005-1011.

Blain, S.W., Montalvo, E. and Massague, J. (1997) Differential interaction of the cyclin-dependent kinase (Cdk) inhibitor p27Kip1 with cyclin A-Cdk2 and cyclin D2-Cdk4. *J. Biol. Chem.*, 272, 25863-25872.

Bodrug, S.E., Warner, B.J., Bath, M.L., Lindeman, G.J., Harris, A.W. and Adams, J.M. (1994) Cyclin D1 transgene impedes lymphocyte maturation and collaborates in lymphomagenesis with the myc gene. *Embo J.*, 13, 2124-2130.

Borg, A., Sandberg, T., Nilsson, K., Johannsson, O., Klinker, M., Masback, A., Westerdahl, J., Olsson, H. and Ingvar, C. (2000) High frequency of multiple melanomas and breast and pancreas carcinomas in CDKN2A mutation-positive melanoma families. *J. Natl. Cancer Inst.*, 92, 1260-1266.

Boyle, J.O., Langenfeld, J., Lonardo, F., Sekula, D., Reczek, P., Rusch, V., Dawson, M.I. and Dmitrovsky, E. (1999) Cyclin D1 proteolysis: a retinoid chemoprevention signal in normal, immortalized, and transformed human bronchial epithelial cells. *J. Natl. Cancer Inst.*, 91, 373-379.

Burd, C.J., Petre, C.E., Morey, L.M., Wang, Y., Revelo, M.P., Haiman, C.A., Lu, S., Fenoglio-Preiser, C.M., Li, J., Knudsen, E.S., Wong, J. and Knudsen, K.E. (2006) Cyclin D1b variant influences prostate cancer growth through aberrant androgen receptor regulation. *Proc. Natl. Acad. Sci. U S A*, 103, 2190-2195.

Cardozo, T. and Pagano, M. (2004) The SCF ubiquitin ligase: insights into a molecular machine. *Nat. Rev. Mol. Cell Biol.*, 5, 739-751.

Carraway, H. and Hidalgo, M. (2004) New targets for therapy in breast cancer: mammalian target of rapamycin (mTOR) antagonists. *Breast Cancer Res*, 6, 219-224.

Casanovas, O., Miro, F., Estanyol, J.M., Itarte, E., Agell, N. and Bachs, O. (2000) Osmotic stress regulates the stability of cyclin D1 in a p38SAPK2-dependent manner. *J. Biol. Chem*, 275, 35091-35097.

Castellano, M., Pollock, P.M., Walters, M.K., Sparrow, L.E., Down, L.M., Gabrielli, B.G., Parsons, P.G. and Hayward, N.K. (1997) CDKN2A/p16 is inactivated in most melanoma cell lines. *Cancer Res*, 57, 4868-4875.

Chen, Y.R., Wang, X., Templeton, D., Davis, R.J. and Tan, T.H. (1996) The role of c-Jun N-terminal kinase (JNK) in apoptosis induced by ultraviolet C and gamma radiation. Duration of JNK activation may determine cell death and proliferation. *J. Biol. Chem.*, 271, 31929-31936.

Cheng, M., Olivier, P., Diehl, J.A., Fero, M., Roussel, M.F., Roberts, J.M. and Sherr, C.J. (1999) The p21(Cip1) and p27(Kip1) CDK 'inhibitors' are essential activators of cyclin D-dependent kinases in murine fibroblasts. *Embo. J.*, 18, 1571-1583.

Choi, Y.H., Lee, S.J., Nguyen, P., Jang, J.S., Lee, J., Wu, M.L., Takano, E., Maki, M., Henkart, P.A. and Trepel, J.B. (1997) Regulation of cyclin D1 by calpain protease. *J. Biol. Chem*, 272, 28479-28484.

Collier, H.O. (1969a) New light on how aspirin works. *Nature*, 223, 35-37.

Collier, H.O. (1969b) A pharmacological analysis of aspirin. *Adv Pharmacol Chemother*, 7, 333-405.

Comstock, C.E., Revelo, M.P., Buncher, C.R. and Knudsen, K.E. (2007) Impact of differential cyclin D1 expression and localisation in prostate cancer. *Br. J. Cancer*, 96, 970-979.

Coqueret, O. (2002) Linking cyclins to transcriptional control. *Gene*, 299, 35-55.

De los Santos, M., Zambrano, A. and Aranda, A. (2007a) Combined effects of retinoic acid and histone deacetylase inhibitors on human neuroblastoma SH-SY5Y cells. *Mol. Cancer Ther.*, 6, 1425-1432.

De los Santos, M., Zambrano, A., Sanchez-Pacheco, A. and Aranda, A. (2007b) Histone deacetylase inhibitors regulate retinoic acid receptor beta expression in neuroblastoma cells by both transcriptional and posttranscriptional mechanisms. *Mol. Endocrinol.*, 21, 2416-2426.

Demarchi, F., Bertoli, C., Copetti, T., Eskelinen, E.L. and Schneider, C. (2007) Calpain as a novel regulator of autophagosome formation. *Autophagy*, 3, 235-237.

Demarchi, F. and Schneider, C. (2007) The calpain system as a modulator of stress/damage response. *Cell Cycle*, 6, 136-138.

Deshaies, R.J. (1999) SCF and Cullin/Ring H2-based ubiquitin ligases. *Annu. Re.v Cell. Dev. Biol.*, 15, 435-467.

Diehl, J.A., Cheng, M., Roussel, M.F. and Sherr, C.J. (1998) Glycogen synthase kinase-3beta regulates cyclin D1 proteolysis and subcellular localization. *Genes Dev.*, 12, 3499-3511.

Diehl, J.A., Zindy, F. and Sherr, C.J. (1997) Inhibition of cyclin D1 phosphorylation on threonine-286 prevents its rapid degradation via the ubiquitin-proteasome pathway. *Genes Dev*, 11, 957-972.

Dmitrieva, N.I. and Burg, M.B. (2005) Hypertonic stress response. *Mutat. Res.*, 569, 65-74.

Dmitrieva, N.I., Burg, M.B. and Ferraris, J.D. (2005) DNA damage and osmotic regulation in the kidney. *Am. J. Physiol. Renal. Physiol.*, 289, F2-7.

Dohner, H., Stilgenbauer, S., Dohner, K., Bentz, M. and Lichter, P. (1999) Chromosome aberrations in B-cell chronic lymphocytic leukemia: reassessment based on molecular cytogenetic analysis. *J. Mol. Med.*, 77, 266-281.

Dokmanovic, M. and Marks, P.A. (2005) Prospects: histone deacetylase inhibitors. *J. Cell. Biochem.*, 96, 293-304.

Dong, J., Peng, J., Zhang, H., Mondesire, W.H., Jian, W., Mills, G.B., Hung, M.C. and Meric-Bernstam, F. (2005) Role of glycogen synthase kinase 3beta in rapamycin-mediated cell cycle regulation and chemosensitivity. *Cancer Res.*, 65, 1961-1972.

Dragnev, K.H., Feng, Q., Ma, Y., Shah, S.J., Black, C., Memoli, V., Nugent, W., Rigas, J.R., Kitareewan, S., Freemantle, S. and Dmitrovsky, E. (2007) Uncovering novel targets for cancer chemoprevention. *Recent Results Cancer Res.*, 174, 235-243.

Dragnev, K.H., Petty, W.J., Shah, S., Biddle, A., Desai, N.B., Memoli, V., Rigas, J.R. and Dmitrovsky, E. (2005) Bexarotene and erlotinib for aerodigestive tract cancer. *J. Clin. Oncol.*, 23, 8757-8764.

Dragnev, K.H., Pitha-Rowe, I., Ma, Y., Petty, W.J., Sekula, D., Murphy, B., Rendi, M., Suh, N., Desai, N.B., Sporn, M.B., Freemantle, S.J. and Dmitrovsky, E. (2004) Specific chemopreventive agents trigger proteasomal degradation of G1 cyclins: implications for combination therapy. *Clin. Cancer Res.*, 10, 2570-2577.

el-Deiry, W.S., Tokino, T., Velculescu, V.E., Levy, D.B., Parsons, R., Trent, J.M., Lin, D., Mercer, W.E., Kinzler, K.W. and Vogelstein, B. (1993) WAF1, a potential mediator of p53 tumor suppression. *Cell*, 75, 817-825.

Elsheikh, S., Green, A.R., Aleskandarany, M.A., Grainge, M., Paish, C.E., Lambros, M.B., Reis-Filho, J.S. and Ellis, I.O. (2007) CCND1 amplification and cyclin D1 expression in breast cancer and their relation with proteomic subgroups and patient outcome. *Breast Cancer Res. Treat.*

Feng, Q., Sekula, D., Muller, R., Freemantle, S.J. and Dmitrovsky, E. (2007) Uncovering residues that regulate cyclin D1 proteasomal degradation. *Oncogene*.

Feuer, R., Mena, I., Pagarigan, R., Slifka, M.K. and Whitton, J.L. (2002) Cell cycle status affects coxsackievirus replication, persistence, and reactivation in vitro. *J. Virol.*, 76, 4430-4440.

Foster, J.S., Henley, D.C., Ahamed, S. and Wimalasena, J. (2001) Estrogens and cell-cycle regulation in breast cancer. *Trends Endocrinol. Metab.*, 12, 320-327.

Freemantle, S.J., Liu, X., Feng, Q., Galimberti, F., Blumen, S., Sekula, D., Kitareewan, S., Dragnev, K.H. and Dmitrovsky, E. (2007) Cyclin degradation for cancer therapy and chemoprevention. *J. Cell Biochem.*, 102, 869-877.

Friedman, E. (2007) Mirk/Dyrk1B in cancer. *J. Cell Biochem.*, 102, 274-279.

Fry, D.W., Harvey, P.J., Keller, P.R., Elliott, W.L., Meade, M., Trachet, E., Albassam, M., Zheng, X., Leopold, W.R., Pryer, N.K. and Toogood, P.L. (2004) Specific inhibition of cyclin-dependent kinase 4/6 by PD 0332991 and associated antitumor activity in human tumor xenografts. *Mol. Cancer Ther.*, 3, 1427-1438.

Fukami-Kobayashi, J. and Mitsui, Y. (1999) Cyclin D1 inhibits cell proliferation through binding to PCNA and cdk2. *Exp. Cell Res.*, 246, 338-347.

Fulda, S. and Debatin, K.M. (2006) Resveratrol modulation of signal transduction in apoptosis and cell survival: a mini-review. *Cancer Detect. Prev.*, 30, 217-223.

Ganiatsas, S., Dow, R., Thompson, A., Schulman, B. and Germain, D. (2001) A splice variant of Skp2 is retained in the cytoplasm and fails to direct cyclin D1 ubiquitination in the uterine cancer cell line SK-UT. *Oncogene*, 20, 3641-3650.

Gera, J.F., Mellinghoff, I.K., Shi, Y., Rettig, M.B., Tran, C., Hsu, J.H., Sawyers, C.L. and Lichtenstein, A.K. (2004) AKT activity determines sensitivity to mammalian target of rapamycin (mTOR) inhibitors by regulating cyclin D1 and c-myc expression. *J. Biol. Chem.*, 279, 2737-2746.

Germain, D., Russell, A., Thompson, A. and Hendley, J. (2000) Ubiquitination of free cyclin D1 is independent of phosphorylation on threonine 286. *J. Biol. Chem.*, 275, 12074-12079.

Gingras, A.C., Raught, B. and Sonenberg, N. (2004) mTOR signaling to translation. *Curr. Top. Microbiol. Immunol.*, 279, 169-197.

Gladden, A.B. and Diehl, J.A. (2005) Location, location, location: the role of cyclin D1 nuclear localization in cancer. *J. Cell Biochem.*, 96, 906-913.

Glaser, K.B. (2007) HDAC inhibitors: Clinical update and mechanism-based potential. *Biochem. Pharmacol.*

Guan, K.L., Jenkins, C.W., Li, Y., Nichols, M.A., Wu, X., O'Keefe, C.L., Matera, A.G. and Xiong, Y. (1994) Growth suppression by p18, a p16INK4/MTS1- and p14INK4B/MTS2-related CDK6 inhibitor, correlates with wild-type pRb function. *Genes Dev.*, 8, 2939-2952.

Guo, Y., Harwalkar, J., Stacey, D.W. and Hitomi, M. (2005a) Destabilization of cyclin D1 message plays a critical role in cell cycle exit upon mitogen withdrawal. *Oncogene*, 24, 1032-1042.

Guo, Y., Yang, K., Harwalkar, J., Nye, J.M., Mason, D.R., Garrett, M.D., Hitomi, M. and Stacey, D.W. (2005b) Phosphorylation of cyclin D1 at Thr 286 during S phase leads to its proteasomal degradation and allows efficient DNA synthesis. *Oncogene*, 24, 2599-2612.

Hannon, G.J. and Beach, D. (1994) p15INK4B is a potential effector of TGF-beta-induced cell cycle arrest. *Nature*, 371, 257-261.

Hay, N. and Sonenberg, N. (2004) Upstream and downstream of mTOR. *Genes Dev.*, 18, 1926-1945.

Hayward, N. (2000) New developments in melanoma genetics. *Curr. Oncol. Rep.*, 2, 300-306.

Herber, B., Truss, M., Beato, M. and Muller, R. (1994) Inducible regulatory elements in the human cyclin D1 promoter. *Oncogene*, 9, 1295-1304.

Hidalgo, M. and Rowinsky, E.K. (2000) The rapamycin-sensitive signal transduction pathway as a target for cancer therapy. *Oncogene*, 19, 6680-6686.

Hirai, H., Roussel, M.F., Kato, J.Y., Ashmun, R.A. and Sherr, C.J. (1995) Novel INK4 proteins, p19 and p18, are specific inhibitors of the cyclin D-dependent kinases CDK4 and CDK6. *Mol. Cell. Biol.*, 15, 2672-2681.

Hiramatsu, Y., Kitagawa, K., Suzuki, T., Uchida, C., Hattori, T., Kikuchi, H., Oda, T., Hatakeyama, S., Nakayama, K.I., Yamamoto, T., Konno, H. and Kitagawa, M. (2006) Degradation of Tob1 Mediated by SCFSkp2-Dependent Ubiquitination. *Cancer Res.*, 66, 8477-8483.

Hitomi, M. and Stacey, D.W. (1999a) Cellular ras and cyclin D1 are required during different cell cycle periods in cycling NIH 3T3 cells. *Mol. Cell Biol.*, 19, 4623-4632.

Hitomi, M. and Stacey, D.W. (1999b) Cyclin D1 production in cycling cells depends on ras in a cell-cycle-specific manner. *Curr. Biol.*, 9, 1075-1084.

Hitomi, M. and Stacey, D.W. (2001) Ras-dependent cell cycle commitment during G2 phase. *FEBS Lett.*, 490, 123-131.

Ho, M.S., Tsai, P.I. and Chien, C.T. (2006) F-box proteins: the key to protein degradation. *J Biomed. Sci.*, 13, 181-191.

Hosokawa, Y., Gadd, M., Smith, A.P., Koerner, F.C., Schmidt, E.V. and Arnold, A. (1997) Cyclin D1 (PRAD1) alternative transcript b: full-length cDNA cloning and expression in breast cancers. *Cancer Lett.*, 113, 123-130.

Hosokawa, Y., Joh, T., Maeda, Y., Arnold, A. and Seto, M. (1999) Cyclin D1/PRAD1/BCL-1 alternative transcript [B] protein product in B-lymphoid malignancies with t(11;14)(q13;q32) translocation. *Int. J. Cancer*, 81, 616-619.

Huang, G., Chen, Y., Lu, H. and Cao, X. (2007) Coupling mitochondrial respiratory chain to cell death: an essential role of mitochondrial complex I in the interferon-beta and retinoic acid-induced cancer cell death. *Cell Death Differ.*, 14, 327-337.

Huang, J.W., Shiau, C.W., Yang, J., Wang, D.S., Chiu, H.C., Chen, C.Y. and Chen, C.S. (2006) Development of small-molecule cyclin D1-ablative agents. *J. Med. Chem.*, 49, 4684-4689.

Huang, J.W., Shiau, C.W., Yang, Y.T., Kulp, S.K., Chen, K.F., Brueggemeier, R.W., Shapiro, C.L. and Chen, C.S. (2005) Peroxisome proliferator-activated receptor gamma-independent ablation of cyclin D1 by thiazolidinediones and their derivatives in breast cancer cells. *Mol. Pharmacol.*, 67, 1342-1348.

Hui, R., Finney, G.L., Carroll, J.S., Lee, C.S., Musgrove, E.A. and Sutherland, R.L. (2002) Constitutive overexpression of cyclin D1 but not cyclin E confers acute resistance to antiestrogens in T-47D breast cancer cells. *Cancer Re.s*, 62, 6916-6923.

Hussussian, C.J., Struewing, J.P., Goldstein, A.M., Higgins, P.A., Ally, D.S., Sheahan, M.D., Clark, W.H., Jr., Tucker, M.A. and Dracopoli, N.C. (1994) Germline p16 mutations in familial melanoma. *Nat. Genet.*, 8, 15-21.

Jang, M., Cai, L., Udeani, G.O., Slowing, K.V., Thomas, C.F., Beecher, C.W., Fong, H.H., Farnsworth, N.R., Kinghorn, A.D., Mehta, R.G., Moon, R.C. and Pezzuto, J.M. (1997) Cancer chemopreventive activity of resveratrol, a natural product derived from grapes. *Science*, 275, 218-220.

Janus, A., Robak, T. and Smolewski, P. (2005) The mammalian target of the rapamycin (mTOR) kinase pathway: its role in tumourigenesis and targeted antitumour therapy. *Cell Mol. Biol. Lett*, 10, 479-498.

Joe, A.K., Liu, H., Suzui, M., Vural, M.E., Xiao, D. and Weinstein, I.B. (2002) Resveratrol induces growth inhibition, S-phase arrest, apoptosis, and changes in biomarker expression in several human cancer cell lines. *Clin. Cancer Re.s*, 8, 893-903.

Johnstone, R.W. (2002) Histone-deacetylase inhibitors: novel drugs for the treatment of cancer. *Nat. Rev. Drug Disco.v*, 1, 287-299.

Kamb, A., Gruis, N.A., Weaver-Feldhaus, J., Liu, Q., Harshman, K., Tavtigian, S.V., Stockert, E., Day, R.S., 3rd, Johnson, B.E. and Skolnick, M.H. (1994a) A cell cycle regulator potentially involved in genesis of many tumor types. *Science*, 264, 436-440.

Kamb, A., Shattuck-Eidens, D., Eeles, R., Liu, Q., Gruis, N.A., Ding, W., Hussey, C., Tran, T., Miki, Y., Weaver-Feldhaus, J. and et al. (1994b) Analysis of the p16 gene (CDKN2) as a candidate for the chromosome 9p melanoma susceptibility locus. *Nat Genet*, 8, 23-26.

Kawamata, N., Chen, J. and Koeffler, H.P. (2007) Suberoylanilide hydroxamic acid (SAHA; vorinostat) suppresses translation of cyclin D1 in mantle cell lymphoma cells. *Blood*, 110, 2667-2673.

Kim, H., Jo, C., Jang, B.G., Oh, U. and Jo, S.A. (2007) Oncostatin M induces growth arrest of skeletal muscle cells in G(1) phase by regulating cyclin D1 protein level. *Cell Signal.*

Kisselev, A.F. and Goldberg, A.L. (2001) Proteasome inhibitors: from research tools to drug candidates. *Chem Biol*, 8, 739-758.

Knudsen, K.E. (2006) The cyclin D1b splice variant: an old oncogene learns new tricks. *Cell Div.*, 1, 15.

Knudsen, K.E., Diehl, J.A., Haiman, C.A. and Knudsen, E.S. (2006) Cyclin D1: polymorphism, aberrant splicing and cancer risk. *Oncogene*, 25, 1620-1628.

Kota, B.P., Huang, T.H. and Roufogalis, B.D. (2005) An overview on biological mechanisms of PPARs. *Pharmacol. Res*, 51, 85-94.

Krieger, S., Gauduchon, J., Roussel, M., Troussard, X. and Sola, B. (2006) Relevance of cyclin D1b expression and CCND1 polymorphism in the pathogenesis of multiple myeloma and mantle cell lymphoma. *BMC Cancer*, 6, 238.

LaBaer, J., Garrett, M.D., Stevenson, L.F., Slingerland, J.M., Sandhu, C., Chou, H.S., Fattaey, A. and Harlow, E. (1997) New functional activities for the p21 family of CDK inhibitors. *Genes Dev.*, 11, 847-862.

Lamb, J., Ramaswamy, S., Ford, H.L., Contreras, B., Martinez, R.V., Kittrell, F.S., Zahnow, C.A., Patterson, N., Golub, T.R. and Ewen, M.E. (2003) A mechanism of cyclin D1 action encoded in the patterns of gene expression in human cancer. *Cell*, 114, 323-334.

Langenfeld, J., Kiyokawa, H., Sekula, D., Boyle, J. and Dmitrovsky, E. (1997) Posttranslational regulation of cyclin D1 by retinoic acid: a chemoprevention mechanism. *Proc. Natl. Acad. Sci. U S A*, 94, 12070-12074.

Langston, J.W. and Irwin, I. (1986) MPTP: current concepts and controversies. *Clin. Neuropharmacol.*, 9, 485-507.

Lee, D.H. and Goldberg, A.L. (1998) Proteasome inhibitors: valuable new tools for cell biologists. *Trends Cell. Biol.*, 8, 397-403.

Lee, M.H., Reynisdottir, I. and Massague, J. (1995) Cloning of p57KIP2, a cyclin-dependent kinase inhibitor with unique domain structure and tissue distribution. *Genes Dev.*, 9, 639-649.

Lee, P., Hur, J., Lee, J., Kim, J., Jeong, J., Kang, I., Kim, S.Y. and Kim, H. (2006) 15,16-dihydrotanshinone I suppresses the activation of BV-2 cell, a murine microglia cell line, by lipopolysaccharide. *Neurochem. Int.*, 48, 60-66.

Lin, D.I., Barbash, O., Kumar, K.G., Weber, J.D., Harper, J.W., Klein-Szanto, A.J., Rustgi, A., Fuchs, S.Y. and Diehl, J.A. (2006) Phosphorylation-dependent ubiquitination of cyclin D1 by the SCF(FBX4-alphaB crystallin) complex. *Mol. Cell*, 24, 355-366.

Lin, D.I., Lessie, M.D., Gladden, A.B., Bassing, C.H., Wagner, K.U. and Diehl, J.A. (2007) Disruption of cyclin D1 nuclear export and proteolysis accelerates mammary carcinogenesis. *Oncogene*.

Lin, S., Huang, H.C., Chen, L.L., Lee, C.C. and Huang, T.S. (2001) GL331 induces down-regulation of cyclin D1 expression via enhanced proteolysis and repressed transcription. *Mol. Pharmacol.*, 60, 768-775.

Lovec, H., Grzeschiczek, A., Kowalski, M.B. and Moroy, T. (1994) Cyclin D1/bcl-1 cooperates with myc genes in the generation of B-cell lymphoma in transgenic mice. *Embo J.*, 13, 3487-3495.

Lu, F., Gladden, A.B. and Diehl, J.A. (2003) An alternatively spliced cyclin D1 isoform, cyclin D1b, is a nuclear oncogene. *Cancer Res.*, 63, 7056-7061.

Lukas, J., Pagano, M., Staskova, Z., Draetta, G. and Bartek, J. (1994) Cyclin D1 protein oscillates and is essential for cell cycle progression in human tumour cell lines. *Oncogene*, 9, 707-718.

Luo, H., Zhang, J., Dastvan, F., Yanagawa, B., Reidy, M.A., Zhang, H.M., Yang, D., Wilson, J.E. and McManus, B.M. (2003) Ubiquitin-dependent proteolysis of cyclin D1 is associated with coxsackievirus-induced cell growth arrest. *J. Virol.*, 77, 1-9.

Ma, Y., Feng, Q., Sekula, D., Diehl, J.A., Freemantle, S.J. and Dmitrovsky, E. (2005) Retinoid targeting of different D-type cyclins through distinct chemopreventive mechanisms. *Cancer Res.*, 65, 6476-6483.

Marks, P.A. and Jiang, X. (2005) Histone deacetylase inhibitors in programmed cell death and cancer therapy. *Cell Cycle*, 4, 549-551.

Matsuoka, S., Edwards, M.C., Bai, C., Parker, S., Zhang, P., Baldini, A., Harper, J.W. and Elledge, S.J. (1995) p57KIP2, a structurally distinct member of the p21CIP1 Cdk inhibitor family, is a candidate tumor suppressor gene. *Genes Dev.*, 9, 650-662.

Matsushime, H., Roussel, M.F., Ashmun, R.A. and Sherr, C.J. (1991) Colony-stimulating factor 1 regulates novel cyclins during the G1 phase of the cell cycle. *Cell*, 65, 701-713.

Mehnert, J.M. and Kelly, W.K. (2007) Histone deacetylase inhibitors: biology and mechanism of action. *Cancer J.*, 13, 23-29.

Mikhailov, A., Shinohara, M. and Rieder, C.L. (2004) Topoisomerase II and histone deacetylase inhibitors delay the G2/M transition by triggering the p38 MAPK checkpoint pathway. *J. Cell Biol.*, 166, 517-526.

Mita, M.M., Mita, A. and Rowinsky, E.K. (2003a) Mammalian target of rapamycin: a new molecular target for breast cancer. *Clin. Breast Cancer*, 4, 126-137.

Mita, M.M., Mita, A. and Rowinsky, E.K. (2003b) The molecular target of rapamycin (mTOR) as a therapeutic target against cancer. *Cancer Biol. Ther.*, 2, S169-177.

Mori, J., Takahashi-Yanaga, F., Miwa, Y., Watanabe, Y., Hirata, M., Morimoto, S., Shirasuna, K. and Sasaguri, T. (2005) Differentiation-inducing factor-1 induces cyclin D1 degradation through the phosphorylation of Thr286 in squamous cell carcinoma. *Exp. Cell Res.*, 310, 426-433.

Morris, H.R., Taylor, G.W., Masento, M.S., Jermyn, K.A. and Kay, R.R. (1987) Chemical structure of the morphogen differentiation inducing factor from Dictyostelium discoideum. *Nature*, 328, 811-814.

Motokura, T. and Arnold, A. (1993) PRAD1/cyclin D1 proto-oncogene: genomic organization, 5' DNA sequence, and sequence of a tumor-specific rearrangement breakpoint. *Genes Chromosomes Cancer*, 7, 89-95.

Motokura, T., Bloom, T., Kim, H.G., Juppner, H., Ruderman, J.V., Kronenberg, H.M. and Arnold, A. (1991) A novel cyclin encoded by a bcl1-linked candidate oncogene. *Nature*, 350, 512-515.

Mukhopadhyay, A., Banerjee, S., Stafford, L.J., Xia, C., Liu, M. and Aggarwal, B.B. (2002) Curcumin-induced suppression of cell proliferation correlates with down-regulation of cyclin D1 expression and CDK4-mediated retinoblastoma protein phosphorylation. *Oncogene*, 21, 8852-8861.

Muller, H., Lukas, J., Schneider, A., Warthoe, P., Bartek, J., Eilers, M. and Strauss, M. (1994) Cyclin D1 expression is regulated by the retinoblastoma protein. *Proc. Natl. Acad. Sci. U S A*, 91, 2945-2949.

Musgrove, E.A. (2006) Cyclins: roles in mitogenic signaling and oncogenic transformation. *Growth Factors*, 24, 13-19.

Newman, R.M., Mobascher, A., Mangold, U., Koike, C., Diah, S., Schmidt, M., Finley, D. and Zetter, B.R. (2004) Antizyme targets cyclin D1 for degradation. A novel mechanism for cell growth repression. *J. Biol. Chem.*, 279, 41504-41511.

Nobori, T., Miura, K., Wu, D.J., Lois, A., Takabayashi, K. and Carson, D.A. (1994) Deletions of the cyclin-dependent kinase-4 inhibitor gene in multiple human cancers. *Nature*, 368, 753-756.

Noh, W.C., Mondesire, W.H., Peng, J., Jian, W., Zhang, H., Dong, J., Mills, G.B., Hung, M.C. and Meric-Bernstam, F. (2004) Determinants of rapamycin sensitivity in breast cancer cells. *Clin. Cancer Res.*, 10, 1013-1023.

Okabe, H., Lee, S.H., Phuchareon, J., Albertson, D.G., McCormick, F. and Tetsu, O. (2006) A Critical Role for FBXW8 and MAPK in Cyclin D1 Degradation and Cancer Cell Proliferation. *PLoS ONE*, 1, e128.

Pagano, M., Theodoras, A.M., Tam, S.W. and Draetta, G.F. (1994) Cyclin D1-mediated inhibition of repair and replicative DNA synthesis in human fibroblasts. *Genes Dev*,. 8, 1627-1639.

Petroulakis, E., Mamane, Y., Le Bacquer, O., Shahbazian, D. and Sonenberg, N. (2006) mTOR signaling: implications for cancer and anticancer therapy. *Br. J. Cancer.*, 94, 195-199.

Petty, W.J., Dragnev, K.H. and Dmitrovsky, E. (2003) Cyclin D1 as a target for chemoprevention. *Lung Cancer*, 41 Suppl 1, S155-161.

Price, M.A., Cruzalegui, F.H. and Treisman, R. (1996) The p38 and ERK MAP kinase pathways cooperate to activate Ternary Complex Factors and c-fos transcription in response to UV light. *Embo J.*, 15, 6552-6563.

Qi, C., Zhu, Y. and Reddy, J.K. (2000) Peroxisome proliferator-activated receptors, coactivators, and downstream targets. *Cell Biochem. Biophys.*, 32 Spring, 187-204.

Qin, C., Burghardt, R., Smith, R., Wormke, M., Stewart, J. and Safe, S. (2003) Peroxisome proliferator-activated receptor gamma agonists induce proteasome-dependent degradation of cyclin D1 and estrogen receptor alpha in MCF-7 breast cancer cells. *Cancer Res.*, 63, 958-964.

Ranade, K., Hussussian, C.J., Sikorski, R.S., Varmus, H.E., Goldstein, A.M., Tucker, M.A., Serrano, M., Hannon, G.J., Beach, D. and Dracopoli, N.C. (1995) Mutations associated with familial melanoma impair p16INK4 function. *Nat. Genet.*, 10, 114-116.

Reinhardt, H.C., Aslanian, A.S., Lees, J.A. and Yaffe, M.B. (2007) p53-deficient cells rely on ATM- and ATR-mediated checkpoint signaling through the p38MAPK/MK2 pathway for survival after DNA damage. *Cancer Cell*, 11, 175-189.

Reis-Filho, J.S., Savage, K., Lambros, M.B., James, M., Steele, D., Jones, R.L. and Dowsett, M. (2006) Cyclin D1 protein overexpression and CCND1 amplification in breast carcinomas: an immunohistochemical and chromogenic in situ hybridisation analysis. *Mod. Pathol.*, 19, 999-1009.

Rigas, J.R. and Dragnev, K.H. (2005) Emerging role of rexinoids in non-small cell lung cancer: focus on bexarotene. *Oncologist*, 10, 22-33.

Roovers, K. and Assoian, R.K. (2000) Integrating the MAP kinase signal into the G1 phase cell cycle machinery. *Bioessays*, 22, 818-826.

Rosenberg, C.L., Motokura, T., Kronenberg, H.M. and Arnold, A. (1993) Coding sequence of the overexpressed transcript of the putative oncogene PRAD1/cyclin D1 in two primary human tumors. *Oncogene*, 8, 519-521.

Rosenberg, C.L., Wong, E., Petty, E.M., Bale, A.E., Tsujimoto, Y., Harris, N.L. and Arnold, A. (1991) PRAD1, a candidate BCL1 oncogene: mapping and expression in centrocytic lymphoma. *Proc. Natl. Acad. Sci. U S A*, 88, 9638-9642.

Russell, A., Thompson, M.A., Hendley, J., Trute, L., Armes, J. and Germain, D. (1999) Cyclin D1 and D3 associate with the SCF complex and are coordinately elevated in breast cancer. *Oncogene*, 18, 1983-1991.

Sabbah, M., Courilleau, D., Mester, J. and Redeuilh, G. (1999) Estrogen induction of the cyclin D1 promoter: involvement of a cAMP response-like element. *Proc. Natl .Acad. Sci. U S A*, 96, 11217-11222.

Schneider, G., Reichert, M., Saur, D., Hamacher, R., Fritsch, R. and Schmid, R.M. (2007) HDAC3 is linked to cell cycle machinery in MiaPaCa2 cells by regulating transcription of skp2. *Cell Prolif.*, 40, 522-531.

Serrano, M., Hannon, G.J. and Beach, D. (1993) A new regulatory motif in cell-cycle control causing specific inhibition of cyclin D/CDK4. *Nature*, 366, 704-707.

Shapiro, G.I. (2006) Cyclin-dependent kinase pathways as targets for cancer treatment. *J. Clin. Oncol.*, 24, 1770-1783.

Sharma, R.A., Gescher, A.J. and Steward, W.P. (2005) Curcumin: the story so far. *Eur. J. .Cancer*, 41, 1955-1968.

Sherr, C.J. and Roberts, J.M. (1999) CDK inhibitors: positive and negative regulators of G1-phase progression. *Genes Dev.*, 13, 1501-1512.

Solomon, D.A., Wang, Y., Fox, S.R., Lambeck, T.C., Giesting, S., Lan, Z., Senderowicz, A.M., Conti, C.J. and Knudsen, E.S. (2003) Cyclin D1 splice variants. Differential effects on localization, RB phosphorylation, and cellular transformation. *J. Biol. Chem.*, 278, 30339-30347.

Sonoda, H., Omi, K., Hojo, K., Nishida, K., Omura, S. and Sugita, K. (1999) Suppression of oncogenic transformation by hypothemycin associated with accelerated cyclin D1 degradation through ubiquitin-proteasome pathway. *Life Sci.*, 65, 381-394.

Soos, T.J., Kiyokawa, H., Yan, J.S., Rubin, M.S., Giordano, A., DeBlasio, A., Bottega, S., Wong, B., Mendelsohn, J. and Koff, A. (1996) Formation of p27-CDK complexes during the human mitotic cell cycle. *Cell Growth Differ.*, 7, 135-146.

Spinella, M.J., Freemantle, S.J., Sekula, D., Chang, J.H., Christie, A.J. and Dmitrovsky, E. (1999) Retinoic acid promotes ubiquitination and proteolysis of cyclin D1 during induced tumor cell differentiation. *J. Biol. Chem.*, 274, 22013-22018.

Sumrejkanchanakij, P., Tamamori-Adachi, M., Matsunaga, Y., Eto, K. and Ikeda, M.A. (2003) Role of cyclin D1 cytoplasmic sequestration in the survival of postmitotic neurons. *Oncogene*, 22, 8723-8730.

Takahashi-Yanaga, F., Mori, J., Matsuzaki, E., Watanabe, Y., Hirata, M., Miwa, Y., Morimoto, S. and Sasaguri, T. (2006) Involvement of GSK-3beta and DYRK1B in differentiation-inducing factor-3-induced phosphorylation of cyclin D1 in HeLa cells. *J. Biol. Chem.*, 281, 38489-38497.

Takahashi-Yanaga, F., Taba, Y., Miwa, Y., Kubohara, Y., Watanabe, Y., Hirata, M., Morimoto, S. and Sasaguri, T. (2003) Dictyostelium differentiation-inducing factor-3 activates glycogen synthase kinase-3beta and degrades cyclin D1 in mammalian cells. *J Biol. Chem.*, 278, 9663-9670.

Tamamori-Adachi, M., Ito, H., Sumrejkanchanakij, P., Adachi, S., Hiroe, M., Shimizu, M., Kawauchi, J., Sunamori, M., Marumo, F., Kitajima, S. and Ikeda, M.A. (2003) Critical role of cyclin D1 nuclear import in cardiomyocyte proliferation. *Circ. Res*, 92, e12-19.

Tashiro, E., Tsuchiya, A. and Imoto, M. (2007) Functions of cyclin D1 as an oncogene and regulation of cyclin D1 expression. *Cancer Sci.*, 98, 629-635.

Tetsu, O. and McCormick, F. (1999) Beta-catenin regulates expression of cyclin D1 in colon carcinoma cells. *Nature*, 398, 422-426.

Thangapazham, R.L., Sharma, A. and Maheshwari, R.K. (2006) Multiple molecular targets in cancer chemoprevention by curcumin. *Aaps. J.*, 8, E443-449.

Thoms, H.C., Dunlop, M.G. and Stark, L.A. (2007) p38-mediated inactivation of cyclin D1/cyclin-dependent kinase 4 stimulates nucleolar translocation of RelA and apoptosis in colorectal cancer cells. *Cancer Res.*, 67, 1660-1669.

Toyoshima, H. and Hunter, T. (1994) p27, a novel inhibitor of G1 cyclin-Cdk protein kinase activity, is related to p21. *Cell*, 78, 67-74.

Tsai, S.L., Suk, F.M., Wang, C.I., Liu, D.Z., Hou, W.C., Lin, P.J., Hung, L.F. and Liang, Y.C. (2007) Anti-tumor potential of 15,16-dihydrotanshinone I against breast adenocarcinoma through inducing G1 arrest and apoptosis. *Biochem Pharmacol*, 74, 1575-1586.

Wakita, K., Minami, M., Venkateswarlu, A., Sharma, V.M., Ramesh, M. and Akahane, K. (2001) Antitumor bicyclic hexapeptide RA-VII modulates cyclin D1 protein level. *Anticancer Drugs*, 12, 433-439.

Wang, C., Fu, M., D'Amico, M., Albanese, C., Zhou, J.N., Brownlee, M., Lisanti, M.P., Chatterjee, V.K., Lazar, M.A. and Pestell, R.G. (2001) Inhibition of cellular proliferation through IkappaB kinase-independent and peroxisome proliferator-activated receptor gamma-dependent repression of cyclin D1. *Mol. Cell Biol.*, 21, 3057-3070.

Wang, T., Xu, J., Yu, X., Yang, R. and Han, Z.C. (2006) Peroxisome proliferator-activated receptor gamma in malignant diseases. *Crit. Rev. Oncol. Hematol*, 58, 1-14.

Weinberg, R.A. (1995) The retinoblastoma protein and cell cycle control. *Cell*, 81, 323-330.

Wiederrecht, G.J., Sabers, C.J., Brunn, G.J., Martin, M.M., Dumont, F.J. and Abraham, R.T. (1995) Mechanism of action of rapamycin: new insights into the regulation of G1-phase progression in eukaryotic cells. *Prog. Cell Cycle Res.*, 1, 53-71.

Vigushin, D.M. and Coombes, R.C. (2002) Histone deacetylase inhibitors in cancer treatment. *Anticancer Drugs*, 13, 1-13.

Vivanco, I. and Sawyers, C.L. (2002) The phosphatidylinositol 3-Kinase AKT pathway in human cancer. *Nat. Rev. Cancer*, 2, 489-501.

Xiong, Y., Hannon, G.J., Zhang, H., Casso, D., Kobayashi, R. and Beach, D. (1993) p21 is a universal inhibitor of cyclin kinases. *Nature*, 366, 701-704.

Yamamoto, M., Tamakawa, S., Yoshie, M., Yaginuma, Y. and Ogawa, K. (2006) Neoplastic hepatocyte growth associated with cyclin D1 redistribution from the cytoplasm to the nucleus in mouse hepatocarcinogenesis. *Mol. Carcinog.*, 45, 901-913.

Yang, K., Guo, Y., Stacey, W.C., Harwalkar, J., Fretthold, J., Hitomi, M. and Stacey, D.W. (2006) Glycogen synthase kinase 3 has a limited role in cell cycle regulation of cyclin D1 levels. *BMC Cell Biol.*, 7, 33.

Yu, F., Yang, G., Zhao, Z., Ji, L., Cao, Y., Bai, L., Lu, F., Fu, H., Huang, B., Li, H., Zhang, J., Yao, L. and Lu, Z. (2007) Apoptosis related protein 3, an ATRA-upregulated membrane protein arrests the cell cycle at G1/S phase by decreasing the expression of cyclin D1. *Biochem. Biophys. Res. Commun.*, 358, 1041-1046.

Yu, Q., Geng, Y. and Sicinski, P. (2001) Specific protection against breast cancers by cyclin D1 ablation. *Nature*, 411, 1017- 1021.

Yu, Z.K., Gervais, J.L. and Zhang, H. (1998) Human CUL-1 associates with the SKP1/SKP2 complex and regulates p21(CIP1/WAF1) and cyclin D proteins. *Proc. Natl. Acad. Sci. U S A*, 95, 11324-11329.

Zhang, Z., Ferraris, J.D., Irarrazabal, C.E., Dmitrieva, N.I., Park, J.H. and Burg, M.B. (2005) Ataxia telangiectasia-mutated, a DNA damage-inducible kinase, contributes to high NaCl-induced nuclear localization of transcription factor TonEBP/OREBP. *Am. J. Physiol. Renal. Physiol.*, 289, F506-511.

Zou, Y., Ewton, D.Z., Deng, X., Mercer, S.E. and Friedman, E. (2004) Mirk/dyrk1B kinase destabilizes cyclin D1 by phosphorylation at threonine 288. *J. Biol. Chem.*, 279, 27790-27798.

Zwijsen, R.M., Wientjens, E., Klompmaker, R., van der Sman, J., Bernards, R. and Michalides, R.J. (1997) CDK-independent activation of estrogen receptor by cyclin D1. *Cell*, 88, 405-415.

In: Oncogene Proteins: New Research
Editors: A. H. Malloy and E.C. Carson, pp.129-159
ISBN 978-1-60456-290-3

Chapter 4

Towards Understanding the Multifunction RET Protooncogene

S. W. Moore
Faculty of Health Sciences, University of Stellenbosch, South Africa

Abstract

The RET *(REarranged during Transfection)* proto-oncogene plays a vital role in controlling cell proliferation, migration, differentiation, and continued cellular survival during development. It is central to the development of the kidney and enteric nervous systems as well as being chiefly responsible for at least four conditions associated with mutations at different sites.

Much progress has been made in identifying the functions of this gene and its ligands GDNF, GFRα and neurturin in Tyrosine kinase activation , particularly as it pertains to its influence in oncogenesis, in the thyroid and other endocrine glands (MEN syndromes) Up-regulation of the RAS-RAF-ERK and PI3K-PDK1-Akt sinaling pathways results. In addition, fusion to other genes (eg RET/PTC), activates the *BRAF, RET*, or *RAS* genes of the MAPK system.

This article examines the significance and functions of RET and its ligands reviewing its oncogenic potential as well as possible molecular targets in therapy.

Introduction

It has become clear over recent years that that receptor tyrosine kinases (RTK) function as oncogenes and are important in the development of cancer. [1]. The RET proto-oncogene *(REarranged during Transfection* ; RET*)* is regarded as a proto-type RTK because of the numbers of rearrangements not infrequently isolated from tumour cells[1] . It remains one of most important proto-oncogenes within the human genome, encoding a tyrosine kinase receptor (RTK) which is a critical signalling component required for normal nervous system and kidney development, as well as for spermatogenesis to take place[2]. In general, tyrosine kinases are composed of an extracellular ligand binding, a transmembrane and an intracellular tyrosine kinase domains. In addition there are usually additional attached sequences of amino acids that may have modifying/regulatory functions. In this review, we describe the current knowledge about the molecular mechanism of RET activation in human neuroendocrine tumors as well as the physiological roles and signal transduction

RET is mainly expressed in cells of neural origin and is involved in cell proliferation, migration and differentiation of cells[3]. Impairment of the RET proto-oncogene is known to play a pivotal role in at least 4 clinical syndromes [4;5]. (Hirschsprung's disease (HSCR), Multiple Endocrine Neoplasia (MEN) Type 2 A and B and Familial Medullary thyroid carcinoma (FMTC)) in a unique "switch on"/ "switch off" manner [6]. Loss-of-function mutations of the RET receptor tyrosine kinase result in Hirschsprung's disease and RET remains the major genetic susceptability gene in this condition[7-9]. In addition, it may be responsible for developmental kidney anomalies.

Similar to other oncogenes, RET upregulates protein tyrosine kinase activation, resulting in "gain of function" and increased uninhibited cellular proliferation, poor cellular differentiation and lack of apoptotic control[10], associated with certain forms of thyroid cancer in humans. [4]. The RET proto-oncogene appears unique, as it is associated the least four clinical conditions [11;11]. These include Hirschsprung's disease (HSCR) in addition to Multiple endocrine neoplasia (MEN) syndromes Type 2 (A and B) and Familial Medullary Thyroid Carcinoma (FMTC) which are associated with mutations at different sites of the same gene.

Not all kinase-activating mutations active in oncogenesis are equivalent in terms of the degree of activation produced but are likely to be subject to modifying influences determining their eventual phenotypic expression. As a result, mutations at certain sites on the RET gene are regarded as high risk for medullary thyroid cancer, whereas others are intermediate and yet others low risk [12]. The site of the genetic variation also tends to determine the phenotypic expression eg. extracellular cysteine-related RET gene variations result in the MEN 2A tumours whereas genetic variation of the intracellular tyrosine-kinase region (especially M918T), account for the majority of the MEN 2B phenotype. In addition, the oncogenic potential of each specific RET protein appears important and similar kinetic behavior has been shown for the longer forms of RET (containing the juxtamembrane domain and C-terminal tail)[13]. These appear to determine the altering of RET autophosphorylation (eg MEN2B), or alternatively the modification of the subcellular distribution of the active kinase but may also induce unusual interactions with other proteins (eg RET/PTC)[10]. As a result, Knowles et al [13] suggested that RET kinase activity is autoregulated by means of an alternative inhibitory mechanism/s as there appears to be no cis-inhibitory mechanism within the intracellular domain of RET.

It has been proposed that activating RET genetic variations not only increase RET kinase activity but may also modify its signaling by linking to a number of other downstream signaling pathways [14]. Phosphotyrosine binding (PTB) domains are components of cytoplasmic docking proteins that bind cell surface receptors through NPXY motifs.[15;16]

Important Structural Sites on RET and their Relationship to Function

The RET (Rearranged during transfection) proto-oncogene consists 21 exons which lie over a 55Kb area at 10q11.2 [17]. Like other RTKs it is made up of at least 3 functional areas [viz: extracellular, trans-membrane and intracellular regions].[18](Figure 1) each of which appear to have specific functions. RET signaling is of vital importance in development and RET knockout mice display major renal defects and fail to survive [19]

The extracellular domain of RET consists of four cadherin-like domains (CLD1-4) followed by a cysteine-rich membrane-related domain adjacent to the transmembranous domain [20]. The intracellular tyrosine kinase domain appears to be extremely important as it appears to control the activation of downstream signalling cascades within the cell.

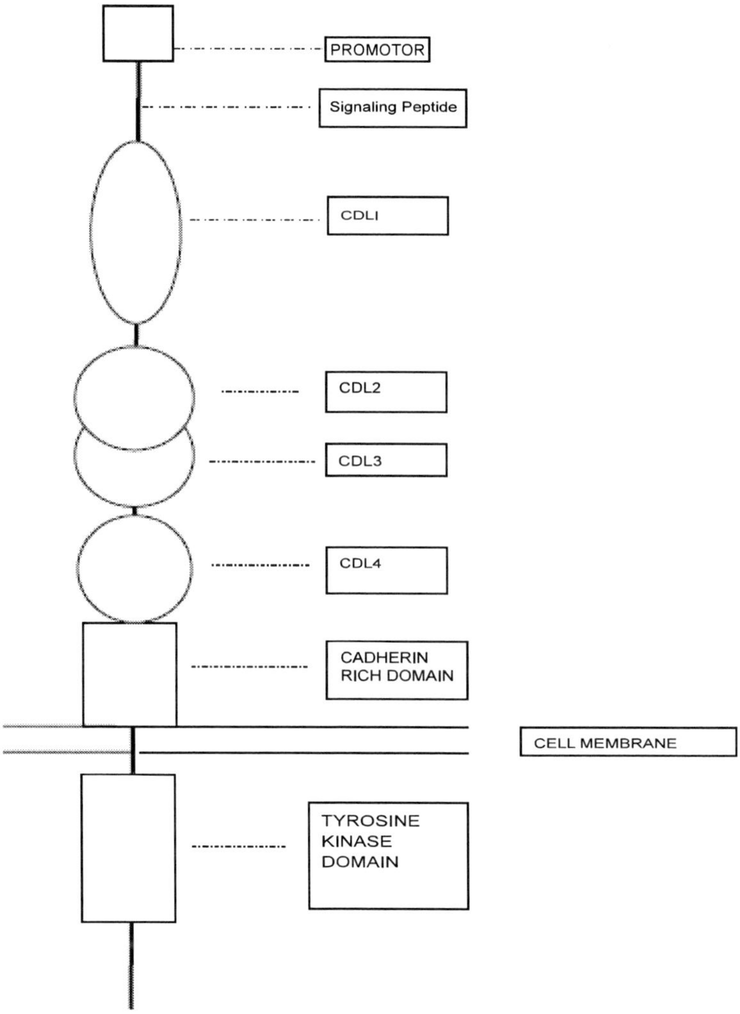

Figure 1. Schematic diagram of the 21exons of the RET proto-oncogene showing the relevant functional areas.[CDL =Cadherin-like domains 1-4].

The four cadherin-like domains are tandemly arranged and lie within the extracellular component of the gene, suggesting a role in cell-cell interaction [21]. This is followed by the vital cysteine-rich domain with critical cysteine residues in codons 609, 618, 620, 630 and 634. This cysteine-rich region lies adjacent to the transmembranous domain and appears an important contact site for both GDNF and GFR alpha 1 molecule. [20]. As a result, it is thought to play a vital role in protein folding and dimerization of the receptor[7]

The function of RET appears to be to transduce those extracellular signals which affect cellular growth and differentiation, this function being performed via the transmembrane tyrosine kinase receptor [22]. The tyrosine kinase receptor of RET does not function in isolation but is activated by binding to a ligand complex formed by the glial cell line-derived neurotrophic factor (GDNF) [23] and its ligand co-receptor, the GDNF-family receptor-alpha (GFR alpha) receptors which are GPI linked thus permitting dimerization [24]. This complex is known to involve lipid rafts via GDNF[25]. GDNF has not only been shown to stimulate the differentiation of neuronal cells during development but also appears to prevent cellular apoptosis protecting against neurodegeneration[26]. Members of the glial cell line-derived neurotrophic factor (GDNF) family of ligands (including GDNF), also include ligands such as neurturin, artemin, and persephin which may also act as ligands and substitute for GDNF in RET signaling[7]. GDNF has also been shown to act independently in the binding of TRKA to nerve growth protein to maintain sympathetic neurons [27]

The binding of RET to GDNF and the specific GFR alpha subtype, leads to known biological responses (Lee RH 2006) and autophosphorylization results at various sites. Phosphorylation of intracellular TK sites further by influences the docking sites for adaptor proteins which leads to activation of critical downstream signalling pathways.

RET tyrosine kinase belongs to the Cadherin super family of genes,(at least 80 members) which also include other classic cadherins, desmogleins, desmocollins, protocadherins, CNRs, Fats, and other transmembrane cadherins [28]. Essentially, cadherins are glycoproteins that mediate calcium-dependent cell-cell adhesion, their normal function being essential for normal development. Aberrent cadherin function has also been implicated in oncogenesis.[29]

Regulatory developmental susceptibility genetic variations [Single nucleic polymorphisms (SNPs)] have been identified within the promotor region of RET in patients with Hirschsprung disease (HSCR).[30] Conserved elements within this region of RET have been shown to have influence the activation or suppression of gene activity in vitro. This critical area of RET has been shown to be conserved in evolution [31] and has led to speculation of possible evolutionary conserving of the promotor region in African as opposed to Chinese[32] and Western[33;34] patients . The RET proximal promoter region has furthermore,been identified as a polypurine (guanine)/polypyrimidine (cytosine)-rich sequence (the G-strand consisting of five consecutive runs of guanines) which is capable of forming intramolecular G-quadruplexes. [35]The formation of these stable secondary G-quadruplex and i-motif structures appears vital to RET basal transcription and suggests that they play a vital role in its transcriptional regulation [35].

This proximal promotor region of RET has specifically been investigated in Hirschsprungs etiology, particularly the role of polymorphisms occurring within the intron 1 region [33;36-38]. A predisposing RET haplotype (A-C-A) representing three identified

SNPs [ie. NT_033985.6:g.975824G>A and NT_033985.6:g.975820C>A plus c.135G>A (a silent polymorphism)], has been proposed as a potential predisposing factor in HSCR pathogenesis(39). It has been found in 62% of HSCR but only 22% of controls[34;40]. and lies in an area approximately -1 bp and -5 bp from the RET transcription start site of the gene. Functional evaluation of this 5' area has shown that the individual genetic variations fail to interfere with RET mRNA splicing or reporter gene transcription.[33]. On the other hand, the 3' untranslated region (UTR) of the gene has been shown to offer protection from HSCR[39]. Real-time RT-PCR, shows that ACA haplotype homozygosity is associated with a decrease in RET gene expression suggesting that instead of single polymorphisms affecting RET expresion, disequilibrium of the ACA haplotype, forms a HSCR susceptibility allele. [33] This would negatively influence RET transcription and result in a reduction of cell surface RET receptor. It may thus act as a low penetrance susceptibility locus or be in linkage disequilibrium (LD) with another RET susceptibility locus(41) It has also been pointed out that these SNPs overlap a TTF-1 binding site and a decrease in TTF-1-activated RET transcription in HSCR has been reported[41]

Genetic variation in the extracellular domain will thus prevent RET expression at the cell surface by affecting processing in the endoplasmic reticulum. [42] which will affect RET transport via Ca++binding sites.

In addition to being of importance in HSCR etiology, RET upregulates protein tyrosine kinase activation, resulting in gain of function which is associated with certain forms of thyroid cancer in humans. RET transportation to the cell surface are likely to be affected by gene variations within the extracellular part. [14] Genetic variation may also impair the correct sequencing of the RET protein allowing.the activation of RET[43] The cysteine molecules are thought to be involved in molecule-molecule connections with other RET molecules (eg GDNF) thus leading to RET receptor dimerization as well as TK activation in the absence of the normal ligand co-factors[44] It has been postulated that in addition to affecting the binding to ligands extracellular domain mutations (eg C620S) may affect polarity due to the unfolding of RET which has been shown to result in activation [45].

One of the problems in the hypothesis that certain commonly occurring genetic variation of RET (S767R and P1039L) have failed to show RET malfunction [46], whereas others (eg R231H) do [26]. This observation raises the role of RET ligands (eg GDNF) in the survival and differentiation of developing neurones. It also raises the case for a second intracellular mechanism [47].

It would appear that gene variations in the intracellular portion of RET impairs its tyrosine kinase activity which results from activation of the tyrosine kinase domain in the intracellular portion of the gene. This tyrosine kinase activation appears to dictate the downstream signalling cascades which have mostly been attributed to 2 signalling pathways. The first is activation of the MAPK and Ras/ERK molecular signalling cascades by a mechanism which involves grb2/mSOS recruitment [8]. The other signals via PI3K and Akt upregulation following recruitment of grb2/gab2 [49].(Figure 2) A number of studies have shown RET interaction with a number of molecules which include phospholipase Cy, Shc, enigma, grb2, /Grb7.10, Src kinase, and RAS-GAP. [50-52].In addition, further RET signaling pathways appear to be partly controlled by Heparan sulfate interactions.(Wang XJ 2007).

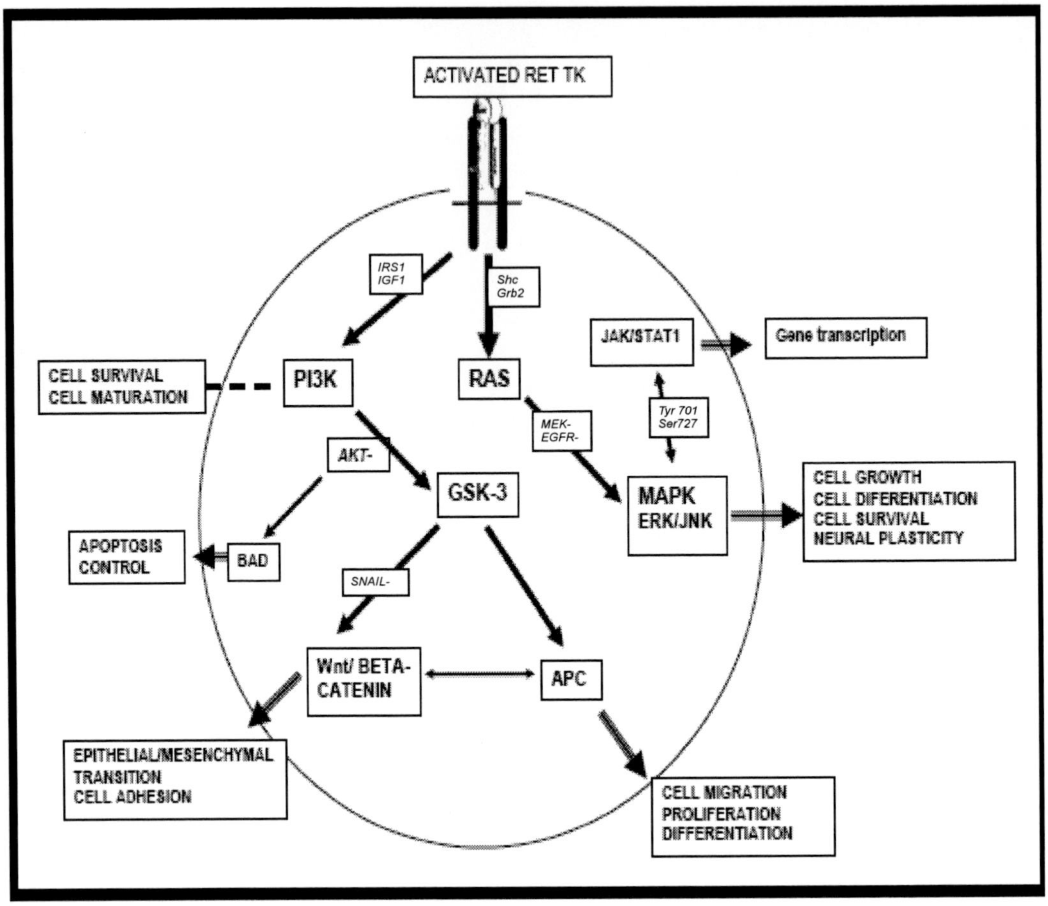

Figure 2. Schematic representation of downstream intracellular pathways and their possible functions folowing RET Tyrosine Kinase activation.

Following activation by the RET/GDNF/GFR alpha complex at least two receptor tyrosine kinase isoforms are encoded which then play a vital role in the development of the neural crest and kidney. [49] These have both been shown to assemble on phosphorylated Tyr-1062 of the molecule (via Shc). [53] One of these leads to activation of the Ras/Erk pathway (via of Grb2/Sos recruitmen. The other activates the PI3K/Akt pathway (via of Grb2/Gab2 recruitment followed by p85(PI3K) and SHP-2 recruitment) and link to phosphorylated Tyr-1096. This latter pathway may represent an alternative signaling pathway to PI3K stimulation. [53] Both Ret isoforms ret 9 and ret 51 appear to bind to Grb2 via different modalities) [49]

Genetic variations such as Polymorphisms (SNP's) are mostly located in exons 11 (G691S), 13 (L769L), 14 (S836S), and 15 (S904S) . It would appear that they are over-represented in sporadic MTC patients (especially from American and European countries) . RET Single nucleotide polymorphisms (SNPs) do not seem to be implicated directly in any tyrosine kinase activation but a number of studies suggest that they could be modifiers of disease susceptibility and clinical phenotype [54;55].

RET Ligands and Activation

In keeping with other growth factors RET signals by means of an activated complex of receptor molecules in order to modulate the response. In the case of RET this complex consists of the Glial cell line-derived neurotrophic factor (GDNF) family ligands, as well as GPI-anchored ligand binding GDNF family receptor alphas (GFRalphas) in addition to RET. Mammals have been shown to have at least 4 different ligand-receptor binding pairs of the GDNF (glial cell line-derived neurotrophic factor) family which all appear to signal via the transmembrane RET receptor tyrosine kinase.

Distinct regions within the Ret molecule that interact with GDNF and GFRalpha have been identified . and include both GDNF and GFR alpha sites. Lack of mature sites in these regions are most likely to result in defective GDNF/GFR alpha binding. [56]The co-receptor GDNF-family receptors alpha1-3 (GFRalpha1-3) have been fairly extensively studied as potential ligands.[56]. The first three cysteine-rich domains of GFRalpha1 appear important with the first stabilizing the interaction between GFRalpha1 and GDNFD1 thus contributing to its optimal function[57;58]. The other two domains (viz: 2 and 3) appear to be chiefly responsible for GDNF binding. More recently, animal studies suggest that GFRalpha4 is also important as it is able to activate RET phosphorylation independently thus inducing downstream signalling cascades independent of its natural ligand persephin. [59]. Mutations in this area have been shown to induce a frame shift thus affecting the encoding transcripts.

In addition to all this other , GRAL (GDNF Receptor Alpha-Like) proteins of unknown function as well as molecules like Gas1 (growth arrest specific 1) have GDNF family receptor (GFR)-like domains [60]. The net effect of these signalling processes would be increased neuronal survival and differentiation.

MEN 2B-RET phenotyopes have been shown to be able to dimerize and autophosphorylate RET in the absence of ligand stimulation suggesting that other complementary molecular mechanisms underlie the MEN 2B phenotype and provide potential targets for effective therapeutics (Gujral TS 2006)

The Ret receptor tyrosine kinase is activated through binding to the GDNF/GFR alpha complex at a number of sites. The binding of the RET/GDNF/GFR alpha complex activates RET and leads to homodimerization. This is then followed by autophosphorylation and activation at the various intracellular tyrosine residues. Mass spectrometric analysis has shown these autophosphorylation sites to be at Tyr(806), Tyr(809), Tyr(900), Tyr(905), Tyr(981), Tyr(1062), Tyr(1090), and Tyr(1096). [61;61] This is not exclusively so as the RET activation reported after the experimental introduction of a cysteine at the 833 position demonstrates [62].

It would appear that the recruitment at the phosphorylated Tyr1062) site in RET is one of the most important in MEN2B etiology as it includes a variety of phosphotyrosine binding (PTB) domain-containing adaptor proteins(which include Shc and Frs2), a vital step in RET receptor function. This step appears to result in Grb2 recruitment and downstream signaling through the Ras and Raf pathways.[53] In addition, Malignant cell growth in MEN2A involves the downstream activation of the STAT3 signaling cascade[63]

Crosstalk with other Signaling Systems

The main concern of those studying the action of RET in HSCR is the lack of consistency and wide variation in RET mutations in cases of HSCR (50% of familial cases and 7-35% in the sporadic form [64]). This strongly suggests alternative pathways or gene-gene interaction.

It would appear that the accumulation of mutations/variations in critical genes contributes towards the resultant Hirschsprung's phenotype. Although it has also become clear that it is the mutations of the c-ret gene that eventually leads to aganglionosis[65], combined associations of RET and EDNRB have been suggested to interact in the development of the enteric nervous system[66-68] . EDNRB mutations have been shown to delay the migration of neural crest cells In a number of mammalian species, resulting in pigmentary/spotted coat as well as the ENS disturbance. Without EDN3/EDNRB signaling there is poor migration of neuroblasts to the distal gut [69] and mutations of both ET-3 [70]and the EDNRB [71]genes have been reported in association with HSCR. Although the specific role of ET-3 in preventing premature neural cell differentiation or controlling neuroblast proliferation is as yet unclear [72], RET and ENDRB have been shown to act in a co-ordinated synergistic manner in both animal models[68]and humans[73]. Upon entering the caecum from the small bowel, the migrating neuroblasts are affected by the highest levels of GDNF and ET-3 suggesting that they are co-ordinately controlled by both EDNRB and RET[66]. Exactly why they then move on to populate the entire colon is unclear but probably depends on GDNF availability as well as another possible trophic factor[72]. Recent gut transplant experiments [66;74] have shown that multipotential neural crest cells from Endothelin receptor B (Ednrb)-deficient gut form neurons as efficiently as did wild-type neural crest stem cells when transplanted into Hirschsprungs animal models.

This co-ordinated interdependent cross-talk between the receptor tyrosine kinases (eg RET) and the G-protein coupled receptors (eg. EDNRB) appear to involve Protein A as a key component [66] . It has more recently become clear that the Sox 10 transcription factor is co-expressed with the RET and EDNRB receptors on developing neurocytes both in the lung [75]and the ENS[76]which suggests interaction. SOX10 is known to interact with Sp1 to activate the EDNRB promotor.[77] It also has been shown to interact with Pax3 leading to activation of the c-ret promoter[76;78]. It is possible that this may result from a reverse effect resulting in reduced levels of endothelin-converting enzyme -1 and endothelin 3 mRNA [as has been reported in the Ret dominant-negative mouse model [79)]. It has also been suggested that variations in the GDNF, EDNRB and EDN3 genes may influence the penetration of RET mutations thus determining the phenotypic expression[80;81] .

RET has been shown to activate a variety of signaling pathways in association with GDNF upregulation [82]. These include the RAS/extracellular signal-regulated kinase (ERK), phosphatidylinositol 3-kinase (PI3K)/AKT, p38 mitogen-activated protein kinase (MAPK) and c-Jun N-terminal kinase (JNK) pathways (Figure 2). Downstream signaling then follows via binding of adaptor proteins to auto-phosphorylated intracellular RET tyrosine residues. Those pathways frequently upregulated include the RAS-RAF-ERK and PI3K/PDK1/Akt signaling cascades [83;84]. In addition, the RAC1/JNK pathway has fairly recently been shown to be regulated by serine phosphorylation at the juxtamembrane region of RET [85].

Mutations of the oncogenes BRAF, RAS and RET/PTC co-exist in up to two thirds of certain thyroid cancers sugesting that they are significant in thyroid cancer oncogenesis along with tumor suppressor genes (TP53) inactivation .

There is some current interest in non-cysteine intracellular sites because of indications of a second intracellular pathway in the oncogenic process relating to MTC. It has been hypothesized that activation of the RAS, PI3K/AKT, and PPARgamma pathways ultimately affects the beta-catenin/ Wnt pathway [86] and may be part of the explanation as to why cancer develops in RET activation

An interesting additional finding is that RET appears to increase heat shock transcription factor HSF1 activation, the major mechanism of stress response induction. [87] suggesting an additional role for RET in the expression of stress response proteins and initiation of the stress response.

The Effect of RET Mutation on RET Protein Function

Alternative splicing of RET results in 3 isoforms RET mutations vary in type. Missense mutations result in the insertion of an amino acid in an abnormal place in the resulting gene product causing it to function in an unregulated way. On the other hand point mutations may introduce a stop codon within the coding sequence resulting in incoplete translation and loss of function. Structural and frame shift mutations also occur which may result in an abnormally functioning protein.

Alternative splicing of the RET pre-mRNA results in several 3' and 5' RET variants being transcribed which are rigidly regulated during kidney development. [88] Those related to the 3' variation transcribe at least 3 isoforms, [viz: RET9, RET51, and RET43]. These contain 9, 51 or 43 amino-acids respectively. 5' variations , on the othrer hand, give rise to RET2/4, RET2/5 and RET2/6 respectively by varying the transcription code.

There is evidence to suggest specific functions for the different protein forms during mammalian embryogenesis [89] and each specific RET protein appears to have different extracellular ligand-binding sites. Experiments on mice expressing either the the human RET9 or RET51 isoform both demonstrated similar severe renal anomalies (bilateral megaureters and multicystic kidneys) when a mutation of the docking Tyr 1015 (Y1015F) was present[90]. Both RET isoforms ret 9 and ret 51 appearing to bind to Grb2 (via different modalities) [49]. On the other hand, renal agenesis occurred with of RET9(Y1062)-mediated AKT/MAPK activation loss, whereas no obvious effects were noted on AKT/MAPK activity and renal development with RET51 mutation. These suggest specific signaling pathways in embryonic kidney development related to individual RET proteins, in keeping with previous reports.

The Role of RET in Cellular Apoptosis

There is some experimental evidence suggesting that the activation of pathways involving protein tyrosine kinases (eg TrkA, RET and EGFR), could induce the cleavage of the beta-amyloid precursor protein, APP, by beta- and gamma-secretases, thus promoting the generation of amyloid peptides. [91] The process of RET-induced apoptosis appears to be caspase dependent and the cleavage sites have been mapped to lie between D707 and D1017 and genetic variation in this area is likely to affect apoptosis.[26]

GDNF and its ligands have been shown to be involved in the survival of cultured neural crest-derived cells in a number of studies [92;93] and promotes survival of developing neurons[94]. Furthermore, animal GDNF disruption models result in aganglionosis and kidney agenesis[95] and RET receptor expression has been shown to induce apoptosis which may be blocked by GDNF[26]. It would, therefore, seem reasonable to hypthesise that in situations beyond ligand availability or ligand (GDNF)blocking (due to genetic variation at critical sites) may result in cells undergoing programmed death (apoptosis). This is paricularly relevant to HSCR etiology where both migration and differentiation and survival of migrating neurones are implicated.

The question is still unresolved although the hypothesis appears attractive at this stage.

Outcomes of RET Signaling

A. The Role of Ret During Development

Neural development is an extraordinarily complex process involving the migration of neural crest cells, their colonizing and differentiation at the required destinations, followed by the development of the synaptic pathways. [96] Signals are mediated by cell surface receptors, extracellular matrix and basement membrane molecules, growth factors and cytokines, governed by a complex array of transcriptional switches. The neural crest cells migrate throughout the developing fetus and develop into a wide variety of cell types including peripheral sensory, autonomic and enteric nervous system glia and neurons as well as melanoctytes, medullary neurons of the adrenal gland, the great vessels of the cardiac system and sections of the cranial musculoskeletal systems[72]. All cells in the ENS appear to derive from vagal or sacral inflow of progenitor cells and diferentiate into at least 14 different cell types to perform functions of motility, sensory response, secretion and blood flow control. It therefore stands to reason that the regulation of neuroblast proliferation, migration, differentiation and maturation is a complicated interactive system involving numerous signaling pathways which act in a spatiotemporal manner. As a result, there are many different stages of neural cell differentiation present within the ENS and cell surface signalling proteins and their intracellular transduction cascades are upregulated or down regulated in a spatio-temporal manner . Myenteric and submucosal plexus formation result from perpendicular migrations of neural crest cells within the gut wall.[97].

The normal development of the pluripotential neural crest cells is essentially controlled by encoding a RET transmembrane receptor kinase during the cell migration stages during early development. RET therefore plays a role in controlling the survival, proliferation, differentiation and migration of the enteric nervous system (ENS) progenitor cells [66;98], the survival and regeneration of sympathetic neural cells [99], as well as controlling normal kidney development [19;100] and spermatogenesis[2]

RET knockout mice display major renal defects and fail to survive [19]. In addition to kidney development, the GDNF-RET receptor complex is involved in the development of neurons within the sympathetic, parasympathetic and enteric nervous systems. [101] It would appear that the RET isoforms have specific developmental roles in this process and Ret9 (but not Ret51) has been shown to be essential for kidney development [102] . The reason for this may be that the binding of this particular isoform, activates the Phosphatidylinositol 3-kinase (PI3K) signalling cascade at this particular site thus activating the relevant downstream signalling cascades. RET gene ablation studies [19] have shown that RET activation results in increased levels of phosphatidylinositol 3-kinase (PI3K) activity and Akt/PKB phosphorylation. PI3K appears essential and the branching morphogenesis which occurs following mesenchymal invasion of the ureteric bud during kidney development via its regulation of GDNF [103;104] In addition, prolonged PI3K and Erk-MAPK signaling, appears critical for tubules to form [102] . This process is affected by the Shank3 adaptor protein in association with recruitment of the Grb2 adaptor protein suggesting a reason why the different Ret isoforms affect the process in different ways. [102]

Important ligands of GDNF 9 in sympathetic nerve development include Retinoic acid[105], Bone morphogenetic protein 2.[106] and Neurturin. Neurturin, (one of the GDNF-related ligands) appears to be an important neurotrophic factor for parasympathetic neuronal development possibly linking with RET via its preferred ligand GFR alpha 2[107] There also appears to be a symbionic relationship between RET and nerve growth factor (NGF) in developing as well as mature sympathetic neurons. NGF has been shown to stimulate phosphorylation and RET tyrosine kinase activation in order to promote neural growth. [108]. NGF also inhibited GDNF-induced Ret degradation by increasing the autophosphorylated Ret on the cell surface whereas GDNF activation led to the increase of activated Ret within the cell.

B.RET and Hirschsprungs Disease

The development of the ENS and migration of enteric neurones requires adequate expression of RETon the cell surface[109]. The precise etiology of loss of function genetic variatons that result in Hirschsprungs disease are more difficult to explain as the associated genetic variations occur throughout the gene[36]. More than 100 individual genetic variations have been described, particularly in the 50% of familial cases with mutations. The percentage of mutations identified in sporadic cases is much lower varies between 7-35% (depending on the series). [7] Current understanding is that a heterozygous coding sequence mutation or a polymorphism within intron 1 leads to defective RET expression [110]. This, in turn, results in loss of function in the resultant RET protein and poor enteric nervous system development.

In keeping with these findings, the majority of Hirschsprungs mutations appear to lie within the RET extracellular portion[111] particularly in the first 3 cadherin-like domains[7;33;42] . A variety of genetic variations have been identified in association with HSCR including missense or point mutations, microdeletions, insertions, and defects in splicing and frame shifts[31].

Although the cadherin domains remain important in HSCR pathogenesis, HSCR mutations are not necessarily confined to the cadherin motif [109]. At least 4 classes of mutations of the extracellular domain have been reported and include defects in glycolization (N-glycolisation) calcium^{++} co-ordination,disulphide bond formation as well as those with no predictable efect. [42]. the majority of these areas prevent in difficulties in protein maturation and folding. It is thought that the most likely mechanism is impaired protein folding resulting in inhibition of the kinase activity but defective ligand binding and/or prevention of receptor maturation are other possibilities [7].

Further investigation of these areas for RET processing and function demonstrate inability to interact with the GDNF/GFR $\propto$ activating complex. As a result processing of RET in the endoplasmic reticulum may be affected. It stand to reason that the important functions that RET loses in this situation are its ability to bind Ca^{++} or disulphide bridge formation. Defective RET with persistence of the 150kDa isoform(as opposed to the normal 170kDa protein) has the tendency to form dimers and hetromeric complex impairing function.[111]. In addition, defects in GDNF/GFR $\propto$ complex binding would also result in impairment in protein folding[42].

Associated congenital anomalies occur in at least 5-32% (mean 21%) of Hirschspurngs patients and certain syndromic phenotypes have been linked to specific genetic sites, indicating underlying genetic associations of the disease and probable gene-gene interaction in its pathogenesis. In certain series approximately 30% of Hirschsprungs disease are syndromic [110]. Those currently known include congenital central hypoventilation syndrome, Shah-Waardenburg (WS4), the Mowat-Wilson syndrome, Bardet-Biedl , cartilage-hair-hypoplasia, Smith-Lemli-Opitz and Goldberg-Sprintzsen syndromes as well as aqueductal stenosis and hydrocephalus [80;110;112]. In these situations, RET appears to act as a modifier particularly in Downs, congenital central hypoventilation and Bardet-Biedl syndromes. A different mechanism via SOX10/EDNRB appears to account for WS4 and the Mowat-Wilson syndromes. [110]

RET therefore can be seen as the most important susceptability gene in Hirschsprungs disease.

C. The Role of Ret in Cancer

RET is closely involved in the development of several human neuroendocrine conditions. Oncogenesis relies on tyrosine kinase activation of growth factors in order to promote cellular angiogenesis in addition to increasing cellular proliferation, invasion and metastases that form an essential part of the oncogenic process. Ultimately, RET variations that confer oncogenic gain of function result in the development of cancer due to uncontrolled cellular proliferation as well as failure to undergo normal differentiation and the loss of apoptotic functions.

The constitutive activation of RET by somatic rearrangement with other partner genes or germ-line mutations initially linked RET with cancer. Demonstration of Tyrosine kinase activity upregulation by the RET/PTC fusion gene in papillary thyroid carcinoma[113] has been shown to be highly significant in the pathogenesis of Papillary thyroid carcinoma, particularly in childhood. Following the Chernobyl disaster (ionizing radiation), papillary thyroid tumors occurring in children show a high prevalence of RET fusion gene rearrangements with at least 11 of these being described (the majority being a RET/PTC3 rearrangement) [114]. Other less frequent rearangements include the RET/PTC1 and some more novel rearrangements (eg RET/PTC5). Further gene analysis has shown a balanced intrachromosomal inversion whereby *RET* receptor tyrosine kinase gene activation occurs following fusion of the RET 3' portion to the 5' portion of various genes [eg *BRAF, RET*, or *RAS* genes of the MAPK system] to form RET/PTC and similar chimeras[115]. As a result, the tyrosine kinase RET domain then becomes controlled by 5' fused regulatory sequences of other genes with dimerization potential.[116] It is interesting that activating RET mutations are also associated with other thyroid carcinomas [117].

In addition to the potential tyrosine kinase activation via fusion genes, there are other genetic variations identified in RET malfunction [13]. As a result RET is now known as a significant etiological factor in human papillary thyroid carcinoma (via fusion genes) as well as multiple endocrine neoplasia (MEN) type 2A and 2B (via direct mutations). The first group are those regions in the extracellular portion of the gene which control protein folding (mutations which result in a non-folded RET protein at the endoplasmic reticulum). Those occuring within the 1st cadherin-like domain in HSCR [20] are thought to influence its pathogenesis [33]. Those affecting one of the 6 cysteine radicals (viz: 609, 611, 618, 620, 630 and 634 positions) affect gene stability and influence RET activity resulting in MEN2A [14;109]. The second group of RET gene variations are those within the terminal tyrosine kinase (KD) intracellular region of the gene (eg exon 16 M918/T mutations in MEN 2B)[14;109]. A third area is emerging in those gene variations which affect the docking sites for downstream signaling pathways[47]. There is a further group of MEN-related genetic variations, associated with non-cysteine areas of gene which raises interesting questions as to the oncogenic mechanism. This latter group appeared to occur more frequently than expected in our own series[118] in keeping with other genetic pools[119].

Activation of the RET protooncogene occurs as a result of RET mutations has been causally linked to both sporadic and familial forms of MEN2 with MTC[120-123]. Oncogenic RET mutations mostly occur de novo and their effect on protein synthesis and function clearly involves a number of downstream molecular signaling pathways [14].

The biological properties of defective RET proteins then appear to determine the particular disease phenotype. It would appear that the oncogenic capability of the various sites of RET genetic variation, result in genetic dysfunction which, in turn, results in the production of mutant proteins which then determine the phenotypic expression. Currently known mechanisms of RET activation include the induction of a disulfide-linked homodimerization particularly in exon 11, (codon 634 point mutations) [43;124]. Where mutations of this area occur, oncogenesis results from the removal of a half of the intermolecular bond which thus allows an abnormal bond with a second mutant molecule to occur. This aberrant disulfide homodimerization will result in a decrease in RET at the plasma membrane [125]. Abnormal disulphide homodimerization of RET is the most likely explanation of gene activation following cysteine-substitution mutations at RET codons 609, 618, 620, 630 and 634. [126;127] This suggested mechanism of action, probably also applies to all gain of function mutations associated with MTC.

It is also clear that a number of additional molecular mechanisms exist in addition to the activating effect of RET mutations to potentiate the enzymatic activity of the RET kinase. The unfolding of RET due to extracellular domain mutations (eg C620S) affects polarity which may result in tyrosine kinase activation [45]. Mutation in the Tyrosine-kinase rich region (eg. in 95% of MEN 2B patients) , on the other hand, alters the substrate specificity of RET tyrosine kinase and appears to induce a different set of downstream signaling genes from that carrying the MEN 2A mutation. This appears to function via the sic protein [128] which is known to induce downstream activation of the Ras/mitogen – activated protein- and P13K/Akt signalling cascades [129;130]. It may well have bearing on the speed and aggressiveness of the oncogenic process as well as influencing the penetrance of the MEN 2 phenotype. Both the phosphorylated and non-phosphorylated forms have been shown to have a pre-organized activation loop and are competent to bind ATP and substrate [13].

In addition to the primary causes of RET activation, other possible mechanisms include the autophosphorylation of RET (eg. MEN2B) as well as the modifying the subcellular distribution of the active kinase[10] as well as defective downstream signalling. The role of downstream signaling systems implicates up-regulation of the RAS-RAF-ERK and PI3K-PDK1-Akt signaling pathways[131;132]. In addition, the multidocking RET C-tail is vital to downstream signaling due to the number of signaling molecules that interact there[eg. Shc, FRS2, IRS1, Gab1/2, and Enigma.]

Additional factors influencing gene penetration and the signalling pathways involved in metasatasis and invasion are less clear, but it would appear that factors such as gene methylation may play an epigenetic role by altering the phenotype without affecting the genotype.

RET Genotype - Phenotype Associations in MEN 2

There appear to be fairly consistent associations between genotype and phenotype with MEN 2A [133;134] with the type of amino acid substitution encoded at the cysteine radicals in the extracellular RET domain having fairly consistent transforming power [135;136]. MEN 2B, is also associated with a point mutation (Methionine→threonine) in exon 16 (M918/T) in 95% of cases[137;138]. It appears, therefore, that the mutation site is one of the most important determining factors of risk.

RET Testing and Inherited Medullary Thyroid Carcinoma (MTC) Risk

Current evidence suggests that almost all cases of MEN 2 syndromes result from gene up regulation due to germline activating mutations in the RET proto-oncogene. Genetic variations in exons 10, 11, 13, 14, 15, and 16 of the RET proto-oncogene have been identified in at least 92-95% of MEN 2 cases presenting with MTC [139;140]. RET-testing has been shown to be vastly superior to Calcitonin in identifying pre-clinical MTC cases with a specificity approaching 95-100% [141].

Activating mutations of RET appear to be of the order of 1:500,000 in the general population[142] and consistent associations exist between RET genotype and phenotype. MEN 2A MTC is mostly associated with variations in the 5 cysteine radicals in the extracellular domain with 95% of MEN 2B patients are associated with a point mutation (Methionine→threonine) in exon 16(M918/T) of the RET proto-oncogene. There is some evidence that oncogenic RET mutations may vary between specific ethnic groups[119].

The highest penetrance for MTC is in the 634 (exon 11) cysteine –related point mutations as well as those occurring in the extracellular tyrosine kinase domain (918 codon variations) Mutations of the high risk RET codons 634 and 618 have been associated with the youngest MTC in MEN2A (3 yrs and 7 yrs respectively)[143]. On the other hand, codons 790, 620 and 611 appear associated with an a lesser (or intermediate risk), and codons 768 [144] and 804 as having a relatively low MTC risk. Patients with phenotypic features resembling MEN2B always require genetic testing in spite of a negative family history because of the high incidence of spontaneous mutations (approximately 50%) [145;146] and the early MTC onset. Offspring of MEN 2B-related MTC have a particularly high risk of an early aggressive carcinoma, the youngest reported case being at 2 months of age [147].

Genetic Screening for Inherited MTC Risk

Carriers of RET variations are mostly heterozygous and carry one mutant and one "wild-type" RET allele.[148]. Pre-clinical genetic analysis of the RET proto-oncogene results in improved risk prediction and permits prophylactic surgery. Research has shown a distinct age-related progression to MTC of RET gene mutation carriers with a clear genotype-phenotrype correlation if located in specific sites [149]. It is therefore clear that genetic screening should occur prior to the onset of any clinical MTC symptoms to allow for adequate early risk assessment and prophylactic management.

The ability to accurately predict the risk of MEN2 and medullary thyroid carcinoma (MTC) by genetic RET proto-oncogene analysis, has resulted in the active follow-up of children of families with RET cysteine mutations (particularly those with exon 10, in 11 and 16 mutations) to identify those at risk of early metastatic tumours and to prevent it by prophylactic thyroidectomy. Genetic screening for RET has been shown to be an extremely sensitive marker in MEN2 syndromes [141], with the majority of mutations relating to the cysteine radicals in exons 10, in 11 and 16.

The prevention and management of inherited MTC thus rests upon RET genetic screening because of potential benefits of prophylactic surgical intervention. Ongoing screening of family members of those with known mutations should be performed as maternal MTC has been reported to occur much later [150-153;153]

Co-Segregation between HSCR and MTC

It is becoming clear that RET should be also be screened in selected cases of HSCR (Particularly in long segment HSCR (L-HSCR) [118]) to identify those at risk of developing coexisting MEN2. The uncommon co-segregation of HSCR and MEN2 in the same patient is a fascinating finding as it involves both gain and loss of function in gthe same patient. Less than 30 cases of MEN2A-HSCR have been reported in the literature[154]. [120;150-153;155-161]. The most common mutation associated with MEN2A/ FMTC-HSCR associations is the 620 missense mutations (mostly C620R[65;162] and occasionally C620S and rarely C620W[162]. This association with codon 620 has led to the concept of the so-called "Janus gene" mutation which like the Roman god of doorways can face in both directions. The effect is not a constant one, on the one hand, MTC has been reported to occur as early as 12 years after surgical correction of HSCR in a patient with this variation [163]. By way of contrast, Fernandez et al [164] described a case where the *RET* C620S mutation caused multiple endocrine neoplasia type 2A (MEN2A) but not Hirschsprung disease (HSCR) in a family cosegregating both phenotypes. It is clear that co-segregation of these genes may lead to both short and long segment aganglionosis [as evidenced by the series reported by Decker et al where 7 out of 13 had L-HSCR and 6 out of 13 S-HSCR [162]. L HSCR predominates in the reported cases, however and should be a potential risk factor for co-segregation of the two conditions. The association between HSCR and MEN2A is not limited to codon 620 cysteine substitutions, however, and a 32 -35% association between HSCR with codon 618 (C 618R2x >C 618S-) missense mutations has been reported in MTC [159]. Other variations associated with HSCR include a 15-20% association with codon 609 (C 609Y mostly) and a 4% association with 611 mutations.

RET and Phaeochromocytoma

Phaeochromocytomas which are related to RET and the MEN2 syndromes appear to have distinct clinical differences, generally appearing at a younger age[165]. Pheochromocytoma occurring in childhood tends to be familial with sporadic pheochromocytoma accounting for less than 1% of childhood hypertension [166]. This suggests a possible increase in RET-related tumours [138]. Although genetic testing of Phaeochromocytoma patients is currently recommended in all patients, it is considered to be essential in young patients, or in multiple or recurrent tumours because of increased risk with RET mutations[167]. being identified in 64% of bilateral tumours in one series[168;169].

Genetic variations (at least 5 different genes) in phaeochromocytoma are not infrequent and are thought to be involved in its pathogenesis [169]. Known genes include RET(MEN 2 associations). [170] but other susceptability genes include VHL gene variations (von Hippel-Lindau (VHL) disease associations) and SDHB, D and C genes as well as neurofibromatosis Type I (NF1) gene mutations. They may also be associated with Hirschsprungs disease [171], an association which further underlines the important potential role RET plays in the pathophysiology of phaeochromocytoma tumours.

The cumulative frequencies of germline mutations in sporadic pheochromocytomas have been estimated to vary between 7.5% to 29% [168]. Takaya et al [172]reported RET overexpression in the majority of sporadic Phaeochromocytomas but the RET mRNA levels appeared to vary considerably between individuals. As a result, he concluded that high level of RET expression may be relevant to the development of sporadic Phaeochromocytomas despite the absence of large structural changes on RET gene on Southern Blot analysis, [172]. This finding was supported by Cho et al[173] but Powers et al [174] failed to identify RET overexpression in the early stages of Phaeochromocytoma development .

Although all subtypes of MEN2 are caused by activating mutations of the RET gene, phaeochromocytoma is mostly associated with the MEN2A and MEN2B phenotypes, and occurs in up to 50% of cases. [138] High risk RET mutations (eg C 634S) appears to carry a particularly significant risk of phaeochromocytoma. It is interesting to note that Phaeochromocytoma usually appears much later than the medullary thyroid carcinomas in most MEN2 cases usually occurring at approximately 30 years of age in other MEN situations.[175]. It does also not appear to be a feature of familial Medullary thyroid carcinoma syndromes where the genetic variations occur less focally being more diffusely distributed throughout the gene. In support of this, there has been a report of a 634 codon RET mutation carrier developing a phaeochromocytoma as early as 5 yrs of age[133]. This suggests the possibility that other modifying factors may have considerable influence on the final phenotype. It would appear that other genetic variations(eg . RET haplotypes) may be prime candidates as potential modifying factors in this regard[176;177].

McWhinney et al [177] reported that intronic single nucleotide polymorphisms in the RET protooncogene are associated with a subset of apparently sporadic pheochromocytomas [177]. In his study, the SNPs IVS1-126, IVS1-1463 as well as a deletion of intron 1 in the RET 5' promotor region were overrepresented in sporadic phaeochromocytomas, suggesting a possible modulatory role. This association was not verified by Fernandez et al[178] , however,

Therapeutic Implications

A. Risk Level and Codon-Directed Surgery

The diagnosis of MEN2 is currently mostly confirmed on the basis of the genetic features although the clinical phenotype remains important. Effective management therefore depends on early diagnosis and the gene carriers can now be identified before any clinical or biochemical abnormalities are present. These children can therefore successfully be offered a prophylactic thyroidectomy to prevent the development of MTC with its associated high rate of metastatic disease.

Machens et al [179] have stratified the MTC risk into three categories according to the mutation-related aggressiveness. Giving rise to a concept of "codon–directed" timing of surgery. Mutations at residues 634 result in a much higher cell-surface *RET* expression than 618 and 620 (and 609 to a much lesser extent) [126;135], suggesting that differences in the mechanism of RET activation may exist. The youngest MTC occurrence in one large series , was at 41yrs, 7yrs, 18yrs, 29yrs, 13yrs, 47yrs, 20yrs , and 15 years in patients with variations in codons 611, 618, 620, 768 , 790, 791, 804 and 891 respectively [180]. A recent long term follow-up study of 46 RET gene carriers categorized the risk to children and young adults (aged 4 - 21 years) into level 1 (low risk) and level 2 mutations (intermediate to high risk)[181]. Level 1 mutations (n=11) appeared to be associated with variations in codons 790, 791, 804 and 891 and had a high incidence of cure. By way of contrast, 5 (14%) of the 35 level 2 patients [mutations of codons 618, 620, 630 and 634], had ongoing disease. On long term follow-up (mean 6.4 yrs), 2 of those with 634 mutations, developed other MEN manifestations (viz: hyperparathyroidism and bilateral Phaeochromocytoma). This latter risk would probably increase with time and long-term follow-up is necessary.

Codon-risk protocols are of considerable value but not completely watertight and the age of MTC onset does not always appear to be consistent. It would appear to be justified to view mutations involving codons 609,768, 790, 791,804 or 891 as well as other *RET* genetic variations as being a 4th risk category representing those of uncertain risk where modifying influences may influence outcome.

Recent Advances in Treatment

Although a number of questions still exist relating to the timing and extent of genetic screening , the significance of specific intracellular *RET* genetic variations as well as the reliability of codon-directed prophylactic surgery, the search is on for molecular targets to regulate metastatic MTC. Recent advances into the significance of RET proto-oncogene signaling and the molecular pathways of RET signal transduction in the development of MTC, and oncogenesis have created new potential treatment modalities with exciting new possibilities for management. Recent therapeutic options include tyrosine kinase inhibition, gene therapy of significant RET mutations, the promotion of cellular death [182], monoclonal antibodies against oncogenic proteins [183] and, as well as nuclease-resistant factors that both recognize and inhibit RET [184].

Of particular current interest in this regard, is the development of RET-targeting tyrosine kinase inhibitors (eg: ZD6474; RPI-1) which have currently entered clinical trials. Among the other functions of these inhibitors is an effect on vascular endothelial growth factor receptors (VEGF) which results in angiogenesis inhibition as well as affecting certain other kinases[182]. Particular RET genetic variations may show varying response to therapeutic agents and recent studies into RET ultrastructure demonstrate that the response is not likely to be uniform depending on where the oncogenic mutation lies [13].

Arising out of this an array of monoclonal antibodies has currently been introduced as targeted treatment in the treatment of head and neck, breast and lung cancers. These novel therapies specifically target receptors or ligands in cellular proliferation and inflammation pathways and prevent phosphorylation by blocking the ATP-binding domain. This results in inhibition of signal transduction required for upregulation of function thus potentially giving a therapeutic response.

Early results with these modalities appear encouraging in metastatic MTC studies. One research group reports a statistically significant 51% tumour inhibition, a 210 % increase in apoptotic cells, a 47% loss of cellularity and a 37% decrease in micro vessel density [182]

Overall, this means that the scientific study of the RET proto-oncogene has resulted novel potential modalities for treatment and management for a number of conditions. Further research is required elucidate and evaluate this novel approach to disease.

References

[1] Porter AC, Vaillancourt RR. Tyrosine kinase receptor-activated signal transduction pathways which lead to oncogenesis. *Oncogene* 17[11 Reviews], 1343-1352. 1998.

[2] Takahashi M, Ritz J, Cooper GM. Activation of a novel human transforming gene, ret, by DNA rearrangement. *Cell* 42[2], 581-588. 1985.

[3] Pachnis V, Mankoo B, Constantini F. Expression of c-ret protooncogene during mouse embryogenesis. *Development* 119[4], 1005--1017. 1993.

[4] Mulligan LM, Kwok JB, Healey CS, Elsdon MJ, Eng C, Gardner E, et al. Germ-line mutations of the RET proto-oncogene in multiple endocrine neoplasia type 2A. *Nature* 363[6428], 458-460. 1993.

[5] Donis-Keller H, Dou S, Chi D, Carlson KM, Toshima K, Lairmore TC, et al. Mutations in the RET proto-oncogene are associated with MEN 2A and FMTC. *Hum. Mol. Genet.* 2[7], 851-856. 1993.

[6] van Heyningen V. Genetics. One gene--four syndromes. *Nature* 367[6461], 319-320. 1994.

[7] Lantieri F, Griseri P, Ceccherini I. Molecular mechanisms of RET-induced Hirschsprung pathogenesis.. *Ann. Med.* 38[1], 11-9. 2006.

[8] Edery P, Lyonnet S, Mulligan L, et al. Mutations of the *RET* proto-oncogene in Hirschsprung's disease. *Nature* 1994; 1994;367:378-80.

[9] Romeo G, Ronchetto P, et al. Point mutations affecting the tyrosine kinase domain of the *RET proto-oncogene* in Hirschsprung's disease. *Nature* 1994;367:378-80.

[10] Santoro M, Melillo RM, Carlomagno F, Fusco A, Vecchio G. Molecular mechanisms of RET activation in human cancer. *Ann. N Y Acad. Sci.* 2002 Jun;963:116-21.

[11] Passarge E. Dissecting Hirschsprung disease. *Nat. Genet.* 2002;1:11-2.

[12] Machens A, Ukkat J, Brauckhoff M, Gimm O, Dralle H. Advances in the management of hereditary medullary thyroid cancer. *Journal of Internal Medicin*e 2005 Jan;257(1):50-9.

[13] Knowles PP, Murray-Rust J, Kjaer S, Scott RP, Hanrahan S, Santoro M, et al. Structure and chemical inhibition of the RET tyrosine kinase domain. *J. Biol. Chem.* 281[44], 33577-33587. 2006.

[14] Carlomagno F, De Vita G, Berlingieri MT, de Franciscis V, Melillo RM, Colantuoni V, et al. Molecular heterogeneity of RET loss of function in Hirschsprung's disease. *EMBO J.* 15[11], 2717-2725. 1996.

[15] Smith MJ, Hardy WR, Murphy JM, Jones N, Pawson T. Screening for PTB domain binding partners and ligand specificity using proteome-derived NPXY peptide arrays. *Mol. Cell Biol.* 2006 Nov;26(22):8461-74.

[16] Rajala RV, Chan MD. Identification of a NPXY motif in growth factor receptor-bound protein 14 (Grb14) and its interaction with the phosphotyrosine-binding (PTB) domain of IRS-1. *Biochemistry* 2005 Jun 7;44(22):7929-35.

[17] Ceccherini I, Bocciardi R, Luo Y, Pasini B, Hofstra R, Takahashi M, et al. Exon structure and flanking intronic sequences of the human RET proto-oncogene. Biochem *Biophys. Res. Commun*.1993 Nov 15;196(3):1288-95.

[18] Takahashi M, Cooper GM. ret transforming gene encodes a fusion protein homologous to tyrosine kinases. *Mol. Cell Biol.* 1987 Apr;7(4):1378-85.

[19] Schuchardt A, D'Agati V, Larsson-Blomberg L, Constanini F, Pachnis V. Defects in the kidney and enteric system of mice lacking the tyrosine- kinase receptor ret. *Nature* 1994;367:380-3.

[20] Kjaer S, Ibanez CF. Identification of a surface for binding to the GDNF-GFR alpha 1 complex in the first cadherin-like domain of RET. *J. Biol. Chem.* 278[48], 47898-47904. 2003.

[21] Takeichi M. Cadherin cell adhesion receptors as a morphogenetic regulator. *Science* 1991 Mar 22;251(5000):1451-5.

[22] Andl CD, Rustgi AK. No One-way Street: Cross-Talk between E-cadherin and Receptor Tyrosine kinase (RTK) Signaling: A Mechanism to Regulate RTK Activity. *Cancer Biol.* Ther 4[1]. 2005.

[23] Jing S, Wen D, Yu Y, et al. GDNF-induced activation of the *RET* protein tyrosine kinase is mediated by *GDNFR-alpha*, a novel receptor for *GDNF*. *Cell* 1996;85:1113-24.

[24] Baloh RH, Tansey MG, Golden JP, Creedon DJ, Heuckeroth RO, Keck CL, et al. TrnR2, a novel receptor that mediates neurturin and GDNF signaling through Ret. *Neuron* 1997 May;18(5):793-802.

[25] Saarma M. GDNF recruits the signaling crew into lipid rafts. *Trends Neurosci.* 2001 Aug;24(8):427-9.

[26] Bordeaux MC, Forcet C, Granger L, Corset V, Bidaud C, Billaud M, et al. The RET proto-oncogene induces apoptosis: a novel mechanism for Hirschsprung disease. *EMBO J.* 19[15], 4056-4063. 2000.

[27] Tsui-Pierchala BA, Milbrandt J, Johnson EM, Jr. NGF utilizes c-Ret via a novel GFL-independent, inter-RTK signaling mechanism to maintain the trophic status of mature sympathetic neurons. *Neuron* 2002 Jan 17;33(2):261-73.

[28] Pincon-Raymond M. [Cadherins: structure and signalisation protein of cell-cell junctions. Introduction and conclusions]. *J. Soc. Biol*.2004;198(4):353-6.

[29] Anders J, Kjar S, Ibanez CF. Molecular modeling of the extracellular domain of the RET receptor tyrosine kinase reveals multiple cadherin-like domains and a calcium-binding site. *J. Biol. Chem.* 2001;276:35808-17.

[30] Grice EA, Rochelle ES, Green ED, Chakravarti A, McCallion AS. Evaluation of the RET regulatory landscape reveals the biological relevance of a HSCR-implicated enhancer. *Hum. Mol. Genet.* 14[24], 3837-3845. 2005.

[31] Kashuk CS, Stone EA, Grice EA, Portnoy ME, Green ED, Sidow A, et al. Phenotype-genotype correlation in Hirschsprung disease is illuminated by comparative analysis of the RET protein sequence. *Proc. Natl. Acad .Sci .U S A.* 102[25], 8949-8954. 2005.

[32] Garcia-Barcelo M, Sham MH, Lee WS, Lui VC, Chen BL, Wong KK, et al. Highly recurrent RET mutations and novel mutations in genes of the receptor tyrosine kinase and endothelin receptor B pathways in Chinese patients with sporadic Hirschsprung disease. *Clin. Chem* .50[1], 93-100. 2004.

[33] Griseri P, Bachetti T, Puppo F, Lantieri F, Ravazzolo R, Devoto M, et al. A common haplotype at the 5' end of the RET proto-oncogene, overrepresented in Hirschsprung patients, is associated with reduced gene expression. *Hum. Mutat.* 25[2], 189-195. 2005.

[34] Griseri P, Pesce B, Patrone G, Osinga J, Puppo F, Sancandi M, et al. A rare haplotype of the RET proto-oncogene is a risk-modifying allele in hirschsprung disease. *Am. J. Hum. Genet.* 2002 Oct;71(4):969-74.

[35] Guo K, Pourpak A, Beetz-Rogers K, Gokhale V, Sun D, Hurley LH. Formation of pseudosymmetrical G-quadruplex and i-motif structures in the proximal promoter region of the RET oncogene. *J. Am. Chem .Soc.* 2007 Aug 22;129(33):10220-8.

[36] Burzynski GM, Nolte IM, Osinga J, Ceccherini I, Twigt B, Maas S, et al. Localizing a putative mutation as the major contributor to the development of sporadic Hirschsprung disease to the RET genomic sequence between the promoter region and exon 2. *Eur. J. Hum. Genet.* 2004 Aug;12(8):604-12.

[37] Lantieri F, Griseri P, Puppo F, Campus R, Martucciello G, Ravazzolo R, et al. Haplotypes of the human RET proto-oncogene associated with Hirschsprung disease in the Italian population derive from a single ancestral combination of alleles. *Ann. Hum. Genet.* 70[Pt 1], 12-26. 2006.

[38] de Pontual L, Pelet A, Trochet D, Jaubert F, Espinosa-Parrilla Y, Munnich A, et al. Mutations of the RET gene in isolated and syndromic Hirschsprung's disease in human disclose major and modifier alleles at a single locus. *J. Med. Genet.* 43[5], 419-423. 2006.

[39] Griseri P, Lantieri F, Puppo F, Bachetti T, Di DM, Ravazzolo R, et al. A common variant located in the 3'UTR of the RET gene is associated with protection from Hirschsprung disease. *Hum. Mutat.* 2007 Feb;28(2):168-76.

[40] Fitze G, Appelt H, Konig IR, Gorgens H, Stein U, Walther W, et al. Functional haplotypes of the RET proto-oncogene promoter are associated with Hirschsprung disease (HSCR). *Hum. Mol. Genet* .12[24], 3207.-3214. 2003.

[41] Garcia-Barcelo M, Ganster RW, Lui VC, Leon TY, So MT, Lau AM, et al. TTF-1 and RET promoter SNPs: regulation of RET transcription in Hirschsprung's disease. *Hum. Mol. Genet.* 14[2], 191-204. 2005.

[42] Kjaer S, Ibanez CF. Intrinsic susceptibility to misfolding of a hot-spot for Hirschsprung disease mutations in the ectodomain of RET. *Hum. Mol. Genet.* 12[17], 2133-2144. 2003.

[43] Asai N, Iwashita T, Matsuyama M, Takahashi M. Mechanism of activation of the ret proto-oncogene by multiple endocrine neoplasia 2A mutations. *Mol. Cell Biol.* 15[3], 1613-1619. 1995.

[44] Carlomagno F, Salvatore G, Cirafici AM, De Vita G, Melillo RM, de Franciscis V, et al. The different RET-activating capability of mutations of cysteine 620 or cysteine 634 correlates with the multiple endocrine neoplasia type 2 disease phenotype. *Cancer Res.* 57[3], 391-395. 1997.

[45] Carlomagno F, Anaganti S, Guida T, Salvatore G, Troncone G, Wilhelm SM, et al. BAY 43-9006 inhibition of oncogenic RET mutants. *J. Natl. Cancer Inst.* 2006 Mar 1;98(5):326-34.

[46] Pelet A, Geneste O, Edery P, Pasini A, Chappuis S, Atti T, et al. Various mechanisms cause RET-mediated signaling defects in Hirschsprung's disease. *J. Clin. Invest.* 1998 Mar 15;101(6):1415-23.

[47] Geneste O, Bidaud C, De Vita G, Hofstra RM, Tartare-Deckert S, Buys CH, et al. Two distinct mutations of the RET receptor causing Hirschsprung's disease impair the binding of signalling effectors to a multifunctional docking site. *Hum. Mol. Genet* .8[11], 1989-1999. 1999.

[48] Arighi E, Borrello MG, Sariola H. RET tyrosine kinase signaling in development and cancer. *Cytokine. Growth Factor Rev.* 16[4-5], 441-467. 2005.

[49] Alberti L, Borrello MG, Ghizzoni S, Torriti F, Rizzetti MG, Pierotti MA. Grb2 binding to the different isoforms of Ret tyrosine kinase. *Oncogene* 1998 Sep 3;17(9):1079-87.

[50] Santoro M, Carlomagno F, Romano A, Bottaro DP, Dathan NA, Grieco M, et al. Activation of RET as a dominant transforming gene by germline mutations of MEN2A and MEN2B. *Science* 267[5196], 381-383. 1995.

[51] Arighi E, Alberti L, Torriti F, Ghizzoni S, Rizzetti MG, Pelicci G, et al. Identification of Shc docking site on Ret tyrosine kinase. *Oncogene* 1997 Feb 20;14(7):773-82.

[52] (52) Lorenzo MJ, Gish GD, Houghton C, Stonehouse T, Pawson T, Ponder B, et al. RET alternate splicing influences the interaction of activated RET with the SH2 and PTB domains of Shc, and the SH2 domain of Grb2. *Oncogene* 1997;14:763-71.

[53] Besset V, Scott RP, Ibanez CF. Signaling complexes and protein-protein interactions involved in the activation of the Ras and phosphatidylinositol 3-kinase pathways by the c-Ret receptor tyrosine kinase. *J. Biol. Chem.* 2000 Dec 15;275(50):39159-66.

[54] da Rocha AP, Magalhaes PK, Maia AL, Maciel LM. [Genetic polymorphisms: implications in the pathogenesis of medullary thyroid carcinoma]. *Arq. Bras. Endocrinol. Metabol* .2007 Jul;51(5):723-30.

[55] Borrego S, Sáez ME, Ruiz A, et al. Specific polymorphisms in the RET proto-oncogene are over-represented in patients with Hirschsprung disease and may represent loci modifying phenotypic expression. *J. Med. Genet.* 1999;36:771-4.

[56] Amoresano A, Incoronato M, Monti G, Pucci P, de F, V, Cerchia L. Direct interactions among Ret, GDNF and GFRalpha1 molecules reveal new insights into the assembly of a functional three-protein complex. *Cell Signal.* 2005 Jun;17(6):717-27.

[57] Virtanen H, Yang J, Bespalov MM, Hiltunen JO, Leppanen VM, Kalkkinen N, et al. The first cysteine-rich domain of the receptor GFRalpha1 stabilizes the binding of GDNF. *Biochem. J.* 2005 May 1;387(Pt 3):817-24.

[58] Airaksinen MS, Titievsky A, Saarma M. GDNF family neurotrophic factor signaling: four masters, one servant? *Mol. Cell Neurosci.* 1999 May;13(5):313-25.

[59] Yang J, Runeberg-Roos P, Leppanen VM, Saarma M. The mouse soluble GFRalpha4 receptor activates RET independently of its ligand persephin. *Oncogene* 2007 May 31;26(26):3892-8.

[60] Li Z, Wang B, Wu X, Cheng SY, Paraoan L, Zhou J. Identification, expression and functional characterization of the GRAL gene. *J. Neurochem* . 2005.

[61] Kawamoto Y, Takeda K, Okuno Y, Yamakawa Y, Ito Y, Taguchi R, et al. Identification of RET autophosphorylation sites by mass spectrometry. *J. Biol. Chem.* 2004 Apr 2;279(14):14213-24.

[62] Cranston AN, Carniti C, Oakhill K, Radzio-Andzelm E, Stone EA, McCallion AS, et al. RET is constitutively activated by novel tandem mutations that alter the active site resulting in multiple endocrine neoplasia type 2B. *Cancer Res.* 2006 Oct 15;66(20):10179-87.

[63] Schuringa JJ, Wojtachnio K, Hagens W, Vellenga E, Hofstra R, Kruijer W. MEN2A-RET-induced cellular transformation by activation of STAT3. *Oncogene* 20[38], 5350-5358. 2001.

[64] Solari V, Ennis S, Yoneda A, Wong L, Messineo A, Hollwarth ME, et al. Mutation analysis of the RET gene in total intestinal aganglionosis by wave DNA fragment analysis system. *J. Pediatr. Surg.* 38[3], 497.-501. 2003.

[65] Attie T, Pelet A, Edery P, et al. Diversity of *RET* proto-oncogene mutations in familial and sporadic Hirschsprung disease. *Hum. Mol. Genet* .1995;4:1381-6.

[66] Barlow A, de Graaff E, Pachnis V. Enteric nervous system progenitors are coordinately controlled by the G protein-coupled receptor EDNRB and the receptor tyrosine kinase RET. *Neuron* 40[5], 905-916. 2003.

[67] Kruger GM, Mosher JT, Tsai YH, Yeager KJ, Iwashita T, Gariepy CE, et al. Temporally distinct requirements for endothelin receptor B in the generation and migration of gut neural crest stem cells. *Neuron* 2003 Dec 4;40(5):917-29.

[68] McCallion AS, Stames E, Conlon RA, Chakravarti A. Phenotype variation in two-locus mouse models of Hirschsprung's disease:Tissue-specific interaction between Ret and Ednrb. *Proc. Natl. Acad. Sci .USA* 2003;100(4):1826-31.

[69] Jacobs-Cohen RJ, Payette RF, Gershon MD, Rothman TP. Inability of neural crest cells to colonise the presumptive aganglionic bowel in ls/ls mutant mice. Requirement for a permissive environment. *J. Comp. Neurol.* 1987;255:425-38.

[70] Baynash AG, Hosoda K, Giaid A, et al. Interaction of endothelin-3 with endothelin-B receptor is essential for development of epidermal melanocytes and enteric neurons. *Cell* 1994;79:1277-85.

[71] Hosoda K, Hammer RE, Richardson JA, et al.(. Targeted and natural (piebald-lethal) mutations of endothelin-B receptor gene produce megacolon associated with spotted coat color in mice. *Cell* 1994;79:1267-76.

[72] Heuckeroth RO. Finding your way to the end: a tale of GDNF and endothelin-3. Neuron 2003 Dec 4;40(5):871-3.

[73] Carrasquillo MM, McCallion AS, Puffenberger EG, Kaschuk CS, No N, Chakravarti A. Genome-wide association study as well as the study of mouse models help to identify

the interaction between RET and EDNRB pathways in Hirschsprung disease. *Nature Genet.* 2002;32:237-44.

[74] Mosher JT, Yeager KJ, Kruger GM, Joseph NM, Hutchin ME, Dlugosz AA, et al. Intrinsic differences among spatially distinct neural crest stem cells in terms of migratory properties, fate determination, and ability to colonize the enteric nervous system. *Dev. Biol.* 2007 Mar 1;303(1):1-15.

[75] Burns AJ, Delalande JM. Neural crest cell origin for intrinsic ganglia of the developing chicken lung. *Dev. Biol.* 277[1], 63-79. 2005.

[76] Chan KK, Wong CK, Lui VC, Tam PK, Sham MH. Analysis of SOX10 mutations identified in Waardenburg-Hirschsprung patients: Differential effects on target gene regulation. *J. Cell Biochem.* 90[3], 573.-585. 2003.

[77] Yokoyama S, Takeda K, Shibahara S. SOX10, in combination with Sp1, regulates the endothelin receptor type B gene in human melanocyte lineage cells. *FEBS J.* 273[8], 1805-1820. 2006.

[78] Lang D, EJA. Sox10 and Pax3 physically interact to mediate activation of a conserved c-RET enhancer. *Hum. Mol. Genet*.12[8], 937-945. 2003.

[79] Vohra BP, Planer W, Armon J, Fu M, Jain S, Heuckeroth RO. Reduced endothelin converting enzyme-1 and endothelin-3 mRNA in the developing bowel of male mice may increase expressivity and penetrance of Hirschsprung disease-like distal intestinal aganglionosis. *Dev. Dyn.* 2007 Jan;236(1):106-17.

[80] Amiel J, Lyonnet S. Hirschsprung disease, associated syndromes, and genetics: a review. *J. Med. Genet.* 38[11], 729-739. 2001.

[81] Lui VC, Leon TY, Garcia-Barcelo MM, Ganster RW, Chen BL, Hutson JM, et al. Novel RET Mutation Produces a Truncated RET Receptor Lacking the Intracellular Signaling Domain in a 3-Generation Family with Hirschsprung Disease. *Clin. Chem* .51[8], 1552.-1554. 2005.

[82] Neff F, Noelker C, Eggert K, Schlegel J. Signaling pathways mediate the neuroprotective effects of GDNF. *Ann. N Y Acad. Sci.* 2002 Nov;973:70-4.

[83] Melillo RM, Castellone MD, Guarino V, De F, V, Cirafici AM, Salvatore G, et al. The RET/PTC-RAS-BRAF linear signaling cascade mediates the motile and mitogenic phenotype of thyroid cancer cells. *J. Clin. Invest.* 2005 Apr;115(4):1068-81.

[84] Vasko VV, Saji M. Molecular mechanisms involved in differentiated thyroid cancer invasion and metastasis. *Curr. Opin. Oncol.* 2007 Jan;19(1):11-7.

[85] Asai N, Fukuda T, Wu Z, Enomoto A, Pachnis V, Takahashi M, et al. Targeted mutation of serine 697 in the Ret tyrosine kinase causes migration defect of enteric neural crest cells. *Development* 2006 Nov;133(22):4507-16.

[86] Abbosh PH, Nephew KP. Multiple signaling pathways converge on beta-catenin in thyroid cancer. *Thyroid* 15[6], 551-561. 2005.

[87] Myers SM, Mulligan LM. The RET receptor is linked to stress response pathways. *Cancer Res.* 2004 Jul 1;64(13):4453-63.

[88] Le HH, Charlet-Berguerand N, De F, V, Thermes C. 5'-End RET splicing: absence of variants in normal tissues and intron retention in pheochromocytomas. *Oncology* 2002;63(1):84-91.

[89] de GE, Srinivas S, Kilkenny C, D'Agati V, Mankoo BS, Costantini F, et al. Differential activities of the RET tyrosine kinase receptor isoforms during mammalian embryogenesis. *Genes. Dev.* 2001 Sep 15;15(18):2433-44.

[90] Jain S, Encinas M, Johnson EM, Jr., Milbrandt J. Critical and distinct roles for key RET tyrosine docking sites in renal development. *Genes. Dev.* 2006 Feb 1;20(3):321-33.

[91] Minopoli G, Passaro F, Aloia L, Carlomagno F, Melillo RM, Santoro M, et al. Receptor- and non-receptor tyrosine kinases induce processing of the amyloid precursor protein: role of the low-density lipoprotein receptor-related protein. *Neurodegener Dis.* 2007;4(2-3):94-100.

[92] Heuckeroth RO, Lampe PA, Johnson EM, Milbrandt J. Neurturin and GDNF promote proliferation and survival of enteric neuron and glial progenitors in vitro. *Dev. Biol.* 1998 Aug 1;200(1):116-29.

[93] Taraviras S, Marcos-Gutierrez CV, Durbec P, Jani H, Grigoriou M, Sukumaran M, et al. Signalling by the RET receptor tyrosine kinase and its role in the development of the mammalian enteric nervous system. *Development* 1999 Jun;126(12):2785-97.

[94] Henderson CE, Phillips HS, Pollock RA, Davies AM, Lemeulle C, Armanini M, et al. GDNF: a potent survival factor for motoneurons present in peripheral nerve and muscle. *Science* 1994 Nov 11;266(5187):1062-4.

[95] Sanchez MP, Silos-Santiago I, Frisen J, He B, Lira SA, Barbacid M. Renal agenesis and the absence of enteric neurons in mice lacking GDNF. *Nature* 1996 Jul 4;382(6586):70-3.

[96] Gariepy CE. Intestinal motility disorders and development of the enteric nervous system. *Pediatr. Res.* 49[5], 605-613. 2001.

[97] Burns AJ, Thapar N. Advances in ontogeny of the enteric nervous system. *Neurogastroenterol. Motil.* 2006 Oct;18(10):876-87.

[98] Musholt PB, Musholt TJ, Goodfellow PJ, Zehnbauer BA, Wells SA, Moley JF. "Cold" single-strand conformational variants for mutation analysis of the *RET* proto-oncogene. *Surgery* 122[363], 371. 1997.

[99] Sariola H, Saarma M. Novel functions and signalling pathways for GDNF. *J. Cell Sci.* 116[Pt 19], 3855-3862. 2003.

[100] Costantini F, Shakya R. GDNF/Ret signaling and the development of the kidney. *Bioessays* 28[2], 117-127. 2006.

[101] Ichihara M, Murakumo Y, Takahashi M. RET and neuroendocrine tumors. *Cancer Lett.* 204[2], 197-211. 2004.

[102] Schuetz G, Rosario M, Grimm J, Boeckers TM, Gundelfinger ED, Birchmeier W. The neuronal scaffold protein Shank3 mediates signaling and biological function of the receptor tyrosine kinase Ret in epithelial cells. *J. Cell Biol.* 2004 Dec 6;167(5):945-52.

[103] Tang MJ, Cai Y, Tsai SJ, Wang YK, Dressler GR. Ureteric bud outgrowth in response to RET activation is mediated by phosphatidylinositol 3-kinase. *Dev. Biol.* 2002 Mar 1;243(1):128-36.

[104] Segouffin-Cariou C, Billaud M. Transforming ability of MEN2A-RET requires activation of the phosphatidylinositol 3-kinase/AKT signaling pathway. *J. Biol. Chem.* 2000 Feb 4;275(5):3568-76.

[105] Doxakis E, Davies AM. Retinoic acid negatively regulates GDNF and neurturin receptor expression and responsiveness in embryonic chicken sympathetic neurons. *Mol. Cell Neurosci.* 2005 Aug;29(4):617-27.

[106] Thang SH, Kobayashi M, Matsuoka I. Regulation of glial cell line-derived neurotrophic factor responsiveness in developing rat sympathetic neurons by retinoic acid and bone morphogenetic protein-2. *J. Neurosci* 2000 Apr 15;20(8):2917-25.

[107] Wanigasekara Y, Airaksinen MS, Heuckeroth RO, Milbrandt J, Keast JR. Neurturin signalling via GFRalpha2 is essential for innervation of glandular but not muscle targets of sacral parasympathetic ganglion neurons. *Mol. Cell Neurosci.* 2004 Feb;25(2):288-300.

[108] Pierchala BA, Tsui CC, Milbrandt J, Johnson EM. NGF augments the autophosphorylation of Ret via inhibition of ubiquitin-dependent degradation. *J. Neurochem.* 2007 Mar;100(5):1169-76.

[109] Iwashita T, Murakami H, Asai N, Takahashi M. Mechanism of ret dysfunction by Hirschsprung mutations affecting its extracellular domain.. *Hum. Mol. Genet.* 5[10], 1577-1580. 1996.

[110] de PL, Pelet A, Clement-Ziza M, Trochet D, Antonarakis SE, ttie-Bitach T, et al. Epistatic interactions with a common hypomorphic Ret allele in syndromic Hirschsprung disease. *Hum. Mutat*. 2007 Mar 30.

[111] Cosma M, Cardone M, Carlomagno F, Colantuoni V. Mutations in the extracellular domain cause *RET* loss of function by a dominant negative mechanism. *Mol. Cell Biol.* 9 1998;18 :3321-32.

[112] Moore SW. The contribution of associated congenital anomalies in understanding Hirschsprung's disease. *Pediatr. Surg. Int.* 2006 Apr;22(4):305-15.

[113] Grieco M, Santoro M, Berlingieri MT, Melillo RM, Donghi R, Bongarzone I, et al. PTC is a novel rearranged form of the ret proto-oncogene and is frequently detected in vivo in human thyroid papillary carcinomas. *Cell* 1990 Feb 23;60(4):557-63.

[114] Rabes HM, Klugbauer S. Molecular genetics of childhood papillary thyroid carcinomas after irradiation: high prevalence of RET rearrangement. *Recent Results Cancer Res.* 1998;154:248-64.

[115] Adeniran AJ, Zhu Z, Gandhi M, Steward DL, Fidler JP, Giordano TJ, et al. Correlation between genetic alterations and microscopic features, clinical manifestations, and prognostic characteristics of thyroid papillary carcinomas. *Am. J. Surg. Pathol.* 2006 Feb;30(2):216-22.

[116] Klugbauer S, Rabes HM. The transcription coactivator HTIF1 and a related protein are fused to the RET receptor tyrosine kinase in childhood papillary thyroid carcinomas. *Oncogene* 1999 Jul 29;18(30):4388-93.

[117] Nikiforov YE, Rowland JM, Bove KE, Monforte-Munoz H, Fagin JA. Distinct pattern of ret oncogene rearrangements in morphological variants of radiation-induced and sporadic thyroid papillary carcinomas in children. *Cancer Res*. 1997 May 1;57(9):1690-4.

[118] Moore SW, Appfelstaedt J, Zaahl MG. Familial medullary carcinoma prevention, risk evaluation, and RET in children of families with MEN2. *J. Pediatr. Surg.* 2007 Feb;42(2):326-32.

[119] Pinna G, Orgiana G, Riola A, Ghiani M, Lai ML, Carcassi C, et al. RET Proto-Oncogene in Sardinia: V804M Is the Most Frequent Mutation and May Be Associated with FMTC/MEN-2A Phenotype. *Thyroid* 2007 Feb;17(2):101-4.

[120] Mulligan LM, Eng C, Attie T, Lyonnet S, Marsh DJ, Hyland VJ, et al. Diverse phenotypes associated with exon 10 mutations of the RET proto-oncogene. *Hum. Mol. Genet*.3[12], 2163-2167. 1994.

[121] Edery P, Eng C, Munnich A, Lyonnet S. RET in human development and oncogenesis. *Bioessays* 19[5], 389-395. 1997.

[122] Hofstra RM, Landsvater RM, Ceccherini I, Stulp RP, Stelwagen T, Luo Y, et al. A mutation in the RET proto-oncogene associated with multiple endocrine neoplasia type 2B and sporadic medullary thyroid carcinoma. *Nature* 367[6461], 375 --376. 1994.

[123] DeLellis RA. Pathology and genetics of thyroid carcinoma. J Surg Oncol 2006 Dec 15;94(8):662-9.

[124] Santoro M, Grieco M, Melillo RM, Fusco A, Vecchio G. Molecular defects in thyroid carcinomas: role of the RET oncogene in thyroid neoplastic transformation. *Eur. J. Endocrinol.* 133[5], 513-522. 1995.

[125] Arighi E, Popsueva A, Degl'Innocenti D, Borrello MG, Carniti C, Perala NM, et al. Biological effects of the dual phenotypic Janus mutation of ret cosegregating with both multiple endocrine neoplasia type 2 and Hirschsprung's disease. *Mol. Endocrinol.* 18[4], 1004-1017. 2004.

[126] Chappuis-Flament S, Pasini A, De Vita G, Segouffin-Cariou C, Fusco A, Attie T, et al. Dual effect on the RET receptor of MEN 2 mutations affecting specific extracytoplasmic cysteines. *Oncogene* 17[22], 2851-2861. 1998.

[127] Takahishi M, Iwashita T, Santoro M, Lyonnet S, Lenoir GM, Billaud M. Co-segregation of MEN2 and Hirschsprung's disease: the same mutation with both gain and loss of function. *Hum. Mut.* 13[4], 331-336. 1999.

[128] Pelicci G, Troglio F, Bodini A, Melillo RM, Pettirossi V, Coda L, et al. The neuron-specific Rai (ShcC) adaptor protein inhibits apoptosis by coupling Ret to the phosphatidylinositol 3-kinase/Akt signaling pathway. *Mol. Cell Biol.* 2002 Oct;22(20):7351-63.

[129] Salvatore D, Melillo RM, Monaco C, Visconti R, Fenzi G, Vecchio G, et al. Increased in vivo phosphorylation of ret tyrosine 1062 is a potential pathogenetic mechanism of multiple endocrine neoplasia type 2B. *Cancer Res.* 61[4], 1426-1431. 2001.

[130] Wong A, Bogni S, Kotka P, de Graaff E, D'Agati V, Costantini F, et al. Phosphotyrosine 1062 is critical for the in vivo activity of the Ret9 receptor tyrosine kinase isoform. *Mol. Cell Biol.* 25[21], 9661-9673. 2005.

[131] Vasko VV, Saji M. Molecular mechanisms involved in differentiated thyroid cancer invasion and metastasis. *Curr .Opin. Oncol.* 2007 Jan;19(1):11-7.

[132] Hou P, Liu D, Shan Y, Hu S, Studeman K, Condouris S, et al. Genetic alterations and their relationship in the phosphatidylinositol 3-kinase/Akt pathway in thyroid cancer. *Clin. Cancer Res .*2007 Feb 15;13(4):1161-70.

[133] Brandi ML, Gagel RF, Angeli A, Bilezikian JP, Beck-Peccoz P, et al. Guidelines for diagnosis and therapy of MEN type 1 and type 2. *J. Clin. Endocrinol. Metab.* 86[12], 5658-5671. 2001.

[134] Yip L, Cote GJ, Shapiro SE, Ayers GD, Herzog CE, Sellin RV, et al. Multiple endocrine neoplasia type 2: evaluation of the genotype-phenotype relationship. *Arch. Surg*.138[4], 409-416. 2003.

[135] Ito S, Iwashita T, Asai N, Murakami H, Iwata Y, Sobue G, et al. Biological properties of Ret with cysteine mutations correlate with multiple endocrine neoplasia type 2A, familial medullary thyroid carcinoma, and Hirschsprung's disease phenotype. *Cancer Res.* 57[14], 2870-2872. 1997.

[136] Dralle H, Ukkat J, Brauckhoff M, Gimm O, Machens A. Consortia experiences - can we cure MTC? *Journal of Internal Medicine* 2004 Jun;255(6):698-9.

[137] Brooks AS, Breuning MH, Osinga J, vd Smagt JJ, Catsman CE, Buys CH, et al. A consanguineous family with Hirschsprung disease, microcephaly, and mental retardation (Goldberg-Shprintzen syndrome). *J. Med. Genet.* 36[6], 485-489. 1999.

[138] Gujral TS, Mulligan LM. Molecular implications of RET mutations for pheochromocytoma risk in multiple endocrine neoplasia 2. *Ann. N Y Acad. Sci.* 1073, 234-240. 2006.

[139] Borrego S, Fernandez RM, Dziema H, Niess A, Lopez-Alonso M, Antinolo G, et al. Investigation of germline GFRA4 mutations and evaluation of the involvement of GFRA1, GFRA2, GFRA3, and GFRA4 sequence variants in Hirschsprung disease. *J. Med. Genet.* 40[3], e18. 2003.

[140] Fitze G, Saeger HD, Roesner D, Schackert HK. Management of multiple endocrine neoplasia syndrome type 2 families in association with rare germline mutations of the RET proto-oncogene. *Klin. Padiatr.* 216[5], 270-276. 2004.

[141] Mulligan LM, Marsh DJ, Robinson BG, Schuffenecker I, Zedenius J, Lips CJ, et al. Genotype-phenotype correlation in multiple endocrine neoplasia type 2: report of the International RET Mutation Consortium. *J. Intern. Med*.238[4], 343-346. 1995.

[142] Russo A, Zanna I, Tubiolo C, Migliavacca M, Bazan V, Latteri MA, et al. Hereditary common cancers: molecular and clinical genetics. *Anticancer. Res.* 20[6C], 4841-4851. 2000.

[143] Machens A, Gimm O, Hinze R, Hoppner W, Boehm BO, Dralle H. Genotype-phenotype correlations in hereditary medullary thyroid carcinoma: Oncological features and biochemical properties. *Journal of Clinical Endocrinology and Metabolism* 2001 Mar;86(3):1104-9.

[144] Antinolo G, Marcos I, Fernandez RM, Romero M, Borrego S. A novel germline point mutation, c.2304 G-->T, in codon 768 of the RET proto-oncogene in a patient with medullary thyroid carcinoma. *Am. J. Med.Genet*.2002 Jun 1;110(1):85-7.

[145] Torre M, Martucciello G, Ceccherini I, Lerone M, Aicardi M, Gambini C, et al. Diagnostic and therapeutic approach to multiple endocrine neoplasia type 2B in pediatric patients. *Pediatr. Surg.Int*.18[5-6], 378-383. 2002.

[146] Carlson KM, Bracamontes J, Jackson CE, Clark R, Lacroix A, Wells SA Jr, et al. Parent-of-origin effects in multiple endocrine neoplasia type 2B. *Am. J. Hum. Genet.* 55[6], 1076-1082. 1994.

[147] Unruh A, Fitze G, Janig U, Bielack S, Lochbuhler H, Coerdt W. Medullary thyroid carcinoma in a 2-month-old male with multiple endocrine neoplasia 2B and symptoms

of pseudo-Hirschsprung disease: a case report. *J. Pediatr. Surg* .2007 Sep;42(9):1623-6.

[148] Machens A, Dralle H. Very early manifestation of hereditary medullary thyroid cancer in carriers of intracellulsar-domain RET mutations. *J. Pediatr. Surg.* 42, 1153. 2007.

[149] Machens A, Dralle H. Genotype-phenotype based surgical concept of hereditary medullary thyroid carcinoma. *World J. Surg.* 2007 May;31(5):957-68.

[150] Caron P, Attie T, David D, Amiel J, Brousset F, Roger P, et al. C618R mutation in exon 10 of the RET proto-oncogene in a kindred with multiple endocrine neoplasia type 2A and Hirschsprung's disease. *J. Clin. Endocrinol. Metab.* 81[7], 2731-2733. 1996.

[151] Peretz H, Luboshitsky R, Baron E, Biton A, Gershoni R, Usher S, et al. Cys 618 Arg mutation in the RET proto-oncogene associated with familial medullary thyroid carcinoma and maternally transmitted Hirschsprung's disease suggesting a role for imprinting. *Hum. Mutat.* 10[2], 155-159. 1997.

[152] Romeo G, Ceccherini I, Celli J, Priolo M, Betsos N, Bonardi G, et al. Association of multiple endocrine neoplasia type 2 and Hirschsprung disease. *J. Intern. Med.* 243[6], 515-520. 1998.

[153] Sijmons RH, Hofstra RM, Wijburg FA, Links TP, Zwierstra RP, Vermey A, et al. Oncological implications of RET gene mutations in Hirschsprung's disease. *Gut* 43[4], 542-547. 1998.

[154] Dvorakova S, Dvorakova K, Malikova M, Skaba R, Vlcek P, Bendlova B. A novel Czech kindred with familial medullary thyroid carcinoma and Hirschsprung's disease. *J. Pediatr. Surg.* 40[6], e1-6. 2005.

[155] Verdy M, Weber AM, Roy CC, Morin CL, Cadotte M, Brochu P. Hirschsprung's disease in a family with multiple endocrine neoplasia type 2. J. *Pediatr. Gastroenterol. Nutr.* 1[4], 603-607. 1982.

[156] Borst MJ, VanCamp JM, Peacock ML, Decker RA. Mutational analysis of multiple endocrine neoplasia type 2A associated with Hirschsprung's disease. *Surgery* 117[4], 386-391. 1995.

[157] Blank RD, Sklar CA, Dimich AB, LaQuaglia MP, Brennan MF. Clinical presentations and RET protooncogene mutations in seven multiple endocrine neoplasia type 2 kindreds. *Cancer* 78[9], 1996-2003. 1996.

[158] Borrego S, Eng C, Sanchez B, Saez ME, Navarro E, Antinolo G. Molecular analysis of the ret and GDNF genes in a family with multiple endocrine neoplasia type 2A and Hirschsprung disease. *J. Clin. Endocrinol. Metab.* 83[9], 3361-3364. 1998.

[159] Decker RA, Peacock ML. Occurrence of MEN 2a in familial Hirschsprung's disease: a new indication for genetic testing of the RET proto-oncogene. *J. Pediatr. Surg.* 33[2], 207-214. 1998.

[160] Inoue K, Shimotake T, Inoue K, Tokiwa K. Mutational analysis of the RET proto-oncogene in a kindred with multiple endocrine neoplasia type 2A and Hirschsprung's disease. *J. Pediatr. Surg.* 34[10], 1552-1554. 1999.

[161] Pasini B, Rossi R, Ambrosio MR, Zatelli MC, Gullo M, Gobbo M, et al. RET mutation profile and variable clinical manifestations in a family with multiple endocrine neoplasia type 2A and Hirschsprung's disease. *Surgery* 131[4], 373-381. 2002.

[162] Decker RA, Peacock ML, Watson P. Hirschsprung disease in MEN 2A: increased spectrum of RET exon 10 genotypes and strong genotype-phenotype correlation. *Hum. Mol. Genet.* 7[1], 129-134. 1998.

[163] Sasaki Y, Shimotake T, Go S, Iwai N. Total thyroidectomy for hereditary medullary thyroid carcinoma 12 years after correction of Hirschsprung's disease. *Eur. J. Surg.* 167[6], 467-469. 2001.

[164] Fernandez RM, Antinolo G, Eng C, Borrego S. The RET C620S mutation causes multiple endocrine neoplasia type 2A (MEN2A) but not Hirschsprung disease (HSCR) in a family cosegregating both phenotypes. *Hum. Mutat.* 22[5], 412-415. 2003.

[165] Yip L, Lee JE, Shapiro SE, Waguespack SG, Sherman SI, Hoff AO, et al. Surgical management of hereditary pheochromocytoma. J Am Coll Surg 198[4], 525-534. 2004.

[166] Bhansali A, Rajput R, Behra A, Rao KL, Khandelwal N, Radotra BD. Childhood sporadic pheochromocytoma: clinical profile and outcome in 19 patients. *J. Pediatr. Endocrinol. Metab.* 19[5], 749-756. 2006.

[167] Mannelli M, Simi L, Gagliano MS, Opocher G, Ercolino T, Becherini L, et al. Genetics and biology of pheochromocytoma. *Exp. Clin. Endocrinol. Diabetes.* 2007 Mar;115(3):160-5.

[168] Korpershoek E, Petri BJ, Van Nederveen FH, Dinjens WN, Verhofstad AA, De Herder WW, et al. Candidate gene mutation analysis in bilateral adrenal pheochromocytoma and sympathetic paraganglioma. *Endocr. Relat. Cancer* 2007 Jun;14(2):453-62.

[169] De Krijger RR, Petri BJ, Van Nederveen FH, Korpershoek E, De Herder WW, De Muinck Keizer-Schrama SM, et al. Frequent genetic changes in childhood pheochromocytomas. Ann N Y Acad Sci 2006 Aug;1073:166-76.

[170] Eng C, Crossey PA, Mulligan LM, Healey CS, Healey CS, Houghton C, et al. Mutations in the RET proto-oncogene and the von Hippel-Lindau disease tumour suppressor gene in sporadic and syndromic phaeochromocytomas. *J. Med. Genet.* 32[12], 934-937. 1995.

[171] Schocket E, Telok HA. Aganglionic megacolon, phaeochromocytoma, megaloureter and neurofibromatosis. *Am. J. Dis. Child* 1957;94:185-91.

[172] Takaya K, Yoshimasa T, Arai H, Tamura N, Miyamoto Y, Itoh H, et al. The RET proto-oncogene in sporadic pheochromocytomas. *Intern. Med.* 35[6], 449-452. 1996.

[173] Cho NH, Lee HW, Lim SY, Kang S, Jung WY, Park CS. Genetic aberrance of sporadic MEN 2A component tumours: analysis of RET. *Pathology* 37[1], 10.-13. 2005.

[174] Powers JF, Brachold JM, Tischler AS. Ret protein expression in adrenal medullary hyperplasia and pheochromocytoma. *Endocr. Pathol.* 14[4], 351-361. 2003.

[175] Machens A, Brauckhoff M, Gimm O, Dralle H. Risk-oriented approach to hereditary adrenal pheochromocytoma. *Pheochromocytoma* 2006;1073:417-28.

[176] Plaza-Menacho I, Burzynski GM, de Groot JW, Eggen BJ, Hofstra RM. Current concepts in RET-related genetics, signaling and therapeutics. *Trends Genet.* 2006 Nov;22(11):627-36.

[177] McWhinney SR, Boru G, Binkley PK, Peczkowska M, Januszewicz AA, Neumann HP, et al. Intronic single nucleotide polymorphisms in the RET protooncogene are associated with a subset of apparently sporadic pheochromocytoma and may modulate age of onset. *J. Clin. Endocrinol. Metab.* 88[10], 4911-4916. 2003.

[178] Fernandez RM, Navarro E, Antinolo G, Ruiz-Ferrer M, Borrego S. Evaluation of the role of RET polymorphisms/haplotypes as modifier loci for MEN 2, and analysis of the correlation with the type of RET mutation in a series of Spanish patients. *Int. J. Mol. Med.* 2006 Apr;17(4):575-81.

[179] Machens A, Schneyer U, Holzhausen HJ, Dralle H. Prospects of remission in medullary thyroid carcinoma according to basal calcitonin level. *J. Clin. Endocrinol. Metab.* 90[4], 2029-2034. 2005.

[180] Gimm O, Ukkat J, Niederle BE, Weber T, Thanh PN, Brauckhoff M, et al. Timing and Extent of Surgery in Patients with Familial Medullary Thyroid Carcinoma/Multiple Endocrine Neoplasia 2A-related RET Mutations Not Affecting Codon 634. *World J. Surg.* 28[12], 1312-1316. 2004.

[181] Frank-Raue K, Buhr H, Dralle H, Klar E, Senninger N, Weber T, et al. Long-term outcome in 46 gene carriers of hereditary medullary thyroid carcinoma after prophylactic thyroidectomy: impact of individual RET genotype. *Eur. J. Endocrinol.* 155[2], 229-236. 2006.

[182] Petrangolini G, Cuccuru G, Lanzi C, Tortoreto M, Belluco S, Pratesi G, et al. Apoptotic cell death induction and angiogenesis inhibition in large established medullary thyroid carcinoma xenografts by Ret inhibitor RPI-1. *Biochem. Pharmacol.* 72[4], 405-414. 2006.

[183] Drosten M, Putzer BM. Gene therapeutic approaches for medullary thyroid carcinoma treatment. *J. Mol. Med.* 81[7], 411-419. 2003.

[184] de Groot JW, Links TP, Rouwe CW, van der Wal JE, Hofstra RM, Plukker JT. Prophylactic thyroidectomy in children who are carriers of a multiple endocrine neoplasia type 2 mutation: description of 20 cases and recommendations based on the literature. *Ned. Tijdschr. Geneeskd.* 150[6], 311-318. 2006.

In: Oncogene Proteins: New Research
Editors: A.H. Malloy and E.C. Carson, pp. 161-176 ISBN 978-1-60456-290-3

Chapter 5

Insulin-Like Growth Factors as Targets for Cancer Treatment

Yang Feng[4] and Dimiter S. Dimitrov
Protein Interactions Group, Nanobiology Program, CCR, NCI-Frederick, NIH, Frederick, MD 21702

Abstract

In recent years we have witnessed the transition in studies of the insulin-like growth facor (IGF) system from its physiological functions to potential roles in cancer therapy. The two IGFs (I and II) and their main receptor, IGF-IR, have long been known to promote cell proliferation and tissue development, particularly prenatal growth. Their role in tumorigenesis is supported by studies from several aspects. IGF-IR function is essential for transformation of mouse embryonic fibroblasts by a variety of oncogenes. Epidemiological data indicate that the circulating IGF-I level is correlated with increased risk for development of colon, prostate, breast, lung and bladder cancers. Elevated plasma IGF-II level has also been found among colon and breast cancer patients. Various strategies have been designed to target components of the IGF system, namely the ligand (IGF-I and II), receptor, and ligand binding proteins, with the receptor and ligand being the main focus. Both small molecule inhibitors and protein-based therapeutics have been developed. Among them, antibodies against the receptor have been actively pursued by many academic groups and commercial entities. High affinity IGF-IR antibodies not only inhibit cell growth in vitro but also reduce tumor size in xenograft animal models. Phase I and II clinical trials of IGF-IR antibodies have been conducted by several companies with promising outcomes so far. The idea of inhibiting the signal from the pathway via targeting its ligand is relatively novel and development of those antibodies is at their early stage. We discuss in this review in detail these therapeutic antibodies currently under investigations and for comparison some small molecule inhibitors of the receptor kinase activity.

4 Corresponding Author: Yang Feng, Ph.D. Nanobiology Program, CCR, NCI-Frederick, NIH, Bldg 469, Rm 139, Frederick, MD 21702-1201, Phone: 301-846-5629; FAX: 301-846-5598, e-mail: yfeng@ncifcrf.gov.

Keywords: *IGF-II, IGF-I, IGF-IR, therapeutic antibody, neutralization, signal transduction, cancer.*

Introduction

More than three decades ago, two polypeptides isolated from human serum were found to possess nonsuppressible insulin-like and cell-growth-promoting activities (NSILA). N-terminal sequencing of the two peptides showed that they share 47% and 57% homology respectively in amino acid to the B-chain of insulin. Therefore, they were named insulin-like growth factor I and II [1]. IGF-I was also designated somatomedin A or C because it mediates the growth promoting effects of somatotropin (or growth hormone) on cartilage and muscles. IGF-IR, the receptor for IGF-I and II, was first identified in 1985 in rat testis and found to have mitogenic activity [2]. IGF-IR shares a similar tetrameric $\alpha2\beta2$ (homodimer of heterodimer) structure with insulin receptor. The IGF-IIR is identical to the mannose-6-phosphate receptor [3]. The system mainly consists of ligands, receptors, and binding proteins (Figure 1), and plays pivotal role in cell proliferation, survival, tumor metastasis and angiogenesis.

Upon ligand binding to the IGF-IR extracellular domain, conformational change in the transmembrane domain of the beta subunit occurs, which leads to trans-phosphorylation of the kinase domain in the two beta chains. The fully-activated tyrosine kinase domain then phosphorylates and activates more tyrosines, which act as the docking site for adaptor proteins such as IRS (insulin receptor substrate), Grb2, SOS and the regulatory subunit of the PI3 kinase. Recruitment of these adaptor proteins eventually activates the PI3 kinase and MAPK cascades. Signals from these pathways activate transcription factors which regulate expression of many genes important for cell growth and proliferation [4]. The level of signal is determined by availability of free ligand, abundance of surface receptor, and intracellular modulator proteins. In order to treat IGF-dependent tumors, interventions at any of these steps should lead to reduction or elimination of the signal from IGF-IR. Protein based therapeutics and small molecule inhibitors have been actively studied to block IGF-IR actions. This review will discuss some of the therapeutics targeting IGF-IR and IGFs with a focus on therapeutic antibodies.

Targeting the IGF-IR

Receptors for IGFs belong to the family of transmembrane receptor tyrosine kinase (RTK). They include IGF-IR, insulin receptor (IR), IR-related receptor (IRR) and IGF-IIR (see Figure 1 where IRR is not shown). IGF-I binds both IGF-IR and IRR with high affinity and mediate physiological response in vivo. IGF-II also binds to IGF-IR, although at slightly lower affinity than IGF-I. IR has two isoforms, IR-A and IR-B, with the former lacking 12-amino acid insert from splicing of exon 11. IR-A functions as high affinity receptor for IGF-II and mediates primary proliferative effects, whereas IR-B mainly regulates metabolic effect through insulin binding [5]. The fourth receptor, IGF-IIR (or mannose-6-phosphate receptor,

M6PR) lacks cytoplasmic tyrosine kinase domain, therefore, it acts as a clearance receptor for IGF-II and modulates the extracellular levels of IGF-II. In breast cancer, loss of IGF-IIR has been observed, which potentially allows more interaction of IGF-II with the functional receptor, IGF-IR [6].

IGF-IR and IR share $\alpha 2\beta 2$ homodimeric structure, with the complete α chain and 195 amino acids of the β chain as extracellular domain. The β chain possesses the transmembrane and cytoplasmic tyrosine kinase domains. The two receptors are highly conserved, especially in the cytoplasmic kinase domain [2]. The normal physiological functions for IR and IGF-IR are glucose metabolism and proliferation regulation, respectively. However, recent studies have shown that there are partially overlapping functions between the two receptors, with IR able to promote growth and IGF-IR to regulate metabolism [7,8]. Adding more complexity to the system, hybrid receptors of IGF-IR/IR-A and IGF-IR/IR-B exist in most cells [9]. They can function as high affinity receptors for IGF-II and IGF-I respectively. (Figure 1). The redundancy of the system implies the important roles that the IGF system plays in growth and development, and it also presents technical challenges to completely inhibit the signal for cancer treatment.

The function of IGF-IR in development is demonstrated by transgenic mouse studies by Efstratiadis et al [10]. Homozygous *igf-ir*(-/-) mice die minutes after birth from respiratory failure.

These mice have muscle hypoplasia and abnormalities in the nervous system, skin and bone development. These results demonstrate that *igf-ir* is an essential gene, and in mice the presence of functional IGF-IR is indispensable for normal embryonic development and survival after birth.

IGF-IR has been shown to be required for malignant transformation of embryonic fibroblasts by SV40 T antigen and H-ras [11]. Transgenic mice with constitutively active IGF-IR develop adenocarcinoma spontaneously in salivary and mammary glands as early as 8 weeks after birth [12]. IGF-IR is universally present in all cancer cell lines tested, including hematologic malignancies and solid tumors, including breast cancer, colon cancer, pancreatic cancer, prostate cancer, liver cancer, Ewing's sarcoma, glioblastoma and neuroblastoma. It is also detected in freshly isolated primary multiple myeloma. Among them, breast cancers seem to have highly expressed IGF-IR [13]. Blocking the activity of IGF-IR with antibodies to IGF-IR, antisense IGF-IR mRNA, or dominant negative form of IGF-IR has been used to inhibit tumor cell growth in vitro and tumor formation in mouse models [14-17]. In cancer cells, IGF-IR promotes tumor growth by inhibition of apoptosis, transformation, metastasis, and induction of angiogenesis through the vascular endothelial growth factor (VEGF). Therefore, IGF-IR has been considered as a prime target for cancer therapeutic by many academic groups and commercial entities. Both antibodies and RTK inhibitors have been developed for this purpose.

Therapeutic antibodies against the IGF-IR A number of IGF-IR antibodies have been developed, and some are in early clinical trials (See Figure 2). One of the first anti-IGF-IR mouse monoclonal antibody αIR3, developed in early 1980s has provided indication that inhibition of IGF-I binding to IGF-IR could inhibit DNA synthesis [18]. αIR3 is among the first IGF-IR antibodies that have been shown to effectively inhibit growth of cancer cells in culture and mouse model [19,20].

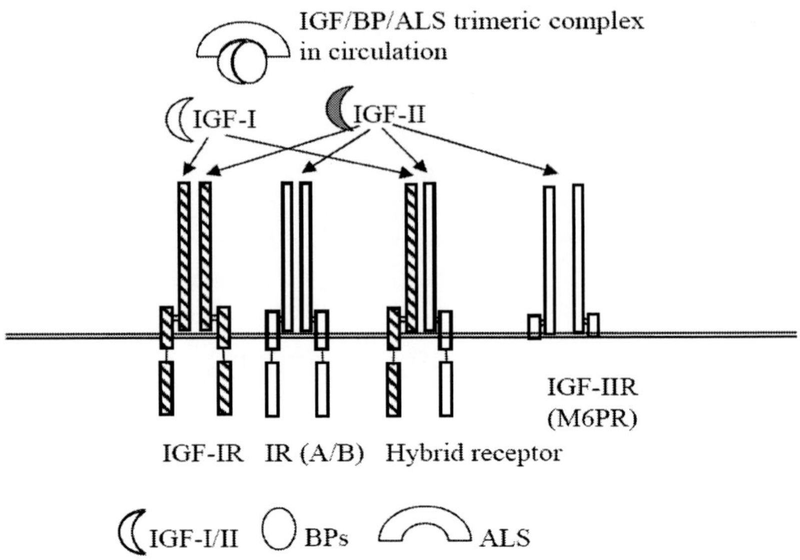

Figure 1. Major components of the IGF system: ligands IGF-I/II, 4 receptors, ALS (acid labile subunit) and binding proteins. Insulin receptor isoform B differs from isoform A by the presence of 12-residual segment from splicing variation. IGF-I binds to IGF-IR and IGF-IR/IR-B hybrid receptor, whereas IGF-II binds to IGF-IR, IR-A and IGF-IR/IR-A hybrid receptor with high affinity. The fourth receptor IGF-IIR lacks cytoplasmic kinase domain and acts as clearance receptor for IGF-II. Majority of IGF-I/II exist as heterotrimolecular complex with one of BPs and ALS. In circulation in rodent, there is <3% of free ligand. There are at least six BPs in human and murine, and together with ALS they control the bioavailability of free ligand to receptors. IGF-independent functions of BPs have also been indentified. Insulin and IRR (insulin receptor-related receptor) are not shown in the diagram.

To date, this antibody is still widely used in many in vitro studies, although it has not been tested as a therapeutic agent. During the past decades, with the advancement of antibody technology, especially humanization of murine antibodies and development of fully human monoclonal antibodies, many potent IGF-IR antibodies have been modified so their efficacies can be tested in preclinical settings and further trialed in phase I and II studies. Antibody CP751,871 was uniquely developed through immunization of Xenomouse with recombinant IGF-IR, and fully human IgG2 was produced through hybridoma technology [21]. With the IgG2 isotype, CP751,871 does not promote ADCC, and its tumor growth inhibition is attributed to the blockade of IGF-IR signal and down-regulation of IGF-IR levels. The lack of ADCC is a desirable feature for IGF-IR antibody because of the expression of IGF-IR in normal cells. Another advantage of CP751,871 is that it recognizes IGF-IR/IR hybrid receptor besides IGF-IR. This binding feature would allow CP751,871 to block signals from IGFs more efficiently in tumor cells expressing both IGF-IR and IR. Phase I trial of CP751,871 has been completed by Pfizer. A total of 36 patients with refractory solid tumors were chosen. Doses at 3~20 mg/kg were found to be safe with minimal side effects [22]. Among 36 patients, three remained stable for more than 6 months at the time of report. Pharmacodynamic studies indicate that CP751,871 targets IGF-IR in granulocytes, tumor cells and tissues involved in regulation of the growth hormone-IGF-I axis. Down regulation of both soluble IGF-IR in circulation and IGF-IR cell surface expression was observed. It should be pointed out that CP751,871 treatment leads to increase in serum IGF-I and IGFBP-3 [23]. CP751,871 is now in Phase II trial of non-small cell lung cancer (NSCLC) with 73 patients [24]. This antibody is at the most advanced stage among all IGF-IR antibodies, and it holds promises for treating IGF-IR driven cancers in both monotherapy and combination therapy.

Ab Name	Nature of Ab	Company	Status	Reference
CP-751,871	Fully human IgG2	Pfizer	Phase II in process	[21-24]
A12	Fully human IgG1	ImClone	Phase I completed	[25,26]
R1507	Fully human IgG1 k	Roche	Phase I completed	[28,30]
AMG479	Fully human IgG	Amgen	Phase I completed	[27,29]
h7C10	Humanized	Merck	Phase I in process	[31]
AVE1642	Humanized	Sanofi-Aventis/ImmunoGen	Phase I in process	[32]

Figure 2. Examples of anti-IGF-IR Antibodies in Development.

A12 is the first IGF-IR antibody reported that has been identified through phage display technology [25]. This fully human IgG1 was selected from a large human naïve phage-displayed Fab library. It is capable of inhibiting MCF-7, BxPC-3 and Colo205 growth in a mouse model. Although A12 is an IgG1 antibody, it does not support ADCC, indicating that its tumoricidal activity is from the blockade of IGF-IR activity. A Phase I trial with A12 has been completed by ImClone Inc. with 21 patients at doses 3~15 mg/kg. Six mg/kg weekly dose was determined to be the Phase II dose. Among 21 patients, 5 experienced stable disease over 4~7 months. No severe side effects were seen [26]. Recently two new IGF-IR antibodies, AMG479 and R1507, have been reported at the 2007 American Society of Clinical Oncology (ASCO) conference. Both antibodies are fully human IgG1. Limited information on preclinical studies for these two antibodies is available in literature [27,28]. A Phase I trial with AMG479 has been completed by Amgen in 33 patients with advanced solid tumors or non-Hodgkin lymphoma. Doses ranged from 1 mg/kg to 20 mg/kg, and most common side effects are thrombocytopenia. Some benefits in patients with Ewing's sarcoma and carcinoid tumor have been observed. Amgen is now conducting a Phase I trial in Ewing's sarcoma and desmoplastic small round cell tumors and Phase IB trial in combination therapy with gemcitabine or panitumumab [29]. A Phase I trial of R1507 is in process by Roche-

Hoffman [30]. Another two antibodies, AVE1642 and H7C10, both as humanized IgG, are in Phase I trials currently [31,32]. These two antibodies, in their murine form and later humanized form, have shown potent IGF-IR antagonist activity and suppressed tumor growth in mouse models. Phase I results are expected in late 2008.

Targeting IGF-IR with high affinity antibodies is an effective method. Fully human and humanized monoclonal antibodies allow longer half-life and are less likely to elicit immune response. Antibodies also have advantage over RTK inhibitors that they do not cross the blood-brain barrier, since IGF signal is also important for the nervous system function even in adults [33].

Treatment with IGF-IR antibodies is accompanied by down-regulation of IGF-IR and IR in tumor cells expressing both receptors. The down-regulation of IR occurs through formation of IGF-IR/IR hybrid receptor and is generally minimized in cells with low levels or no IGF-IR. It is important to know whether IGF-IR is co-expressed with IR in cells which are involved in glucose homeostasis. There has been a concern that IGF-IR antibody therapy may interfere with glucose metabolism, thus will not be appropriate for long term treatment. Outcomes from the four completed Phase I trials of IGF-IR antibodies seem to be encouraging on this aspect. Among the four trials, only in the A12 trial, 4 out of 21 patients experienced hyperglycemia (2 patients at grade 3/4, and 2 in grade 1/2) [26]. In the CP751,871 trial, hyperglycemia was found as one of the minor side effects [22]. Other two Phase I trials did not report hyperglycemia as a major side effect [29,30]. However, we may have to wait till larger size clinical studies over longer term to see the true effects of these antibodies on glucose metabolism. Up-regulation of IGF-I and IGFBPs has been reported with IGF-IR antibody treatment [23]. This may result from feedback mechanism of IGF-I to hGH from reduced IGF-IR activity since IGF-I is under control of the hypothalamus-pituitary-liver axis. Alternatively, consequence from down-regulation of IGF-IR provides lower amounts of receptor for IGF-I binding and internalization. Because only IGF-I has been observed to increase, not IGF-II, the first mechanism is more likely.

Small molecule inhibitors Inhibitors of IGF-IR have been attempted since a decade ago. However, due to the high homology of the IGF-IR with the IR, it is challenging to develop inhibitors that are specific to the IGF-IR. For IGF-IR inhibitors reported so far, mechanisms of inhibition include substrate competitive binding and ATP competitive binding. AG1024 and AG1034 are among the first IGF-IR tyrosine kinase inhibitors (tyrphostins) reported [34]. They are synthetic molecules derived from benzylidene malononitrile nucleus, which resembles the phenolic group on tyrosine. AG1024 and 1034 are capable of inhibiting receptor autophosphorylation and cell proliferation. Although they are not IGF-IR specific, they do show slight preference for IGF-IR. Another inhibitor AG538 and its bioisosteres represent substrate competitive binding inhibitors of IGF-IR and IR [35]. An IGF-IR-specific inhibitor, cyclolignan picropodophyllotoxin (cyclolignan PPP), was identified from plants [36]. Cyclolignan PPP exclusively inhibits phosphorylation of Y1136 of IGF-IR beta subunit and Akt and promotes apoptosis. In a melanoma xenograft model, cyclolignan PPP causes tumor regression and inhibits metastasis [37]. BMS-536924 and BMS-554417 have equal potency against IGF-IR and IR with in vivo antitumor activity [38,39]. A newly reported small molecule inhibitor, PQIP, displays a 14 fold selectivity of IGF-IR over IR. PQIP is 1,3-disubstituted-8-amino-imidazopyrazine derivative. It abolishes IGF signals and activities of phosphorylated ERK and Akt, and inhibits growth of IGF-IR dependent tumor in a mouse model. PQIP also blocks IGF-II/IGF-IR autocrine loop which could be a choice for treating IGF-II-dependent tumors [40].

Recently, two inhibitors, NVP-AEW541 and NVP-ADW742 (Novartis Parma AG), have been reported to have much higher potency to IGF-IR than to IR (at least 27 folds in cellular tests) [13,41]. The two inhibitors belong to the pyrrolo[2,3-*d*]pyrimidine class, and inhibit growth of a number of tumor cell types in vitro. Multiple myeloma cells are particularly sensitive to these inhibitors. In mouse xenograft model, survival benefits have been shown with NVP-AEW 541 or NVP-ADW742 alone. In addition, NVP-ADW-742 maintains antitumor activity when given orally to SCID/NOD mice with multiple myeloma. Therefore, NVP-AEW541 and NVP-ADW742 may prove to be good candidate for further development.

Because IR is involved in IGF-II signaling, it is not clear whether an IGF-IR-specific inhibitor will be sufficient to block all of IGF- signals in clinic. In such cases, inhibitors targeting at both IR and IGF-IR may be considered, as long as care is taken to monitor the glucose metabolism of patients. All of IGF-IR inhibitors mentioned here have shown anti-tumor activity in vitro and in animal models, yet they have not been tested in clinic. It is important that safety and similar efficacy be observed in patients.

The Ligands as Targets for Cancer Therapy

Insulin, IGF-I and II are the ligands of the IGF system. IGF-I (79 a.a.) and IGF-II (67 a.a.) are both single polypeptide in their mature forms. Unlike insulin, their B-chain and A-chain are connected by the C-domain in IGFs (Figure 3). There is a D-domain at the end of the polypeptide, which is not found in insulin. The B-chain of IGFs is responsible for binding to IGF-IR and IGF binding proteins, whereas the A-chain also contributes to the interaction with IGF-IR [42].

The expression and function of the IGFs show different profiles between rodents and humans. In mice, IGF-II is expressed at high levels during embryogenesis, and reduced rapidly after birth except for the choroids plexus and leptomeninges in adult brain. IGF-I in murine is low during the prenatal period, increases after birth, peaks at puberty and declines with the age. This pattern is not seen in human, where both ligands are detectable in plasma throughout their lives. In general, in humans there is higher level of total IGF-II than IGF-I in plasma.

Because of the pro-growth and anti-apoptosis signals IGFs transduce, the ligands have been linked to tumorigenesis and studied as targets for cancer treatment. This link was validated with the epidemiological data demonstrating that men in the highest quartile of serum IGF-I levels have higher risk of prostate cancer compared with men in the lowest quartile [44]. These results from this 1998 epidemiological study have inspired many groups to investigate in depth the possible role of IGFs in tumorigenesis. The later studies have revealed somewhat complex role for IGF-I and unexpected indications for IGF-II as an important player in tumor development. In 1994 and 1995, two groups reported increased frequency of tumor in transgenic mice expressing IGF-II [45,46]. Rogler et al used a major urinay protein promoter, and high amount of IGF-II was made in liver and pituitary gland and secreted into blood. Although high levels of IGF-II was maintained in blood, most tumors developed only after 18 months of age among the test mice. The predominant tumors belong to liver cancer and leukemia [46]. Mammary gland tumor, frequently induced by high levels of IGF-II, was not found. Bates et al used a beta-lactoglobulin promoter, which allowed locally high expression of IGF-II in mammary gland of transgenic mice, and they observed mammary tumors much earlier, at 4~6 months [45]. While plasma IGF-II levels in mice of this study were higher than these in normal mice, they were not as high as these in mice of the Rogler study. Yet tumors were detected at much earlier time and at the organ where IGF-II was highly expressed. The variations in tumor occurrence timeline in these two studies underscore the notion that the most relevant IGF levels in tumor promotion may be tissue levels rather than plasma levels [42]. These results also indicate that autocrine and paracrine mechanisms of IGF-II play a critical role in tumor growth.

```
IGF-II     ~~~~ayrpse tlcggelvdt lqfvcgdrgf yfsrpasrvs ..rrs..rgi
IGF-I    lltftssttagpe tlcgaelvda lqfvcgprgf yfnkptgygs sirrapqtgi
insulin    ~~~~~~fvnq hlcgshlvea lylvcgergf fy........ ....tpktgi
                        B                            C

IGF-II   veeccfrscd lalletycat pakse
IGF-I    vdeccfrscd lrrlemycap lkpaks~~~~~
insulin  veqcctsics lyqlenycn~ ~~~~~
                  A              D
```

Figure 3. Alignment of insulin, IGF-I and II polypeptides. Insulin B chain and A chain are underlined. C and D domains are labeled. Sources of protein sequences are insulin (AAA59172), IGF-I (CAA40092) and IGF-II (P01344) from NCBI protein database.

Both IGF-I and -II are synthesized in liver and bones, and secreted into the circulation. However, only IGF-I is known to be tightly regulated by human Growth Hormone (hGH) through hypothalamus-pituitary-liver axis [43].

Both IGF-I and -II are produced by stromal elements of normal breast tissues. However, in malignant breast cancers IGF-II, not IGF-I, is the predominant type of IGFs.

In colorectal cancer patients, there are increases in both IGF-II and IGFBP-3, whereas IGF-I level did not change compared with healthy volunteers [47]. Similarly, in patients with Ewing sarcoma, serum IGF-I is reduced [48]. Studies in cultured lung and breast cancer cells have found that although most tested cell lines have both IGF-I and IGF-II mRNA, only IGF-II protein, particularly long IGF-II, was expressed by these cells [49]. In highly proliferative glioblastomas, IGF-II and phosphoinositide 3-kinase regulatory subunit 3 (PI3KR3) are over-expressed. Signaling from IGF-II/PI3KR3 promotes the growth of high grade human glioblastoma independent of EGFR amplification [50]. Mutations in the IGF-IR or in the ligands have not been reported. Overexpressed IGF-IR still needs ligands to be active. For reasons described above, IGFs, especially IGF-II, present a window to suppress signals from cascading to downstream targets.

Methods to Reduce Signaling from Igfs

Given the critical role of IGFs in cell proliferation and tumorigenesis, a number of therapeutic agents have been attempted to reduce their levels. These include endocrine small molecule antagonists targeting the GH (growth hormone) and GHRH (growth hormone release hormone) [51], and monoclonal antibodies specific to IGFs. For IGF-I, a good example is pegvisomant (brand name Somavert), a recombinant human GH, which has been used to treat acromegaly, through binding to GH receptor and blocking the effect of native GH. In acromegaly patients and healthy volunteers, pegvisomant significantly reduces IGF-I levels [52]. In addition, it has been demonstrated to reduce xenograft MCF-7 tumor size in mice [53]. However, to date pegvisomant has only been used in acromegaly patients, and how it will benefit cancer patients remains to be tested. Another IGF-I lowering drug OncoLAR, a long lasting IGF-I nona-peptide analog [54] is in phase III trial with neuroendocrine tumors. Similar drug for IGF-II has not been reported.

Antibodies to IGFs have been reported recently. A fully-human IGF-II specific antibody is capable of abolishing IGF-II induced activation of IGF-IR and inhibiting growth of cells in vitro. It also inhibits cell motility [55]. A rat monoclonal antibody recognizing human IGF-I, -II, and mouse IGF-II has been shown to inhibit IGFs-induced phosphorylation of IGF-IR in cultured cell, and suppress tumor growth in SCID mice [56]. These ligand-specific antibodies are still at their early stages compared with the receptor antibodies. Compared to receptor-specific antibodies they could have at least two advantages. Firstly, ligand antibodies shift the equilibration of free and binding protein-bound ligand in circulation and further change the balance of free ligand across the vascular barrier. Therefore, they have potential to reduce intra-tumor ligand level without penetrating to the solid tumors. Secondly, most antibodies targeting the IGF-IR down-regulate the receptor levels through internalization and degradation, and in many cases they also down-regulate the insulin receptor levels, which could potentially lead to type II diabetes. The ligand-specific antibodies leave both receptors intact. These antibodies hold promising future in cancer treatment. Compared with other therapies such as recombinant pegvisomant, fully human antibody is able to reach high plasma concentration and still has relatively mild side effects.

The fate of the ligand-antibody complex in circulation is a subject worth investigating. The stable nature of fully human antibody may prolong the half life of antibody-bound IGFs. Whether this will affect the benefit of antibody treatment remains to be determined. To this end, studies on an anti-angiogenesis antibody, Bevacizumab, may shed light to it [57]. In kidney cancer patients treated with Bevacizumab, a humanized VEGF antibody, which have clinical benefits, the levels of VEGF (not distinguishing free and antibody-bound forms) in circulation are actually higher than these in untreated people in a dose-dependent fashion. However, with the presence of antibody bound-VEGF in circulation, the overall short-term progression-free survival is improved with Bevacizumab treatment. The steady state of VEGF-Bevacizumab makes it possible for some VEGF to dissociate from the antibody-antigen complex and become available to its receptor. We have hypothesized that if the antibody is engineered in such a way that it accelerates the clearance of ligand-antibody complex from circulation without shortening the half life of the free antibodies, the efficacy of the antibody treatment may be greater.

IGF Binding Proteins as Possible Target

Over the years IGF binding proteins (IGFBPs) have been expanded into a sixteen-member superfamily with six of the BPs relatively better characterized. Among the six well-characterized BPs, BP3 is the most abundant binding protein in serum. It carries 75% or more serum IGFs in trimeric complex with IGFs and acid labile subunit. The ternary complex is confined to circulation, whereas the free BPs and binary complex of BP and IGFs are able to pass the vasculature barrier [58]. Mature IGFBPs have 216 to 289-amino acid residues, and are found only extracellularly. They have higher affinity ($\sim 10^{-10}$ M) to IGFs than IGF-IR does, while they have very low affinity to insulin. The main physiological functions of IGFBPs are to sequester IGFs from their receptors and prolong their plasma half life. They have growth inhibitory effects by competitively binding IGFs and preventing them from activating IGF-IR. High ratio of IGF-I to IGFBP3 increases the risk for the development of several cancers, such as breast, colon, prostate and lung [44]. Preclinical studies indicate that interventions that augment IGFBP3 function inhibit tumor growth in xenograft model [59]. Yee et al has reported that BP1 inhibits IGF-I-induced proliferation and motility of breast cancer cells in vivo [60]. When specific inhibition of IGF-IR is required, BPs may be an option as insulin is not bound by BPs. Increased IGF-IR signaling is often associated with developing of Trastuzumab (a recombinant humanized monoclonal antibody directed against epidermal growth factor receptor 2) resistance. IGFBP3 may be useful to treat Trastuzumab-resistant breast cancers where IGF-IR is up-regulated. In addition, recombinant BP3 displays synergistic antitumor activity with Trasuzumab [59]. Thus, a combination therapy of IGFBP and Trastuzumab may be beneficial.

By using IGF-I analogue with reduced binding affinity to BPs, Clemmons et al found that BP2, BP4 and BP5 all inhibit IGF-I-induced DNA synthesis and migration of vascular smooth muscle cells. BP5 also stimulate migration of vascular smooth muscle cells in an IGF-I-independent manner [61]. Such IGFs-independent activities of BPs have been repeatedly observed. IGFBP3 is reported to have IGF-IR-independent growth inhibition in mouse fibroblast cell lines [62,63]. They are capable of interacting with integrin and mediating cell motility [64]. Recently, several groups have found that IGFBPs regulate survival of stem cells [65]. IGFBPs are known to have complex post-translational modification such as glycosylation, phosphorylation and proteolysis. Changes in glycosylation and/or proteolysis status of BPs affect their stability in circulation [66]. With the complexity of IGFBP functions and challenge to produce large quantity of IGFBP, we may still be years away from utilizing them as therapeutics.

Conclusion

It has been well defined that signals from the IGF pathway are involved in tumorigenesis. Now the task ahead of us is to find ways to block the signals efficiently in cancer patients, without harming the normal cells. Both antibodies and small molecular inhibitors have such potentials. The IGF-IR antibody by Pfizer CP751,871 is at the most advanced stage of clinical studies among the other IGF-IR antibodies, and it has ability to inhibit the hybrid receptors and does not support ADCC. CP751,871 may be the model of antibodies that target at IGF-IR. The two IGF-IR- specific inhibitors NVP-AEW541 and NVP-ADW742 have shown tumoricidal effect in animal models. It is hoped that they will prevail in clinical trials.

Due to the complexity of the IGF system (multiple ligands and each activates multiple receptors), it will be difficult to completely abolish signals from this pathway with monotherapy. Thus, the future of IGF-IR therapy may be in combination therapy. For example, treatment with mTOR inhibitor rapamycin in human rhabdomyosarcoma cell lines induces IGF-IR-dependent activation of Akt [67]. However, simultaneous treatment with rapamycin and h7C10, an IGF-IR antibody, result in additive inhibition of growth and survival in these cells. It has been well documented that in patients who develop resistance to Trastuzumab, up-regulation of IGF-IR is observed. Adding IGF-IR antibodies to Trastuzumab therapy may delay or overcome resistance.

Like all "designer" drugs, it is important to identify tumors that rely on IGF signaling, and only these tumors will benefit from IGF therapy. A number of in vitro studies have provided us with vast amount of information relating cancer cell growth to up-regulated signals from IGF system. Whether the same parameters could be applied to patients remains to be determined. We are fortunate that there are multiple ongoing clinical studies of IGF-IR antibodies. Information from these clinical studies should be helpful.

References

[1] Rinderknecht E, Humbel RE: Polypeptides with Nonsuppressible Insulin-Like and Cell-Growth Promoting Activities in Human-Serum - Isolation, Chemical Characterization, and Some Biological Properties of Forms 1 and 2. *Proc. Natl. Acad. Sci. U S A* 1976, 73:2365.

[2] Adams TE, Epa VC, Garrett TP, Ward CW: Structure and function of the type 1 insulin-like growth factor receptor. *Cell Mol. Life Sci.* 2000, 57:1050-1093.

[3] Morgan DO, Edman JC, Standring DN, Fried VA, Smith MC, Roth RA, Rutter WJ: Insulin-like growth factor II receptor as a multifunctional binding protein. *Nature* 1987, 329:301-307.

[4] LeRoith D, Roberts CT, Jr.: The insulin-like growth factor system and cancer. *Cancer Lett.* 2003, 195:127-137.

[5] LeRoith DaRJ, C.T.: The insulin-like growth factor system and cancer. *Cancer Letters* 2003, 195:127-137.

[6] Chappell SA, Walsh T, Walker RA, Shaw JA: Loss of heterozygosity at chromosome 6q in preinvasive and early invasive breast carcinomas. *Br. J. Cancer* 1997, 75:1324-1329.

[7] Di Cola G, Cool MH, Accili D: Hypoglycemic effect of insulin-like growth factor-1 in mice lacking insulin receptors. *J. Clin. Invest.* 1997, 99:2538-2544.

[8] Kim JJ, Accili D: Signalling through IGF-I and insulin receptors: where is the specificity? *Growth Horm IGF Res.* 2002, 12:84-90.

[9] Pandini G, Frasca F, Mineo R, Sciacca L, Vigneri R, Belfiore A: Insulin/insulin-like growth factor I hybrid receptors have different biological characteristics depending on the insulin receptor isoform involved. *J. Biol. Chem.* 2002, 277:39684-39695.

[10] Liu JP, Baker J, Perkins AS, Robertson EJ, Efstratiadis A: Mice Carrying Null Mutations of the Genes Encoding Insulin-Like Growth Factor-I (Igf-1) and Type-1 Igf Receptor (Igf1R). *Cell* 1993, 75:59.

[11] Sell C, Dumenil G, Deveaud C, Miura M, Coppola D, DeAngelis T, Rubin R, Efstratiadis A, Baserga R: Effect of A Null Mutation of the Insulin-Like Growth-Factor-I Receptor Gene on Growth and Transformation of Mouse Embryo Fibroblasts. *Mol. Cell Biol.* 1994, 14:3604.

[12] Carboni JM, Lee AV, Hadsell DL, Rowley BR, Lee FY, Bol DK, Camuso AE, Gottardis M, Greer AF, Ho CP, et al.: Tumor development by transgenic expression of a constitutively active insulin-like growth factor I receptor. *Cancer Res.* 2005, 65:3781-3787.

[13] Mitsiades CS, Mitsiades NS, McMullan CJ, Poulaki V, Shringarpure R, Akiyama M, Hideshima T, Chauhan D, Joseph M, Libermann TA, et al.: Inhibition of the insulin-like growth factor receptor-1 tyrosine kinase activity as a therapeutic strategy for multiple myeloma, other hematologic malignancies, and solid tumors. *Cancer Cell.* 2004, 5:221.

[14] Reinmuth N, Liu W, Fan F, Jung YD, Ahmad SA, Stoeltzing O, Bucana CD, Radinsky R, Ellis LM: Blockade of insulin-like growth factor I receptor function inhibits growth and angiogenesis of colon cancer. *Clin. Cancer Res.* 2002, 8:3259-3269.

[15] Resnicoff M, Coppola D, Sell C, Rubin R, Ferrone S, Baserga R: Growth inhibition of human melanoma cells in nude mice by antisense strategies to the type 1 insulin-like growth factor receptor. *Cancer Res.* 1994, 54:4848-4850.

[16] Sachdev D, Hartell JS, Lee AV, Zhang X, Yee D: A dominant negative type I insulin-like growth factor receptor inhibits metastasis of human cancer cells. *J. Biol. Chem.* 2004, 279:5017-5024.

[17] Seely BL, Samimi G, Webster NJ: Retroviral expression of a kinase-defective IGF-I receptor suppresses growth and causes apoptosis of CHO and U87 cells in-vivo. *BMC Cancer* 2002, 2:15.

[18] Flier JS, Usher P, Moses AC: Monoclonal-Antibody to the Type-I Insulin-Like Growth-Factor (Igf-I) Receptor Blocks Igf-I Receptor-Mediated Dna-Synthesis - Clarification of the Mitogenic Mechanisms of Igf-I and Insulin in Human-Skin Fibroblasts. *Proc. Natl. Acad. Sci. U S A* 1986, 83:664.

[19] Scotlandi K, Benini S, Nanni P, Lollini PL, Nicoletti G, Landuzzi L, Serra M, Manara MC, Picci P, Baldini N: Blockage of insulin-like growth factor-I receptor inhibits the growth of Ewing's sarcoma in athymic mice. *Cancer Res.* 1998, 58:4127-4131.

[20] Zia F, Jacobs S, Kull F, Jr., Cuttitta F, Mulshine JL, Moody TW: Monoclonal antibody alpha IR-3 inhibits non-small cell lung cancer growth in vitro and in vivo. *J Cell Biochem. Suppl.*.1996, 24:269-275.

[21] Cohen BD, Baker DA, Soderstrom C, Tkalcevic G, Rossi AM, Miller PE, Tengowski MW, Wang F, Gualberto A, Beebe JS, et al.: Combination therapy enhances the

inhibition of tumor growth with the fully human anti-type 1 insulin-like growth factor receptor monoclonal antibody CP-751,871. *Clin. Cancer Res.* 2005, 11:2063.

[22] Haluska P, Shaw H, Batzel GN, Molife LR, Adjei AA, Yap TA, Roberts ML, Gualberto A, de Bono JS: Phase I dose escalation study of the anti-IGF-1R monoclonal antibody CP-751,871 in patients with refractory solid tumors. *J Clin Oncol 2007 ASCO Annual Meeting Proceedings Part I.* 2007, 25:3586.

[23] Pollak MN, Lacy MQ, Lipton A, Demers L, Leitzel K, de Bono JS, Yin D, Roberts L, Sharma A, Gualberto A: Pharmacodynamic properties of the anti-IGF-IR monoclonal antibody CP-751,871 in cancer patients. *J. Clin. Oncology, 2007 ASCO Annual Meeting Proceedings Part I.* 2007, 25:3587.

[24] Karp DD, Paz-Ares LG, Blakely LJ, Kreisman H, Eisenberg PD, Cohen RB, Garland L, Langer CJ, Melvin CL, Gualberto A: Efficacy of the anti-insulin like growth factor I receptor (IGF-IR) antibody CP-751871 in combination with paclitaxel and carboplatin as first-line treatment for advanced non-small cell lung cancer (NSCLC). *J Clin Oncol 2007 ASCO Annual Meeting Proceedings Part I.* 2007, 25:7506.

[25] Burtrum D, Zhu Z, Lu D, Anderson DM, Prewett M, Pereira DS, Bassi R, Abdullah R, Hooper AT, Koo H, et al.: A fully human monoclonal antibody to the insulin-like growth factor I receptor blocks ligand-dependent signaling and inhibits human tumor growth in vivo. *Cancer Res.* 2003, 63:8912.

[26] Higano CS, Yu EY, Whiting SH, Gordon MS, LoRusso P, Fox F, Katz TL, Roecker JM, Schwartz MA: A phase I, first in man study of weekly IMC-A12, a fully human insulin like growth factor-I receptor IgG1 monoclonal antibody, in patients with advanced solid tumors. *J. Clin. Oncol. 2007 ASCO Annual Meeting Proceedings Part I.* 2007, 25:3505.

[27] Beltran P, Lu J, cajuilis E, Chung Y, Tsai M, Chen C, Deshpande R, Kendall R, Radinsky R, Calzone F: AMG 479, a fully human anti IGF-i receptor monoclonal antibody, enhances the response of established colon and pancreatic xenografts to chemotherapeutic agents. *Eur. J. Cancer Suppl.* 2006, 4:64.

[28] Schnitzer T, Keuenkele K, Rebers F, Van Vugt M, Klein C, Lanzendoerfer M, Mundigl O, Parren P, van de Winkle J, Schumacher R: Characterization of a recombinant, fully human monoclonal antibody directed against the human insulin-like growth factor receptor. *Eur. J. Cancer Suppl.* 2006, 4:66.

[29] Tolcher AW, Rothenberg ML, Rodon J, Delbeke D, Patnaik A, Nguyen L, Young F, Hwang Y, Haqq C, Puzanov I: A phase I pharmacokinetic and pharmacodynamic study of AMG 479, a fully human monoclonal antibody against insulin-like growth factor type 1 receptor (IGF-1R), in advanced solid tumors. *J Clin Oncol 2007 ASCO Annual Meeting Proceedings Part I.* 2007, 25:3002.

[30] Rodon J, Patnaik A, Stein M, Tolcher A, Ng C, Dias C, Kurman M, Greig G, Kurzrock R, Rubin R: A phase I study of q3W R1507, a human monoclonal antibody IGF-1R antagonist in patients with advanced cancer. *J. Clin. Oncology, 2007 ASCO Annual Meeting Proceedings Part I.* 2007, 25:3590.

[31] Goetsch L, Gonzalez A, Leger O, Beck A, Pauwels PJ, Haeuw JF, Corvaia N: A recombinant humanized anti-insulin-like growth factor receptor type I antibody

(h7C10) enhances the antitumor activity of vinorelbine and anti-epidermal growth factor receptor therapy against human cancer xenografts. *Int. J. Cancer* 2005, 113:316.

[32] Maloney EK, McLaughlin JL, Dagdigian NE, Garrett LM, Connors KM, Zhou XM, Blattler WA, Chittenden T, Singh R: An anti-insulin-like growth factor I receptor antibody that is a potent inhibitor of cancer cell proliferation. *Cancer Res.* 2003, 63:5073-5083.

[33] Bondy CA, Cheng CM: Signaling by insulin-like growth factor 1 in brain. *Eur.J. Pharmacol.* 2004, 490:25-31.

[34] Parrizas M, Gazit A, Levitzki A, Wertheimer E, LeRoith D: Specific inhibition of insulin-like growth factor-1 and insulin receptor tyrosine kinase activity and biological function by tyrphostins. *Endocrinology* 1997, 138:1427-1433.

[35] Blum G, Gazit A, Levitzki A: Development of new insulin-like growth factor-1 receptor kinase inhibitors using catechol mimics. *J. Biol. Chem.* 2003, 278:40442.

[36] Vasilcanu D, Girnita A, Girnita L, Vasilcanu R, Axelson M, Larsson O: The cyclolignan PPP induces activation loop-specific inhibition of tyrosine phosphorylation of the insulin-like growth factor-1 receptor. Link to the phosphatidyl inositol-3 kinase/Akt apoptotic pathway. *Oncogene* 2004, 23:7854-7862.

[37] Girnita A, All-Ericsson C, Economou MA, Astrom K, Axelson M, Seregard S, Larsson O, Girnita L: The insulin-like growth factor-I receptor inhibitor picropodophyllin causes tumor regression and attenuates mechanisms involved in invasion of uveal melanoma cells. *Clin. Cancer Res.* 2006, 12:1383-1391.

[38] Haluska P, Carboni JM, Loegering DA, Lee FY, Wittman M, Saulnier MG, Frennesson DB, Kalli KR, Conover CA, Attar RM, et al.: In vitro and in vivo antitumor effects of the dual insulin-like growth factor-I/insulin receptor inhibitor, BMS-554417. *Cancer Res.* 2006, 66:362.

[39] Wittman M, Carboni J, Attar R, Balasubramanian B, Balimane P, Brassil P, Beaulieu F, Chang CY, Clarke W, Dell J, et al.: Discovery of a 1H-benzoimidazol-2-yl)-1H-pyridin-2-one (BMS-536924) inhibitor of insulin-like growth factor I receptor kinase with in vivo antitumor activity. *J. Medicinal Chem.* 2005, 48:5639.

[40] Ji QS, Mulvihill MJ, Rosenfeld-Franklin M, Cooke A, Feng L, Mak G, O'Connor M, Yao Y, Pirritt C, Buck E, et al.: A novel, potent, and selective insulin-like growth factor-I receptor kinase inhibitor blocks insulin-like growth factor-I receptor signaling in vitro and inhibits insulin-like growth factor-I receptor dependent tumor growth in vivo. *Mol. Cancer Ther.* 2007, 6:2158-2167.

[41] Garcia-Echeverria C, Pearson MA, Marti A, Meyer T, Mestan J, Zimmermann J, Gao JP, Brueggen J, Capraro HG, Cozens R, et al.: In vivo antitumor activity of NVP-AEW541 - A novel, potent, and selective inhibitor of the IGF-IR kinase. *Cancer Cell* 2004, 5:231.

[42] Rechler MM, Nissley SP: *Insulin-like growth factors in "Peptide Growth Factors and Their Receptors"*, vol 95. Edited by Sporn MB, Roberts AB. Berlin: Springer-Verlag; 1990.

[43] Daughaday WH, Rotwein P: Insulin-Like Growth Factor-I and Factor-Ii - Peptide, Messenger Ribonucleic-Acid and Gene Structures, Serum, and Tissue Concentrations. *Endocr. Re.v* 1989, 10:68.

[44] Chan JM, Stampfer MJ, Giovannucci E, Gann PH, Ma J, Wilkinson P, Hennekens CH, Pollak M: Plasma insulin-like growth factor I and prostate cancer risk: A prospective study. *Science* 1998, 279:563-566.
[45] Bates P, Fisher R, Ward A, Richardson L, Hill DJ, Graham CF: Mammary-Cancer in Transgenic Mice Expressing Insulin-Like Growth-Factor-Ii (Igf-Ii). *Br. J. Cancer* 1995, 72:1189.
[46] Rogler CE, Yang DY, Rossetti L, Donohoe J, Alt E, Chang CJ, Rosenfeld R, Neely K, Hintz R: Altered Body-Composition and Increased Frequency of Diverse Malignancies in Insulin-Like Growth Factor-Ii Transgenic Mice. *J Biol. Chem.* 1994, 269:13779.
[47] Elatiq F, Garrouste F, Remaclebonnet M, Sastre B, Pommier G: Alterations in Serum Levels of Insulin-Like Growth-Factors and Insulin-Like Growth-Factor-Binding Proteins in Patients with Colorectal-Cancer. *Int. J. Cancer* 1994, 57:491.
[48] Toretsky JA, Steinberg SM, Thakar M, Counts D, Pironis B, Parente C, Eskenazi A, Helman L, Wexler LH: Insulin-like growth factor type 1 (IGF-1) and IGF binding protein-3 in patients with ewing sarcoma family of tumors. *Cancer* 2001, 92:2941.
[49] Quinn KA, Treston AM, Unsworth EJ, Miller MJ, Vos M, Grimley C, Battey J, Mulshine JL, Cuttitta F: Insulin-like growth factor expression in human cancer cell lines. *J. Biol. Chem.* 1996, 271:11477.
[50] Soroceanu L, Kharbanda S, Chen R, Soriano RH, Aldape K, Misra A, Zha J, Forrest WF, Nigro JM, Modrusan Z, et al.: Identification of IGF2 signaling through phosphoinositide-3-kinase regulatory subunit 3 as a growth-promoting axis in glioblastoma. *Proc. Natl. Acad. Sci.U S A* 2007, 104:3466-3471.
[51] Braczkowski R, Schally AV, Plonowski A, Varga JL, Groot K, Krupa M, Armatis P: Inhibition of proliferation in human MNNG/HOS osteosarcoma and SK-ES-1 Ewing sarcoma cell lines in vitro and in vivo by antagonists of growth hormone-releasing hormone - Effects on insulin-like growth factor II. *Cancer* 2002, 95:1735.
[52] Yin D, Vreeland F, Schaaf LJ, Millham R, Duncan BA, Sharma A: Clinical pharmacodynamic effects of the growth hormone receptor antagonist pegvisomant: implications for cancer therapy. *Clin. Cancer Res* 2007, 13:1000-1009.
[53] Divisova J, Kuiatse I, Lazard Z, Weiss H, Vreeland F, Hadsell DL, Schiff R, Osborne CK, Lee AV: The growth hormone receptor antagonist pegvisomant blocks both mammary gland development and MCF-7 breast cancer xenograft growth. *Breast Cancer Res. Treat.* 2006, 98:315-327.
[54] Khanna C, Prehn J, Hayden D, Cassaday RD, Caylor J, Jacob S, Bose SM, Hong SH, Hewitt SM, Helman LJ: A randomized controlled trial of octreotide pamoate long-acting release and carboplatin versus carboplatin alone in dogs with naturally occurring osteosarcoma: evaluation of insulin-like growth factor suppression and chemotherapy. *Clin. Cancer Res.* 2002, 8:2406-2412.
[55] Feng Y, Zhu Z, Xiao X, Choudhry V, Barrett JC, Dimitrov DS: Novel human monoclonal antibodies to insulin-like growth factor (IGF)-II that potently inhibit the IGF receptor type I signal transduction function. *Mol. Cancer Ther.*2006, 5:114-120.
[56] Goya M, Miyamoto S, Nagai K, Ohki Y, Nakamura K, Shitara K, Maeda H, Sangai T, Kodama K, Endoh Y, et al.: Growth inhibition of human prostate cancer cells in human adult bone implanted into nonobese diabetic/severe combined immunodeficient mice

by a ligand-specific antibody to human insulin-like growth factors. *Cancer Res.* 2004, 64:6252.

[57] Yang JC, Haworth L, Sherry RM, Hwu P, Schwartzentruber DJ, Topalian SL, Steinberg SM, Chen HX, Rosenberg SA: A randomized trial of bevacizumab, an anti-vascular endothelial growth factor antibody, for metastatic renal cancer. *New Engl. J. Medicine* 2003, 349:427.

[58] Firth SM, Baxter RC: Cellular actions of the insulin-like growth factor binding proteins. *Endocr. Rev.* 2002, 23:824-854.

[59] Jerome L, Alami N, Belanger S, Page V, Yu QN, Paterson J, Shiry L, Pegram M, Leyland-Jones B: Recombinant human insulin-like growth factor binding protein 3 inhibits growth of human epidermal growth factor receptor-2-overexpressing breast tumors and potentiates Herceptin activity in vivo. *Cancer Res.* 2006, 66:7245.

[60] Zhang X, Yee D: Insulin-like growth factor binding protein-1 (IGFBP-1) inhibits breast cancer cell motility. *Cancer Res.* 2002, 62:4369-4375.

[61] Hsieh T, Gordon RE, Clemmons DR, Busby WH, Duan CM: Regulation of vascular smooth muscle cell responses to insulin-like growth factor (IGF)-I by local IGF-binding proteins. *J. Biol. Chem.* 2003, 278:42886.

[62] Butt AJ, Fraley KA, Firth SM, Baxter RC: IGF-binding protein-3-induced growth inhibition and apoptosis do not require cell surface binding and nuclear translocation in human breast cancer cells. *Endocrinology* 2002, 143:2693-2699.

[63] Valentinis B, Bhala A, DeAngelis T, Baserga R, Cohen P: The human insulin-like growth factor (IGF) binding protein-3 inhibits the growth of fibroblasts with a targeted disruption of the IGF-I receptor gene. *Mol. Endocrinol.* 1995, 9:361-367.

[64] Maile LA, Imai Y, Clarke JB, Clemmons DR: Insulin-like growth factor I increases alpha(v)beta(3) affinity by increasing the amount of integrin-associated protein that is associated with non-raft domains of the cellular membrane. *J. Biol. Chem.* 2002, 277:1800.

[65] Chang KH, Chan-Ling T, McFarland EL, Afzal A, Pan H, Baxter LC, Shaw LC, Caballero S, Sengupta N, Li Calzi S, et al.: IGF binding protein-3 regulates hematopoietic stem cell and endothelial precursor cell function during vascular development. *Proc. Natl. Acad. Sci. U S A* 2007, 104:10595-10600.

[66] Clemmons DR, Sleevi M, Busby WH: Recombinant, nonglycosylated human insulin-like growth factor-binding protein-3 (IGFBP-3) is degraded preferentially after administration to type II diabetics, resulting in increased endogenous glycosylated IGFBP-3. *J. Clin. Endocrinol. Metab.* 2005, 90:6561.

[67] Wan X, Harkavy B, Shen N, Grohar P, Helman LJ: Rapamycin induces feedback activation of Akt signaling through an IGF-1R-dependent mechanism. *Oncogene* 2007, 26:1932-1940.

In: Oncogene Proteins: New Research
Editors: A. H. Malloy and E.C. Carson, pp.177-193 ISBN 978-1-60456-290-3

Chapter 6

The Role of the eEF1A Family in Human Cancers

B. Scaggiante[(1)], B. Dapas[(1)], L. Cristiano[(1)] and G. Grassi[(1,2)]

[1.] Department of Biochemical, Biophysical and Macromolecular Chemistry, University of Trieste, via Giorgeri 1, Trieste, Italy

[2.] Department of Clinical, Morphological and Technological Sciences, Division of Internal Medicine; University Hospital of Cattinara, strada di Fiume 447, Trieste, Italy

Abstract

Among the *EEF1A* family, two constitutive and actively transcribed genes are known: *EEF1A1* on chromosome 6q14 and *EEF1A2* on chromosome 20q13.3. eEF1A carries aa-tRNA as a ternary complex eEF1A-GTP-aa-tRNA allowing to correct positioning of the aa-tRNA in the A site of the ribosome. In mammalian, eEF1A1 is ubiquitously expressed, whereas eEF1A2 is developmentally regulated, being its expression switched on in adult life in specialised tissues such as skeletal muscle, cardiomyocytes and nervous system . The expression of *EEF1A2* in tissues other than skeletal muscle, heart and brain has been associated to cancer development and aggressiveness, playing a role in ovarian and in breast cancers. Analogously to *EEF1A2*, over-expression of *EEF1A1* has been proposed to be related to cell proliferation and cancer development. Another transcript belonging to this family has been discovered and proposed for tumour transformation: the prostate tumour-inducing gene-1, alias PTI-1, a truncated form of eEF1A1, lacking of 67 aa at NH_2-end.

In our observations an isoform of eEF1A1 and a de-regulation of eEF1A1 and eEF1A2 expression has been found in human haematopoietic cancer cells, but not in the normal counterpart. In hepatocarcinoma cells we document the increased mRNA levels of *EEF1A* genes compared to normal liver and the higher nuclear-enriched/cytoplasm ratio of eEF1A protein in the less differentiated hepatoma cancer cells. Differently, in prostate cancer lines we observed a dramatic up-regulation of *EEF1A2* and the concomitant down regulation of *EEF1A1*, compared to non cancer prostate cells. In prostate cancer cells our data are in favour of a minor role for PTI-1 to the onset of cancer, while a relation with the de-regulation of eEF1A genes was found. PTI-1 mRNA was indeed detected in some hematopoietic and colonadenocarcinoma cell lines, but its role in *in vivo* human tumors remains to be defined. These observations were discussed with the other available data about the involvement of eEF1A family components in human cancers. In conclusion, the up and/or down-regulation of the expression levels of the two eEF1A genes, as well as post-translational modifications of the eEF1A protein, may play a role in human tumorigenesis and cancer progression but in a different manner, depending on tumour type.

1. Introduction

1.1. eEF1A Canonical Functions

The eukaryotic elongation factor 1A (eEF1A, formerly EF-1α) proteins belong to the G-protein superfamily and play a key role in translation, carrying aminoacyl-tRNA (aa-tRNA) as eEF1A-GTP-aa-tRNA ternary complex to the A site of the ribosome (Abbot and Proud, 2004). Numerous other proteins, such as TCTP ("*translationally controlled tumour protein*" also known as p23), ARS ("aminoacyl-tRNA synthetase") and eEF3, participate to this mechanism. TCTP is a tubulin-binding protein that binds to eEF1A-GDP thus preventing the recruitment of the deacylated t-RNA to the translational machinery. This allows a correct tRNA channelling (Cans et al., 2003). ARS acts on the tRNA to produce acylated-tRNA (Petrushenko et al, 2002). The elongation factor eEF3 aids eEF1A to release aa-tRNA to the ribosome (Anand et al., 2006) (Figure 1).

Two actively transcribed genes are known in the human genome for eEF1A: the first, *EEF1A1*, maps on chromosome 6q14.1 and the second, *EEF1A2*, on chromosome 20q13.3. The two paralogues proteins share 78% coding sequence and 92% protein homology (Lund et al, 1996). The two 5' and 3'-UTR regions are highly divergent between the two genes. Moreover, *EEF1A2* lacks the TATA box and the 5' terminal oligopyrimidine tract (TOP) sequence (Bischoff et al, 2000). The two forms have the same *in vitro* translation activity except for a 7 fold higher dissociation rate constant for GDP showed by eEF1A1 (Kahns at al., 1998) and little or no affinity for eEF1B guanine nucleotide exchange factors demonstrated for eEF1A2 (Mansilla et al. 2002).

In mammalians, eEF1A1 is a protein ubiquitously expressed with the exception of heart, brain and skeletal muscle, where during terminal differentiation eEF1A2 is produced (Knudsen et al., 1993). More recently, eEF1A2 expression was found in normal human pancreas islets, in the base of the gut crypts and in specific region of the lung. Notably, in these tissues eEF1A2 was found to be expressed by specific cells that often have a secretory function (Newbery et al., 2007). In normal cells eEF1A is one of the most abundant proteins after actin, accounting for 1-2% of total cellular protein (Condeelis, 1995).

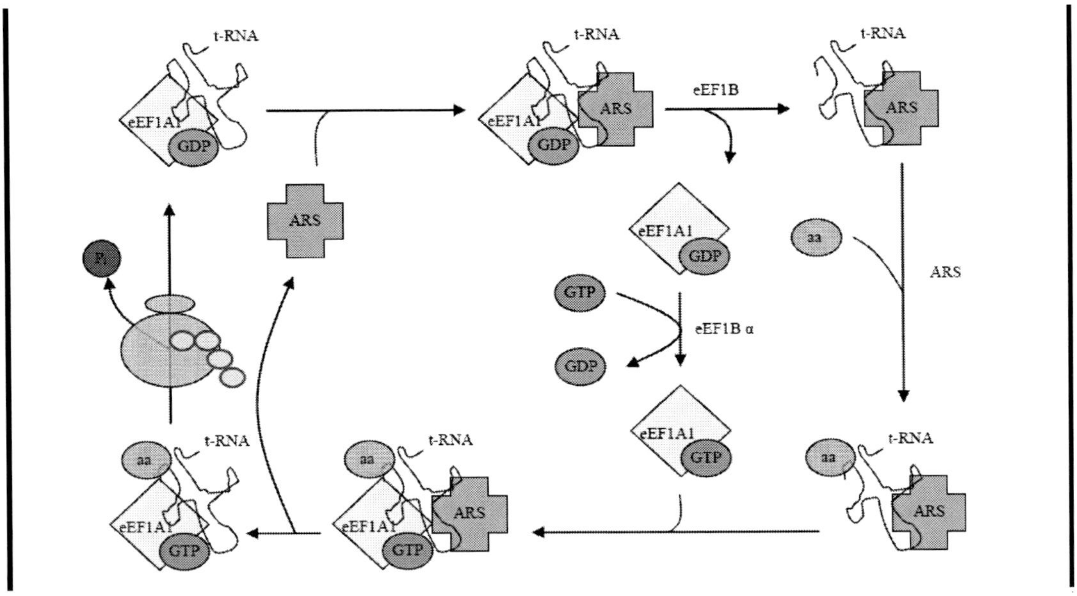

Figure 1. Representative schematic chart for eEF1A1 and some interactive protein partners involved in his canonical function. ARS= aminoacyl t-RNA synthase; eEF1Bα = elongation factor 1B α .

1.2. eEF1A Non Canonical Functions

The role of eEF1As is not confined to protein translation and thus they now appear as multifunctional proteins, especially eEF1A1. Studies in yeast on two closely related eEF1A human genes, i.e. TEF1 and TEF2, confirm that eEF1A1 serves as a direct cytoskeletal modulator that reorganizes microtubules during cell cycle, by bundling of actin filaments and by severing of microtubules (Condeelis et al, 1995; Bektas et al, 1994; Gross and Kinzy, 2005). This function is achieved by an actin-binding site physically separated from the site binding the aa-tRNA (Gross and Kinzy, 2005). The Rho/Rho-kinase pathway modulates eEF1A1 binding to F-actin by phosphorylation (Izawa et al. 2000). Recently, it has been discovered that eEF3 binds to the actin site domain of eEF1A (Anand et al., 2006) and the competition between eEF3 and actin can play a critical role in the modulation of protein translation versus cytoskeletal organization. Moreover, eEF1A1 controls translation fidelity by binding nascent polypeptide chains with an uncorrected folding (Caldas et al., 1998; Hotokezaka et al.2002) and by exhibiting chaperone-like activity to maintain the correct conformation of the macromolecules of the translational compartment (Lukash et al., 2004). eEF1A1 also plays a role in the recognition and degradation of damaged proteins by binding ubiquitinated nascent proteins and by promoting their delivery to the proteosome through the interaction with Rpt1 (Chuang et al., 2005). The eEF1A1 protein is up-regulated by p53 in the process of apoptosis mediated by modification of the cytoskeletal microtubules (Kato et al. 1997; Duttaroy et al. 1998). The silencing of eEF1A1 in Chinese hamster ovary and in rat cardiomyoblast cells under lipotoxic injury confers resistance to oxidative and endoplasmic reticulum (ER) stress, suggesting that eEF1A1 mediates ER-stress induced cell death through a mechanism involving actin cytoskeleton changes (Borradaile et al., 2006). On the contrary,

Talapatra et al., showed that enforced expression of eEF1A1 in a non-transformed murine pro-B cell line confers protection from growth factors withdrawal and ER-stress leading to apoptosis resistance (Talapatra et al., 2002). Thus opposite functions of eEF1A1 on apoptosis can be observed in cells under different stress conditions. Furthermore, eEF1A1 was found to be a co-activator of heat-shock transcription factor 1 (HSF1) suggesting that in consequence of heat shock response, eEF1A1 is released from the collapsed cytoskeleton and/or it becomes available as consequence of translational shutdown, to trigger heat-shock response by binding to HSF1 (Shamovsky et al, 2006). Recently, the human homologue of *mdm2* gene product (HDM2) has been found to bind to eEF1A1 in the cytoplasm of the normal diploid WI38 lung cell line, without altering the translation. HDM2 interacts with several tumour suppressors and it recognizes the transactivation domain of p53 leading to inhibition of p53-mediated transcriptional activation and targeting p53 for ubiquitination. Interestingly, the interaction site of HDM2 for eEF1A1 overlaps the transactivation domain of p53, suggesting that the formation of eEF1A1-HDM2 complex prevents HDM2 from interacting with p53 in the cytoplasm (Frum et al., 2007). Because eEF1A1 is a co-activator of HSF1 (Shamosvsky et al., 2006) it has been hypothesized that HDM2 regulates heat shock response of the cells by binding to eEF1A1 (Frum et al., 2007).

The physiological role of eEF1A2 protein, besides its involvement in translation, is less defined in specialized tissue in adult life. The ectopic over-expression of eEF1A2 protects adult muscle cells from caspase-3 induced apoptosis, whereas eEF1A1 shows apoptotic properties (Ruest et al., 2002; Potter et al. 1998). eEF1A2, as well as eEF1A1, is able to regulate the recycling of M4 muscarinic acetylcholine receptors in adult neurons (McClatchy et al., 2006). Recently, both forms of eEF1A were found to play an as yet un-defined role in peroxisome biology in yeast (Kiel et al. 2007). In mouse tissues, eEF1A2 protein has the unique ability to interact with peroxiredoxin I, potentiating the resistance to oxidative stress-induced cell death and promoting the raise of Akt survival factor (Chang and Wang, 2007). However, eEF1A2 was also found to promote apoptosis in a rat embryonic heart cell line under oxidative stress (Chen et al., 2000). A physiological role in the regulation of PI4K activity has been demonstrated for eEF1A2 as a direct activator of phosphatidylinositol 4-Kinase β (PI4KIIIβ) (Jeganathan and Lee, 2007). Finally, many different potential partners of eEF1A2 have been identified although not yet systematically explored (Chang and Wang, 2007).

Together, the presented data indicate that eEF1A1/2 participate to the regulation of cell survival and growth in numerous pathways by their ability to specifically interact with many proteins and to rapidly raise their protein level either by transcriptional and post-transcriptional events. We believe that their role as pro-apoptotic or anti-apoptotic factors may depend on the phenotype of the cell, on the long-lived or short-lived cell behaviours, as well as on the type of cellular damage.

2. eEF1A in Oncogenesis

Although the role of eEF1A1 in oncogenesis is still under debate, eEF1A2 is recognized as an important oncogene (Thorton et al., 2003) due to its intrinsic ability to modulate kinase signalling pathways (Panasyuk G et al., 2007; Jeganathan and Lee, 2007; Lau et al., 2006; Chang and Wang, 2007). The first evidence of the tumorigenic potential of eEF1A2 came in 2002 when Anand et al. observed that eEF1A2 was expressed in 30% of ovarian tumours but not in the corresponding normal tissue (Anand et al., 2002). Moreover, wild-type expression of eEF1A2 in NIH-3T3 fibroblasts led to cell transformation (Anand et al., 2002). Later, eEF1A2 was found to be highly expressed in 30-60% of human ovary, breast and lung tumours (Lee, 2003; Li et al., 2005; Tomlison et al., 2005; Kulkarni et al., 2006). Interestingly, the over-expression of eEF1A2 was also observed in the majority of human cancer cell lines (Joseph et al., 2004) supporting the hypothesis that the "switch on" of *EEF1A2* gene expression is a common mechanism in tumorigenesis. In other cases, the over-expression of eEF1A2 was related to the differentiation grade of the cancer cells: eEF1A2 transcription levels dramatically increased in an undifferentiated human liver cancer cell line compared to other differentiated hepatoma cells and its over-expression did not confer any increased apoptosis resistance (Grassi et al., 2007).

2.1. eEF1A2 in Ovarian Carcinomas

In primary tumours, the most extensive studies about eEF1A2 are referred to ovarian and breast cancers. With regard to ovarian carcinomas, whereas a large proportion of clear cell carcinoma over-expressed eEF1A2, only a smaller proportion of serous, endometrioid and mucinous tumour cells showed an over-expression of eEF1A2 (Tomlinson et al., 2007). In contrast with previous results from Anand et al. (Anand et al., 2002), Tomlinson et al. (Tomlinson et al., 2007), observed some expression of the gene also in the normal ovary tissue. Interestingly, the authors also showed that not all tumours over-expressing eEF1A2 mRNA display a rise in eEF1A2 protein content. This suggests that in ovarian carcinomas, the over-expression of eEF1A2 does not necessarily lead to the production of a significantly increased quantity of the protein. Notably, a similar behaviour was observed by us with regard to eEF1A1/2 expression in hepatoma cell lines. Here, a strike difference in the mRNA levels evaluated in hepatic cell lines with different phenotypes did not result in a corresponding difference at the protein level (Grassi et al., 2007). It is possible that the increase in *EEF1A1/2* transcription levels provides to the tumour cells an un-translated reservoir useful for a rapid rise of the protein level required for a survival or invasion purposes in response to the surrounding stimuli.

In ovarian cancers the amplification of *EEF1A2* gene did not relate with its over-expression and, notably, gene amplification was found in samples that did not express eEF1A2. Moreover, no activating mutations in the coding sequence have been found in the cases of *EEF1A2* gene over-expression. Finally, the methylation status of the gene was found to be unrelated to the expression level (Tomlison et al. 2007). All these findings suggest that the inappropriate expression of trans-activating factors may be responsible for the over-expression of *EEF1A2* (Tomlinson et al. 2007). In this *scenario* it is interesting to note that in immortalized normal ovarian surface epithelial cells (IOSE) the transfection of the transcription factor ZNF217, frequently amplified in ovarian cancers, increased the expression of *EEF1A2* (Li et al., 2007). This observation further favours the idea that eEF1A2 over-expression may occur *via* a trans-acting factor expressed by tumour cells. In agreement with this, in hepatoma and in prostate adenocarcinoma cell lines, we detected a modest or no significant amplification of *EEF1A2* gene, paralleled by a clear rise in *EEF1A2* mRNA levels (Grassi et al. 2007, Scaggiante et al. unpublished results).

The overall results of Tomlinson et al. (Tomlinson et al. 2007) in ovarian cancers are indicative of the fact that eEF1A2 contributes to the grade of tumour aggressiveness but probably not to the onset of tumorigenesis. This hypothesis is sustained by our results on the hepato-carcinoma cell lines where an inverse relation between *EEF1A2* overexpression and the tumour differentiation grade was found (Grassi et al. 2007).

2.2. eEF1A2 in Breast Carcinomas

The first evidence of eEF1A involvement in breast carcinoma, came from the observation of increased eEF1A levels in breast cancer tissues compared to benign tumour tissues. In particular, eEF1A was related to an increased level of ubiquitin-conjugating enzymes but it was not possible to distinguish between eEF1A1 and 2 forms (Chen and Madura, 2005). Proteosome activity is often elevated in human cancers (Zhang et al., 2004) and eEF1A has been involved in proteosome-mediated degradation of ubiquitinated proteins (Chuang et al. 2005). Thus, it seemed conceivable that eEF1A over-expression could take part to this mechanism, although the contribution of each of the two forms of eEF1A is not known. Tomlison et al. (Tomlinson et al. 2005) demonstrated that eEF1A2 is barely detectable in normal breast, whereas it is over-expressed at different levels in two-third of primary breast tumours, especially those estrogen-positive (ER). This leads to the conclusion that eEF1A2 is a putative oncogene in breast cancer. Notably, there was no significant correlation between eEF1A2 expression level and tumour grade or lymph node positivity. Moreover, the tendency to a positive association with the presence of wild type p53 and eEF1A2 over-expression was observed (Tomlinson et al. 2005).

The fact that the ER positive tumours (usually luminal tumours) have a generally better survival than ER negative ones (usually basal tumours) (Sorlie et al, 2001) suggested that the high eEF1A2 level in ER positive breast tumours might be dependent on the activation of p53 wild type gene. In fact, *EEF1A2* displays a p53 binding site sequence that is shared by *EEF1A1* gene (Kato et al.1997; Tomlinson et al. 2005). Thus the presence of functional p53 could contribute to the up-regulation of eEF1A2 but also could suppress other oncogenes useful for cancer development. Different results were obtained by Kulkarni et al. (Kulkarni et al., 2007) that observed an inverse correlation between eEF1A2 expression and p53 levels, although the up-regulation of eEF1A2 positively correlated with patients'survival. The discrepancy between the two works could be due to the higher number of tumour samples analyzed by Kulkarni with respect to Tomlinson and to the different statistical methods employed. Intriguingly, Joseph et al. (2004) did not detect significant over-expression of eEF1A2 in breast cancer tissues by cDNA array, whereas significant over-expression was found in lung, rectum, kidney and ovary cancer tissues. It cannot be exclude that such discrepancy might be related to the choice of the threshold set between normal and tumour samples, to the lack of information about tumour classification. Indeed, Tomlinson et al. and

Kulkarni et al. found higher eEF1A2 expression in the luminal breast tumour tissues (Tomlinson et a., 2005; Kulkarni et al., 2007). Moreover, the over-expression of eEF1A2 in breast cancer tissues by other authors (Tomlinson et al. 2005; Kulkarni et al., 2007) was proved both at mRNA and protein levels. In conclusion, the over-expression of eEF1A2 plays a significant role in the breast primary tumours although the molecular basis for the protective effect of eEF1A2 in breast cancers remains to be elucidated.

2.3. eEF1A2 Mechanisms in Oncogenesis

The exact mechanism of action of eEF1A2 in tumour cells has not yet been elucidated. A relation-ship between wild-type p53 and eEF1A2 has been found, as above reported, in breast tumour cell and in ovarian clear cell carcinomas (Tomlinson 2005 and 2007) as well as between eEF1A2 and stress response genes in ovarian clear cell carcinomas (Schwartz et al., 2002). In particular it is noteworthy that in NIH-3T3 cells the contemporary over-expression of eEF1A2 and Prdx1 confers to cells a high resistance to apoptosis (Chang and Wang, 2007). Interestingly, clear cell carcinomas expressed p53 and tend to up-regulate stress response genes: it cannot be excluded that eEF1A2 might be up-regulated by p53 and that its rise might relate to the activation of survival stress response genes such as peroxiredoxin 1(Chang and Wang, 2007). eEF1A2 activates the Akt serine/threonine kinase, in a P13K-dependent manner through PI(3,4,5)P cascade signal. Moreover, eEF1A2 can play a role in oncogenesis through PI4K activation as demonstrated in MCF-7 breast cancer cells, where the silencing of eEF1A2 was shown to decrease PI4K activity (Jeganathan and Lee, 2007). The rise in PI4P level in the cells by the over-expressed eEF1A2 can lead to cell growth de-regulation, actin reorganization and remodelling of cytoskeletal functions (Jeganathan and Lee, 2007). This, in turn, can increase cell migration by filipodia formation, thus suggesting a role for eEF1A2 in metastatic invasion (Amiri et al., 2006; Li et al., 2005). Furthermore, filipodia formation seems to be stimulated by eEF1A2 in a P13K-independent manner too (Amiri et al., 2006). Moreover, it has been demonstrated that eEF1A2 has a higher level of tyrosine phosphorylation capacity with respect to eEF1A1. As a consequence, a greater potential to participate to phosphotyrosine-mediated signalling pathways (Panasyuk et al., 2007) has been attributed to eEF1A2 compared to eEF1A1. Finally, a de-regulation of the p16 and p27 cell cycle inhibitors was found in hepatic cell lines over-expressing eEF1A1/2 and this was related to an increased ability of the cells to survive under suboptimal growth conditions (Grassi et al., 2007).

2.4. eEF1A1 in Oncogenesis

Whereas ectopic over-expression of eEF1A2 leads to tumorigenic transformation of NIH-3T3 cells and gives rise to tumours in xenografted nude mice (Anand et al., 2002), it is less clear whether the over-expression of eEF1A1 might possess tumorigenic potential *per se*. The role of eEF1A1 in cancer is thus less defined and controversial, but many evidences assign to eEF1A1 a role in promoting and/or supporting tumorigenesis. This feature is likely to be due to its behaviour of multifunctional protein (Ejiri S, 2002), to its 5' TOP sequence essential for the high activity of the promoter also acting as a cis-regulatory element to increase mRNA translation upon mitogenic stimuli (Bischoff et al., 2000; Shibui-Nihei A et al; 2003). In 1992, de-regulation of *EEF1A1* was shown to increase the transforming potential of 3-methylcholanthrene and UV in rodent cell lines (Tatsuka et al., 1992) and its over-expression was associated to drug-resistance in head and neck cancer cell lines (Johnsson et al., 2000). In a human acute promyelocytic cell line, the all-trans-retinoic acid, a growth inhibitor and differentiation agent, induced a significant reduction of eEF1A1 protein (Harris et al., 2004). Working with the histone deacetylase inhibitor sodium butyrate, the modulation of both eEF1A1 and eEF1A2 proteins was demonstrated in MCF-7 breast cancer cell line suggesting that both forms are protective against apoptosis (Goncalves et al. 2005). Moreover, modification of the expression pattern of eEF1A1 have been identified in esophageal squamous cell cancer (down regulation) (Zhang et al., 2004) and in tongue cancer with metastatic progression (Zhou et al., 2006). Furthermore, microarray and immuno-histochemistry analysis of eEF1A1 revealed that in the infiltrating edges of breast tumours this protein is up-regulated compared to the tumour core (Zu et al. 2003). This suggests that eEF1A1 can play an important role in cancer invasion likely by its moonlighting functions, especially the one involving cytoskeletal organization. eEF1A1 has been recently identified as a binder of Akt2 and Fascin, an actin bundling protein (Lau et al., 2006; Chang and Wang 2007) but it significance in the modulation of cellular functions is not yet clarified. In hepatoma cell lines we found in low differentiation grade cells, a striking over-expression of eEF1A1 with respect to more differentiated hepatoma cells (Grassi et al., 2007). Over-expression of eEF1A1 in non-transformed murine pro-B cells confers selective resistance to apoptosis induced by endoplasmic reticulum stress, promoting long-term viability (Talapatra et al., 2002). Elevated levels of eEF1A1 expression were found also in some hematopoietic cell lines and in primary leukemia (Scaggiante et al., unpublished results), thus suggesting a role for eEF1A1 in haematopoietic tumour development and progression. Moreover, the interaction between eEF1A1 and HDM2 may play a role in oncogenesis in cells over-expressing HDM2, by regulating actin depolymerization (Frum et al, 2007). It should be noted that in human cancer cell lines of different tumour origin we found that both eEF1A1/2 forms are de-regulated. In particular, in hepatoma cells, the over-expression of eEF1A1/2 correlates with the differentiation grade of the tumour and with the ability of the cells to proliferate under non-optimal growing conditions, escaping cell cycle arrest. Moreover, in the more aggressive cells we found a redistribution of eEF1A1/2 forms from the cytoplasm to the cytoskeletal-nuclear fraction (Grassi et al., 2007). The elevated eEF1A1/2 levels could sustain proliferation also by the ability to form complex with ZRP1, a molecule essential for cell viability and cellular growth (Gangwani et al. 1998; Mishra et al. 2007). Moreover, in ovarian and breast carcinomas over-expressing eEF1A2, no difference in eEF1A1 expression levels was seen between normal and tumour tissues (Anand et al., 2002; Tomlinson et al., 2005). However, in human prostate cancer cell lines the level of eEF1A1 was found to decrease with respect to that of non-tumorigenic cell lines (Scaggiante et al., unpublished results) whereas in hepatoma cells the expression of eEF1A1 increase with respect to normal liver (Grassi et al., 2007). Thus this suggests that eEF1A1 expression and more likely the ratio in the expression of eEF1A1/2 forms might affect cancer cell behaviours in some tumours.

2.5. eEF1A1 Isoforms and Oncogenesis

The eEF1A1 protein has some protein homologues with different functions. Interestingly, a CTCL tumour antigen HD-CL-08 of cutaneous T-cell lymphoma has a high degree of homology with eEF1A1 lacking 77 amino acids at NH2-end (Hartmann et al., 2004). An isoform of eEF1A1 named human cervical cancer suppressor-3 (CCS-3) shares eEF1A1 homology with the lack of 101 aminoacid at NH2-terminal region. CCS-3 interacts with the transcription factor PLZF that regulates cell cycle progression. CCS-3 expression is reduced in tumour tissues and transient over-expression of CCS-3 in a cervical cancer cell line inhibits cell growth and induces apoptosis (Rho et al, 2006).

The Prostate Inducing gene-1, sharing 97.7 % identity with eEF1A1 but with a deletion of 67 aminoacid at NH2-end, behaves as an oncogene (Shen et al., 1995) and its forced expression causes cell transformation (Sun et al. 1997). The PTI-1, known as eukaryotic elongation factor 1A-like-14, was initially identified as a gene present in prostate cancer cells and in invasive prostate adenocarcinoma primary tumours, but not in normal cells (Shen et al.1995). Its cDNA consists of 2123 bp with a 5'-UTR sharing partial or total homology with the 23S rRNA of *Mycoplasma hyopneumoniae, Mycoplasma hyorhinis, E.coli* and a coding sequence sharing 99.7 % identity with eEF1A1 (Mansilla, Clark et al.2005; Sun, Katz et al., 1977). The PTI-1 was initially described as a dominant oncogene involved in neoplastic transformation. In this regard, co-transfection of the PTI-1 cDNA into CREF-Trans 6 cells made them able to induce tumours in nude mice and anti-sense PTI-1 molecules was shown to revert the malignant phenotype of the transformed cells (Su et al., 1998). PTI-1 transcript was detected in blood samples of patients with invasive prostate adenocarcinoma but not in those with benign adenoma, suggesting that it could derive from prostate cancer cells detached from the tumour and flowing into the blood stream (Su et al., 1998; Sun, Katz et al., 1997). It has been hypothesized that PTI-1 protein could cause translation infidelity by promoting the insertion of wrong amino acid into the nascent polypeptide chain (Gopalkrishnan et al., 1999; Pogribna et al. 2006) and that an increased cytosolic stability of the protein could lead to intracellular competition with eEF1A1, further promoting cellular transformation (Mansilla, Clark et al., 2005). However, it seems unlikely that the most abundant constitutive eEF1A1 protein in the cells can be competed by the less abundant PTI-1 (Mansilla et al., 2002; Scaggiante et al., 2008). We think that the PTI-1 protein can likely affect other cellular function such as cytoskeleton organization or it may be involved in pathways that regulate cell cycle, or it could participate to the modulation of gene transcription. The PTI-1 transcript was initially identified in LNCaP and DU-145 human prostate cancer cell lines (Sun et al., 1995) conferring to this transcript a role in the oncogenicity. Nevertheless, some of the data available on PTI-1 could be derived from the relief of *Mycoplasma* genera products or other unknown human cancer sequences or of eEF1A1 protein. In the work of Kan et al. (Kan et al.2006) PTI-1 was detected by representational difference analysis of cDNA (cDNA RDA) in an esophageal cancer cell line and by RT-PCR in different esophageal cancer cell lines and in resected esophageal cancer tissues. However, the RT-PCR was performed on cell lines and tissues using primers that span only in the 5'-UTR in common with the 23S rRNA of *Mycoplasma* and do not include the coding sequence of the PTI-1. Despite its potential role in the tumour development, it remains to be clarified its origin in human tumour cells. The PTI-1 gene was located on chromosome 7 (*GC00U900568*). However, at the beginning of 2005 its sequence has been withdrawn from the human genome bank.

Nowadays it remains under investigation whether PTI-1 can play a role in the natural story of prostate adenocarcinoma. Our results (Scaggiante et al., 2008) demonstrate that PTI-1 mRNA is neither detectable by conventional RT-PCR using different primer pairs nor by using the same procedure described by Sun et al (Sun et al., 1997) in the human prostate cancer DU-145, LNCaP and PC-3 lines and in non-tumorigenic control prostate PZHPV-7, nor is detectable its protein product. This discrepancy might derive from the fact that the cell lines in culture can easily change their behaviours or they can undergo *Mycoplasma* infections without exhibiting cell suffering. Furthermore, we were unable to detect any specific PTI-1 amplicons in prostate tumor samples derived from formalin-fixed archive tissues. We therefore believe that the detection of PTI-1 mRNA by other groups in prostate cells (Sheen et al., 1995, Sun et al., 1997; Gopalkrishnan et al. 1999), might be explained only in terms of the possible biology of PTI-1 mRNA. In particular, our hypothesis is that in human prostate adenocarcinoma the generation of PTI-1 mRNA is possible maybe upon *Mycoplasma* infection as suggested by Mansilla et al. (Mansilla et al. 2005), but it unlikely represents the *sine qua non* condition in prostate tumorigenesis. In fact, we found PTI-1 mRNA in human cell lines positive for *Mycoplasma* infection, casually contaminated in the laboratory but not in the same cells *Mycoplasma*-free (Scaggiante et al.,2008). It is possible that, as a consequence of the genomic instability of cancer cells, PTI-1 can randomly derived in an unknown manner from fusion of eEF1A or of the numerous eEF1A1 retropseudogenes (Madsen et al. 1990) with the 23S rRNA of *Mycoplasma* during cancer progression. Moreover, it cannot be excluded that some particular *Mycoplasma* species could be able to give, in a specific manner, a stable genomic rearrangement for PTI-1. Alternatively, the raise of some specific molecules during cancer transformation or progression, especially in the drug-resistant cells (Bertram et al., 1998; Scaggiante et al., unpublished results), might favour the formation of the PTI-1 transcript in the presence of *Mycoplasma* infection. Additionally, the observation that elevated levels of PTI-1 mRNA were detected in surgical specimens of human pancreatic cancers (Ohnami et al. 1999) suggests that *in vivo,* the frequency of PTI-1 mRNA generation may also depend on the type of cancer cells. Finally, despite the fact that at the beginning of 2005 PTI-1 sequence on chromosome 7 has been withdraw from the human genome bank, we cannot completely exclude that PTI-1-like sequences might be located on other chromosomes and that they could undergo particular recombination to origin the PTI-1 oncogenic product in cancer cells independently from *Mycoplasma* infection.

Another contribution to cancer development of eEF1A1, derives from the post-translational modifications of eEF1A1 itself. In particular, an isoform of eEF1A1 was found to play a role in the hematopoietic T-lymphoblastoid cells. This isoform was present in T-lymphoblasts, but not in normal lymphocytes (Dapas et al., 2003). It was initially identified as a specific interactor of aptameric cytotoxic GT oligonucleotides that inhibit the cellular growth of a variety of human cancer cell lines (Scaggiante et al.1998; Scaggiante et al. unpublished observation) but not of normal human lymphocytes (Scaggiante et al.1998) or human dermal microvascular endothelial cells (Goodchild et al., 2007). Only the nuclear-enriched fraction pool of eEF1A1 protein of cancer cells was shown to display the ability to bind to cytotoxic GT oligomers (Scaggiante et al., 1998; Scaggiante et al., 2006) and to display the more basic eEF1A1 isoform (Dapas et al., 2003; Scaggiante et al. unpublished observations). This suggests that the cancer eEF1A1 isoform associated to the cytoskeleton and/or to the nucleus to specifically influence the tumour development and/or progression. Furthermore, we found that the appearance of this isoform can be induced in normal cells upon phorbol ester treatment. This suggests that it likely derived from post-translation modification of the constitutive eEF1A1, possibly after activation of the kinase cascade signalling pathway (Scaggiante et al, unpublished results). Interestingly, also eEF1A2 might undergo post-translational modifications in cancer cells. In tumour samples where eEF1A2 protein is detected, its molecular size could appear to be slightly different from the native form and this might account for post-translational modifications (Tomlinson et al. 2007).

3. Conclusions

The molecular mechanism of action of the eEF1A family in the *in vivo* genesis and progression of human cancers is very complex involving different eEF1A interactors, many of which still need to be fully elucidated. Undoubtedly, both eEF1A1/2 forms can play pivotal role in cancer biology, with the eEF1A2 being the preferential oncogenic form in those tissues that do not express constitutively this protein. Interestingly, eEF1A2 can be a good prognostic marker for some tumours and a worst prognostic factor for others. The reasons for these opposite behaviours deserve further investigations. We think that eEF1A forms can play an important role in cancer invasiveness likely by their ability to modulate the cytoskeleton organization, thus influencing cell migration, both directly, as in the case of eEF1A1 (Gross and Kinzy, 2005), or indirectly, as in the case of eEF1A2 (Jeganathan and Lee 2007; Amiri et al., 2006). Moreover, the redistribution of eEF1A from cytoplasm to other cellular localizations may also play an important role in defining the tumour phenotype and aggressiveness. We believe that in human tumours the de-regulation of one or both eEF1A forms can contribute to cancer onset and progression due to their ability to interact with molecules involved in different pathways controlling cell proliferation, growth, apoptosis and oxidative-response to stress. Furthermore, the possibility that these factors can undergo to post-translational modifications supporting tumour cell growth (Dapas et al., 2003), underlines the multifactor ability of eEF1A to sustain cancer. The putative dominant oncogene PTI-1 remains a mystery, and its role in human cancer still needs to be defined. However, it might emerge as an oncogenic molecule possibly related to the *Mycoplasma* infection, or to the rearrangement of the many eEF1A1 retropseudogenes favoured by the genomic instability during tumour progression. The role of the CCS-3 oncosupressor protein needs to be further analyzed in human cancers other than cervical cancer and might have the potential to function as an interesting marker for tumour prognosis.

Acknowledgements:

we thanks Prof. G. Manfioletti, University of Trieste, to kindly provide the human prostate cell lines. This work was in part supported by FIRB number RBNE0155LB, "Fondazione Cassa di Risparmio of Trieste" and by the "Fondazione Benefica Kathleen-Foreman Casali of Trieste".

References

Abbott, C.M., and Proud, C.G. (2004). Translation factors: in sickness and in health. *Trends Biochem. Sci.,* 29, 25-31.

Amiri, A., Noei, F., Jeganathan, S., Kulkarni, G., Pinke, D.E., and Lee, J.M. (2007). eEF1A2 activates Akt and stimulates Akt-dependent actin remodeling, invasion and migration. *Oncogene,* 26, 3027-3040.

Anand, N., Murthy, S., Amann, G., Wernick, M., Porter, L.A., Cukier, I.H., Collins, C., Gray, J.W., Diebold, J., Demetrick, D.J., and Lee, J.M. (2002). Protein elongation factor EEF1A2 is a putative oncogene in ovarian cancer. *Nat. Genet.,* 31, 301–305.

Anand, M., Balar, B., Ulloque, R., Gross, S.R., and Kinzy, T.G. (2006). Domain and nucleotide dependence of the interaction between Saccharomyces cerevisiae translation elongation factors 3 and 1A. *J. Biol. Chem.,* 281, 32318-3226.

Bektas, M., Nurten, R., Gurel, Z., Sayers, Z., and Bermek, E. (1994). Interactions of eukaryotic elongation factor 2 with actin: a possible link between protein synthetic machinery and cytoskeleton. *FEBS Lett,* 356, 89-93.

Bertram, J., Palfner, K., Hiddemann, W., and Kneba, M. (1998). Overexpression of ribosomal proteins L4 and L5 and the putative alternative elongation factor PTI-1 in the doxorubicin resistant human colon cancer cell line LoVoDxR. *Eur. J. Cancer,* 34, 731-736.

Borradaile, N.M., Buhman, K.K., Listenberger, L.L., Magee, C.J., Morimoto, E.T.A., Ory, D.S. and Schaffer, J.E. (2006). A critical role for eukaryotic elongation factor 1A-1 in lipotoxic cell death. *Mol. Biol. of Cell,* 17, 770-778.

Bischoff, C., Kahns, S., Lund, A., Jorgensen, H.F., Praestegaard, M., Clark, B.F., and Leffers, H. (2000). The human elongation factor 1 A-2 gene (EEF1A2): complete sequence and characterization of gene structure and promoter activity. *Genomics,* 68, 63-70.

Caldas, T.D., El Yaagoubi, A., and Richarme, G. (1998). Chaperone properties of bacterial elongation factor EF-Tu. *J. Biol. Chem.,* 273, 11478–11482.

Chang, R., and Wang, E. (2007). Mouse translation elongation factor eEF1A-2interacts with Prdx-I to protect cells against apoptotic death induced by oxidative stress. *J. Cell Biochem.,* 100, 267-278.

Chen, E., Proestou, G., Bourbeau, D., and Wang, E. (2000). Rapid up-regulation of peptide elongation factor EF-1alpha protein levels is an immediate early event during oxidative stress-induced apoptosis. *Exp. Cell Res.,* 259, 140-148.

Chen, L., and Madura, K. (2005). Increased proteosome activity, ubiquitin-conjugating enzymes, and eEF1A translation factor detected in breast cancer tissue. *Cancer Res.,* 65,5599-5606.

Chuang, S.M., Chen, L., Lambertson, D., Anand, M., Goss, K.T., and Madura, K. (2005). Proteasome-Mediated Degradation of Cotranslationally Damaged Proteins Involves Translation Elongation Factor 1A. *Mol. Cell Biol.,* 25, 403–413.

Condeelis, J. (1995). Elongation factor 1 alpha, translation and the cytoskeleton. *Trends Biochem. Sci.,* 20, 169-170.

Dapas, B., Tell, G., Scaloni, A., Pines, A., Ferrara, L., Quadrifoglio, F., and Scaggiante B. (2003). Identification of different isoforms of eEF1A in the nuclear fraction of human T-lymphoblastic cancer cell line specifically binding to aptameric cytotoxic GT oligomers. *Eur. J. Biochem.,* 270, 3251-3262.

Duttaroy, A., Bourbeau, D., Wang, X.L., and Wang, E. (1998). Apoptosis rate can be accelerated or decelerated by overexpression or reduction of the level of elongation factor-1α. *Exp. Cell Res.,* 238, 168-176.

Ejiri, S. (2002). Moonlighting functions of polypeptide elongation factor 1:from actin bundling to zinc finger protein R1-associated nuclear localization. *Biosci. Biotechno.l Biochem.,* 66, 1-21.

Frum, R., Busby, S.A., Ramamoorthy, M., Deb, S., Shabanowitz, J., Hunt, D.F., and Deb, S.P. (2007). HDM2-binding partners:interaction with translation elongation factor EF1α. *J. Proteome Res.,* 6, 1410-1417.

Gangwani, L., Mikrut, M., Galcheva-Gargova, Z., and Davis, R.J. (1998). Interaction of ZPR1 with translation elongation factor-1alpha in proliferating cells. *J. Cell Biol.,* 143, 1471-1484.

Grassi, G., Scaggiante, B., Farra, R., Dapas, B., Agostini, F., Baiz, D., Rosso, N., and Tiribelli, C. (2007). The expression levels of the translational factors eEF1A 1/2 correlate with cell growth but not apoptosis in hepatocellular carcinoma cell lines with different differentiation grade. *Biochimie,* 89, 1544-52.

Goodchild, A., King, A., Gozar, M.M., Passioura, T., Tucker, C., and Rivory, L. (2007). Cytotoxic G-rich oligodeoxynucleotides: putative protein targets and required sequence motif. *Nucleic Acids Res.,* 35, 4562-4572.

Gonçalves, J., Malta-Vacas, J., Louis, M., Brault, L., Bagrel, D., Monteiro, C., and Brito, M. (2005). Modulation of translation factor's gene expression by histone deacetylase inhibitors in breast cancer cells. *Clin. Chem. Lab. Med.,* 43, 151-156

Gopalkrishnan, R.V., Su, Z.Z., Goldstein, N.I., and Fisher, P.B. (1999). Translational infidelity and human cancer: role of the PTI-1 oncogene. *Int J Biochem Cell Biol,* 31, 151-162.

Gross, S.R., and Kinzy, T.G. (2005). Translation elongation factor 1A is essential for regulation of the actin cytoskeleton and cell morphology. *Nat. Struct. Mol. Biol.,* 12, 772-778.

Harris, M.N., Ozpolat, B., Abdi, F., Gu, S., Legler, A., Mawuenyega, K.G., Tirado-Gomez, M., Lopez-Berestein, G., and Chen, X. (2004). Comparative proteomic analysis of all-trans-retinoic acid treatment reveals systematic posttranscriptional control mechanisms in acute promyelocytic leukemia. *Blood,* 104, 1314-1323.

Hartmann, T.B., Thiel, D., Dummer, R., Schadendorf, D., and Eichmuller, S. (2004). SEREX identification of a new tumour-associated antigens in cutaneous T-cell lymphoma. *Br. J. Dermatol.,* 150, 252-258.

Hotokezaka, Y., Tobben, U., Hotokezaka, H., Van Leyen, K., Beatrix, B., Smith, D.H., Nakamura, T., and Wiedmann, M. (2002). Interaction of the eukaryotic elongation factor 1A with newly synthesized polypeptides. *J. Biol. Chem.,* 277, 18545–18551.

Jeganathan, S., and Lee, J.M. (2007). Binding of elongation factor eEF1A2 to phosphatidylinositol 4-kinase β stimulates lipid kinase activity and phosphatidylinositol 4-phosphate generation. *J. Biol. Chem.,* 282, 372-380.

Johnsson, A., Zeelenberg, I., Min, Y., Hilinski, J., Berry, C., Howell, S.B., and Los, G. (2000). Identification of genes differentially expressed in association with acquired cisplatin resistance. *Br. J. Cancer,* 83, 1047-1054.

Joseph, P., O'Kernick, C.M., Othumpangat, S., Lei, Y.X., Yuan, B.Z., and Ong, T.M. (2004). Expression profile of eukaryotic translation factors in human cancer tissues and cell lines. *Mol. Carcinog.,* 40, 171-179.

Kahns, S., Lund, A., Kristensen, P., Knudsen, C.R., Clark, B.F., Cavallius, J., and Merrick, W.C. (1998). The elongation factor 1 A-2 isoform from rabbit: cloning of the cDNA and characterization of the protein. *Nucleic Acids Res.,* 26, 1884-1890.

Kan, T., Yamasaki, S., Kondo, K., Teratani, N., Kawabe, A., Kaganoi, J., Meltzer, S.J., Iamamura, M., and Shimada, Y. (2006). A new specific gene expression in squamous cell carcinoma of the esophagus detected using representational difference analysis and cDNA microarray. *Oncology,* 70, 25-33.

Kato, M.V., Sato, H., Nagayoshi, M., and Ikawa, Y. (1997). Upregulation of the elongation factor-1alpha gene by p53 in association with death of an erythroleukemic cell line. *Blood,* 90,1373-1378.

Kiel, J., Titorenko, V.I., van der Klei, I.J., and Veenhuis, M. (2007). Overproduction of translation elongation factor 1-a(eEF1A) suppresses the peroxisome biogenesis defect in a Hansenula polymorpha pex3 mutant via translational read-through. *FEM Yeast. Res.*, 7, 1114-11125.

Knudsen, S.M., Frydenberg, J., Clark, B.F., and Leffers, H. (1993). Tissue-dependent variation in the

expression of elongation factor-1 alpha isoforms: isolation and characterisation of a cDNA encoding a novel variant of human elongation-factor 1 alpha. *Eur. J. Biochem.,* 215, 549-554.

Kulkarni, G., Turbin, D.A., Amiri, A., Jeganathan, S., Andrade-Navarro, M.A., Wu, T.D., Huntsman, D.G., and Lee, J.M. (2007). Expression of protein elongation factor eEF1A2 predicts favorable outcome in breast cancer. *Breast Cancer Res. Treat.,* 102, 3027-3040.

Izawa, T., Fukata, Y., Kimura, T., Iwamatsu, A., Dohi, K., and Kaibuchi, K. (2000). Elongation factor-1 alpha is a novel substrate of rho-associated kinase. *Biochem. Biophys.Res. Commun.,* 278, 72-78.

Lau, J., Castelli, L.A., Lin, E.C., and Macaulay, S.L. (2006). Identification of elongation factor 1alpha as a potential associated binding partner for Akt2. *Mol Cell Biochem,* 286, 17–22.

Lee, J.M. (2003). The role of protein elongation factor eEF1A2 in ovarian cancer. *Reprod. Biol. Endocrinol.,* 1, 69.

Lee, S., Wolfraim, L.A., and Wang, E. (1993). Differential expression of S1 and elongation factor-1 alpha during rat development. *J. Biol. Chem.*, 268, 24453–24459.

Li, R., Wang, H., Bekele, B.N., Yin, Z., Caraway, N.P., Katz, R.L., Stass, S.A., and Jiang, F. (2005). Identification of putative oncogenes in lung adenocarcinoma by a comprehensive functional genomic approach. *Oncogene*, 25, 2628–2635.

Li, P., Maines-Bandiera, S., Kuo, W.L., Guan, Y., Sun, Y., Hills, M., Huang, G., Collins, C.C., Leung, P.C., Gray, J.W., and Auersperg, N. (2007). Multiple roles of the candidate oncogene ZNF217 in ovarian epithelial neoplastic progression. *Int. J. Cancer,* 120, 1863-1873.

Lund, A., Knudsen, S.M., Vissing, H., Clark, B., and Tommerup, N. (1996). Assignment of human elongation factor 1alpha genes: EEF1A maps to chromosome 6q14 and EEF1A2 to 20q13.3. *Genomics,* 36, 359-361.

Lukash, T.O., Turkivska, H.V., Negrutskii, B.S., and El'skaya, A.V. (2004). Chaperone-like activity of mammalian elongation factor eEF1A: renaturation of aminoacyl-tRNA synthetases. *Int. J. Biochem. and Cell Biol.,* 36, 1341-1347.

Madsen, H.O., Poulsen, K., Dahl, O., Clark, B.F.C., and Hjorth, J.P. (1990). Retropseudogenes constitute the major part of the human elongation factor 1α gene family. *Nucleic Acids Res.,* 18, 1513-1516.

Mansilla, F., Friis, I., Jadidi, M., Nielsen, K.M., Clark, B.F., and Knudsen, C.R. (2002). Mapping the human translation elongation factor eEF1H complex using the yeast two-hybrid system. *Biochem J.,* 365, 669-676.

Mansilla, F., Hansen, L.L., Jakobsen, H., Kjeldgaard, N.O., Clark, B.F., and Knudsen, C.R. (2005). Deconstructing PTI-1: PTI-1 is a truncated, but not mutated, form of translation elongatin factor 1A1, eEF1A1. *Biochim. Biophys. Acta,* 1727, 116-124.

McClatchy, D.B., Fang, G., and Levey, A.I. (2006). Elongation factor 1A family regulates the recycling of the M4 muscarinic acetylcholine receptor. *Neurochem. Res.,* 31, 975-88.

Mishra, A.K., Gangwani, L., Davis, R.J., and Lambright, D.G. (2007). Structural insights into the interaction of the evolutionarily conserved ZPR1 domain tandem with eukaryotic EF1A, receptors, and SMN complexes. *Proc Natl. Acad. Sc.i USA,* 104, 13930-13935.

Newbery, H.J., Loh, D.H., O'Donoghue, J.E., Tomlinson, V.A., Chau, Y.Y., Boyd, J.A., Bergmann, J.H., Brownstein, D., and Abbott, C.M. (2007). Translation elongation factor eEF1A2 is essential for post-weaning survival in mice. *J. Biol. Chem.,* 282, 28951-28959.

Ohnami, S., Matsumoto, N., Nakano, M., Aoki, K., Nagasaki, K., Sugimura, T., Terada, M., and Yoshida, T. (1999). Identification of genes showing differential expression in antisense K-ras-transduced pancreatic cancer cells with suppressed tumourigenicity. *Cancer Res.,* 59, 5565-5571.

Rho, S.B., Park, Y.G., Park, K., Lee, S.H., Lee, J.H. (2006). A novel cervical cancer suppressor 3 (CCS-3) interacts with BTB domain of PLZF and inhibits the cell growth by inducing apoptosis. *FEBS Lett.,* 580, 4073-4080.

Panasyuk, G., Nemazanyy, I., Filonenko, V., Negrutskii, B., and El'skaya, A.V. (2007). A2 isoform of mammalian translation factor eEF1A displays increased tyrosine phosphorylation and ability to interact with different signalling molecules. *Int J Biochem Cell Biol.,* Sep 1, [Epub ahead of print].

Petrushenko, Z.M., Budkevich, T.V., Shalak, V.F., Negrutskii, B.S., and El'skaya, A.V. (2002). Novel complexes of mammalian translation elongation factor eEF1A·GDP with uncharged tRNA and aminoacyl-tRNA synthetase. Implications for tRNA channelling. *Eur.J. Biochem.*, 269, 4811–4818.

Pogribna, A.P., Negrutskii, B.S., and Elskaya, A.V. (2006). Removal of part of the eEF1A GTP binding domain induced translation errors in vivo. *Biopolymers and Cell,* 22, 186-191.

Potter, M., Bernstein, A., Lee, J.M. (1998). The wst gene regulates multiple forms of thymocyte apoptosis. *Cell Immunol.,* 188, 111-117.

Samovsky, I., Ivannikov, M., Kandel, E.S., Gershon, D., Nudler, D. (2006). RNA-mediated response to heat shock in mammalian cells. *Nature,* 440, 556-560.

Scaggiante, B., Morassutti, C., Dapas, B., Tolazzi, G., Ustulin, F., and Quadrifoglio, F. (1998). Human cancer cell lines growth inhibition by GTn oligodeoxyribonucleotides recognizing single-stranded DNA-binding proteins. *Eur. J. Biochem.,* 252, 207-215.

Scaggiante, B., Dapas, B., Grassi, G., and Manzini, G. (2006). Interaction of G-rich GT oligonucleotides with nuclear-associated eEF1A is correlated with their antiproliferative effect in haematopoietic human cancer cell lines. *FEBS J.,* 273, 1350-1361.

Scaggiante, B., Bonin, S., Cristiano, L., Siracusano, S., Stanta,G., Dapas, B., Giansante, C., Fiotti, N. and Grassi, G. (2008). Prostate Tumour-Inducing gene-1 analysis in human prostate cancer cells and tissue in relation to *Mycoplasma* infection. *Cancer Invest.,* in press.

Schwartz, D.R., Kardia, S.L., Shedden, K.A., Kuick, R., Michailidis, G., Taylor, J.M., Misek, D.E., Wu, R., Zhai, Y., Darrah, D.M., Reed, H., Ellenson, L.H., Giordano, T.J., Fearon, E.R., Hanash, S.M., and Cho, K.R. (2002). Gene expression in ovarian cancer reflects both morphology and biological behavior, distinguishing clear cell from other poor-prognosis ovarian carcinomas. *Cancer Res.,* 62, 4722-4729.

Shen, R., Su, Z.Z., Olsson, C.A., and Fisher, P.B. (1995). Identification of the human prostatic carcinoma oncogene PTI-1 by rapid expression cloning and differential RNA display. *Proc Natl. Acad. Sci. USA,* 92, 6778-6782.

Shibui-Nihei, A., Ohmori, Y., Yoshida, K., Imai, J., Oosuga, I., Iidaka, M., Suzuki, Y., Mizushima-Sugano, J., Yoshitomo-Nakagawa, K., and Sugano, S. (2003). The 5' terminal oligopyrimidine tract of human elongation factor 1A-1 gene functions as a transcriptional initiator and produces a variable number of Us at the transcriptional level. *Gene,* 311, 137-145.

Sørlie, T., Perou, C.M., Tibshirani, R., Aas, T., Geisler, S., Johnsen, H., Hastie, T., Eisen, M.B., van de Rijn, M., Jeffrey, S.S., Thorsen, T., Quist, H., Matese, J.C., Brown, P.O., Botstein, D., Eystein Lønning, P., and Børresen-Dale, A.L. (2001). Gene expression patterns of breast carcinomas distinguish tumor subclasses with clinical implications. *Proc Natl. Acad. Sci. USA,* 98, 10869-10874.

Sun, Y., Lin, J., Katz, A.E., and Fisher, P.B. (1997). Human prostatic carcinoma oncogene PTI-1 is expressed in human tumour cell lines and prostate carcinoma patient blood samples. *Cancer Res.,* 57, 18-23.

Talapatra, S., Wagner, J.D., and Thompson, C.B. (2002). Elongation factor-1 alpha is a selective regulator of growth factor withdrawal and ER stress-induced apoptosis. *Cell Death Differ.*, 9, 856-861.

Tatsuka, M., Mitsui, H., Wada, M., Nagata, A., Nojima, H., and Okayama, H. (1992). Elongation factor-1 alpha gene determines susceptibility to transformation. *Nature,* 359, 333–336.

Thornton, S., Anand, N., Purcell, D., and Lee J. (2003). Not just for housekeeping: protein initiation and elongation factors in cell growth and tumorigenesis. *J. Mol. Med.,* 81, 536–548.

Tomlinson, V.A., Newbery, H.J., Wray, N.R., Jackson, J., Larionov, A., Miller, W.R. Dixson, J.M., and Abbott C.M. (2005). Translation elongation factor eEF1A2is a potential oncoprotein that is overexpressed in two-thirds of breast tumours. *BMC Cancer*, 5, 113.

Tomlinson, V.A., Newbery, H., Bergmann, J.H., Boyd, J., Scott, D., Wray, N.R., Sellar, G.C., Gabra, H., Graham, A., Williams, A.R.W., and Abbott, C.M. (2007). Expression of eEF1A2 is associated with clear cell histology in ovarian carcinomas: overexpression of the gene is not dependent on modifications at EEF1A2 locus. *B. J. Cancer,* 96,1613-1620.

Zhang, J., Gao, F.L., Zhi, H.Y., Luo, A.P., Ding, F., Wu, M., and Liu, Z.H. (2004). Expression patterns of esophageal cancer deregulated genes in C5BL/6J mouse embryogenesis. *World J. Gastroenterol.,* 10, 1088-1092.

Zhou, X., Temam, S., Oh, M., Pungpravat, N., Huang, B.L., Mao, L., and Wong, D.T. (2006). Global expression-based classification of lymph node metastasis and extracapsular spread of oral tongue squamous cell carcinoma. *Neoplasia,* 8, 925-932.

Zhu, G., Reynolds, L., Crnogorac-Jurcevic, T., Gillett, C.E., Dublin, E.A., Marshall, J.F., Barnes, D., D'Arrigo, C., Van Trappen, P.O., Lemoine, N.R., and Hart, I.R. (2003). Combination of microdissection and microarray analysis to identify gene expression changes between differentially located tumour cells in breast cancer. *Oncogene,* 22, 3742-3748.

In: Oncogene Proteins: New Research
Editors: A. H. Malloy and E.C. Carson, pp. 195-216
ISBN 978-1-60456-290-3

Chapter 7

Thromboxane and its Synthase in Human Lung Cancer

Kin Chung Leung and George G. Chen[5]
Department of Surgery, The Chinese University of Hong Kong, Shatin, New Territories, Hong Kong

Abstract

Thromboxane A2 (TXA_2) is a potent arachidonate metabolite in the cyclooxygenase-2 (COX-2) pathway. Thromboxane synthase (THXS) is a member of cytochrome P450 (CYP) superfamily which was first purified in human platelets and porcine lung. This enzyme catabolizes prostaglandin H2 (PGH_2) into 12-*L*-hydroxy-5,8,10-heptadecatrieonic acid (HTT), malondialdehyde (MDA) and TXA_2 in a ratio of 1:1:1. TXA_2 plays a role in platelet aggregation induction, vasocaonstriction and cell proliferation. Its dysfunction is associated with various pathological conditions including cancers. Although the expression of THXS has not been studied in lung cancer, the increase of THXS has been found in a number of other tumors. The level of TXA_2 in lung cancer has been shown to be higher than that in normal lung cells, but its role in lung cancer development is uncertain. TXA_2 can promote cell migration, angiogenesis, tumor metastasis and cancer proliferation. Therefore, we hypothesize that thromboxane plays a role in lung cancer progression. In this review, we will discuss the structure of THXS, the regulation of THXS expression, the effects of TXA_2 modulations and THXS inhibitors on the proliferation and growth of human lung cancer cells.

5 Corresponding Author: George G Chen, Department of Surgery, The Chinese University of Hong Kong, Prince of Wales Hospital, Shatin, New Territories, Hong Kong. Tel: 852 2632 3934, Fax: 852 2645 0605, email:gchen@cuhk.edu.hk.

Introduction

Lung cancer is the most common cancer and the leading cause of cancer cell death in the world. The high incidence rate of lung cancer is due to the increased tobacco smoking in this decade. 85-90% of the lung cancer patients have been exposed to tobacco carcinogens from smoking [1]. There are mainly classified as two groups namely non-small cell lung cancer (NSCLC) and non-small cell lung cancer (SCLC). SCLC constitutes 20% of all lung cancers while NSCLC accounts for the rest. NSCLC can be further sub-divided into adenocarcinoma, squamous cell carcinoma and large cell carcinoma [2, 3]. Among them adenocarcinoma is the most common type.

Around 4000 chemical compounds have been identified as carcinogens in tobacco smoking. Nitrosamines and polycyclic aromatic hydrocarbons are the two major carcinogens in tobacco smoke. These compounds can promote deoxyribonucleic acid damage, and consequently induce DNA and protein mutations, which may account for the majority of lung cancer cases worldwide [1]. The risk for lung cancer is significantly lower in quitters than current smokers [4]. Collectively, there is ample information to support a positive relationship between tobacco smoking and lung cancer. Although smoking is an identifiable factor that is related to lung cancer, high incidence was also seen in non-smokers indicating that other etiological factors can contribute to lung cancer development. These other etiological factors include environmental, occupational, socioeconomic and genetic factors.

Genetic changes are an important factor that leads to lung cancer development. The alteration can occur in tumor suppressor genes, proto-oncogenes, growth factors, telemerase activity and methylation status of promoters [5]. Principal molecular markers found in lung cancer include p53, Rb, p16, p21, K-ras, htERT, Bcl-2, HER-2 and *c-myc* [6-8] while other markers (Table 1) have been recently identified, which may enable early diagnosis, staging and therapeutic options to lung cancer patients.

Thromboxane, THXS and Thromboxane Receptor

CYP assists the transformation of both exogenous and endogenous chemical carcinogens to activated forms. Therefore CYP is believed to play a prominent role in tumor development [11]. In addition, the difference of drug sensitive between tumor and normal cells is correlated to the expression level of CYP enzymes.

THXS is a member of the CYP superfamily and is assigned as CYP 5A1, whereas it does not performed monooxygenase activity. This enzyme was first purified from human platelets [12] and porcine lungs [13] and *in situ* hybridization found that the gene coded for THXS was located in the long arm of chromosome 7 band q33-q34 [14]. Purified THXS exhibits characteristic of CYP and it functions to cleave prostaglandin H_2 into HTT, MDA and TXA_2 at a ratio of 1:1:1 [12, 15].

TXA_2 is a potent platelet aggregator and vasoconstrictor with a half life of only 37 seconds under physiological conditions and rapidly hydrolyzed to form thromboxane B_2 (TXB_2) [16]. The metabolic pathway for TXA_2 synthesis is shown in Figure 1. It is one of the downstream metabolites of COX and converted from prostaglandin H_2 by THXS [17].

Table 1. List of biomarkers for lung cancer

Biomarkers for lung cancer
CYFRA 21-1, NSE, ProGRP, SCC, Carcinoembryonic Antigen (CEA), Tumor M2-PK, insulin-like growth factor (IGF)-I and IGF-binding proteins
Biomarkers for SCLC
CEA, 1-acid glycoprotein (AGP), neuron-specific enolase (NSE), chromogranin A (ChrA), bombesin-like gastrin-releasing peptide (GRP), and BB isoenzyme of creatine kinase (CK-BB) , stem cell factor (SCF) and c-kit, VEGF, Haptoglobin-α Subunit
Biomarkers for NSCLC
K-ras and c-ErbB-1 (EGFR) and c-ErbB-2 (HER-2/neu)

Source: adapted from references [9] and [10]

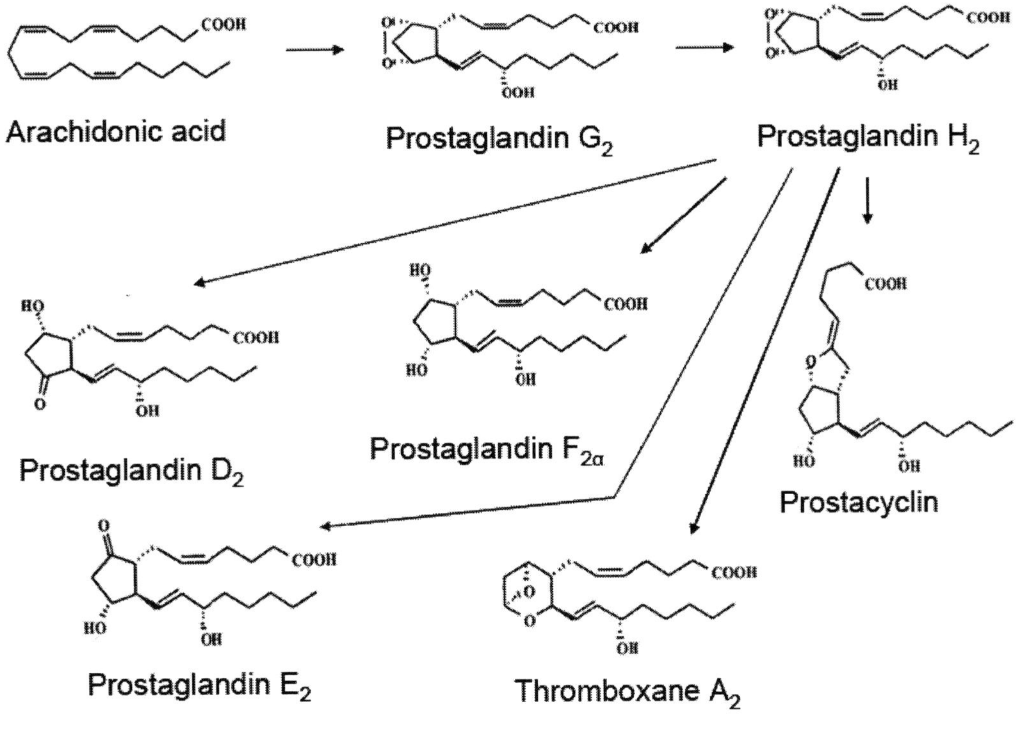

Figure 1. Biosynthesis of thromboxane A_2 from arachidonic acid.

TXA_2 is involved in the pathophysiological process of several diseases such as atherosclerosis, stomach ulcer, preechampsia, bronchial asthma and cardial infarction [18-22]. Furthermore, TXA_2 is also related to metastases of cancer [23, 24]. Generation of TXA_2 depends on the activity of THXS.

THXS mRNA is highly expressed in some human tissues such as peripheral blood leukocytes, prostate gland, spleen, lung and liver but its level is low in kidney, placenta and thymus. TXB_2 is also found in many tissues including lung, spleen, lymphocytes, leukocytes, macrophages, brain and kidney [25]. The expression of THXS is observed in cell lines including NCI-H596, Wi-38, IC21 and HL-60 [26, 27].

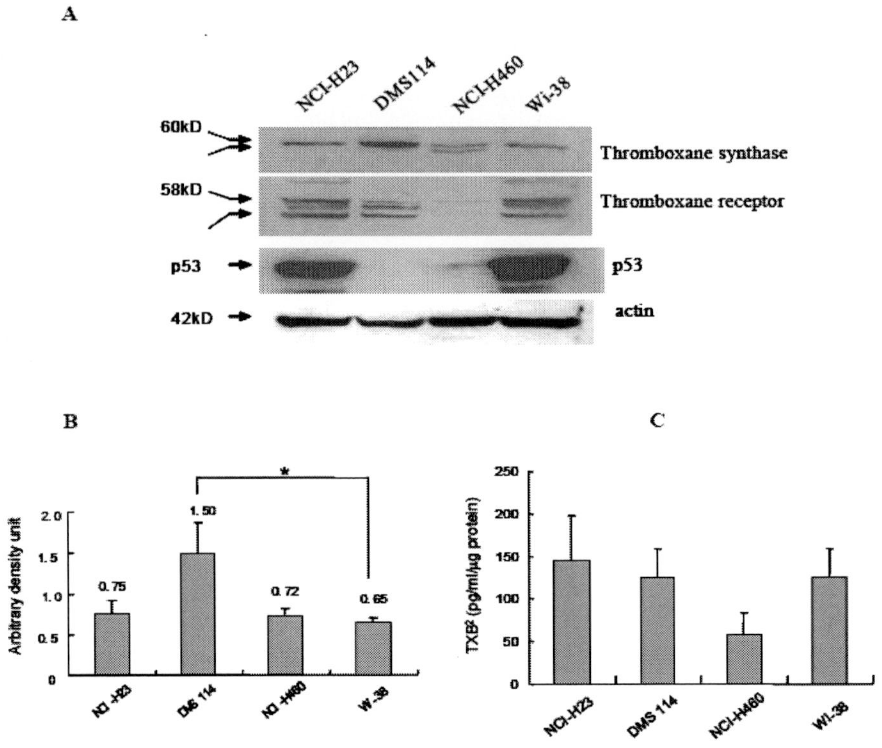

Figure 2. The levels of THXS, thromboxane receptor and TXB_2 in lung cancer cells. **A**) The expression of THXS (60kD and 52kD), thromboxane receptors (58 and 55kD), p53 (53 kD) and actin (42kD) in lung cancer cells NCI-H23, DMS 114 and NCI-H460. Lung fibroblast cells Wi-38 were used as a positive control. B) Protein expression level of THXS in different lung cell lines. Densitometry quantification using Biorad GS-700 imaging densitometer of the data shown in **B**). Average value was obtained from three independent experiments and normalized by actin. * represents statistically significant ($P < 0.05$) compared to Wi-38. **C**) TXB_2 level in culture medium in different lung cancer cells were measured by ELISA. The concentrations were divided by protein concentrations in each cell line to obtain a unit of (pg/ml/μg protein).

Our data showed that THXS and thromboxane receptors were expressed in lung cancer cells and normal lung fibroblast Wi-38. [Figure.2A]. Human lung fibroblast Wi-38 has been documented to express THXS and produce TXA_2 [28, 29], as a result it was used as a positive control in our study. Our experiment demonstrated that all of the lung cancer cell lines, including two NSCLC (NCI-H23 and NCI-H460) and one SCLC (DMS114) expressed the full length 60KD THXS whereas the C terminal truncated form with a 52 KD in size was expressed in NCI-H460. The level of THXS in NCI-H23 and NCI-H460 was higher than Wi-38 while the highest expression level was seen in DMS114 [Figure 2B].

Genomic Structure of THXS

The primary structure and the promoter sequence of THXS have been reported [14, 30, 31]. The gene spans 193 kb in size, which represents the largest CYP gene, and consists of 13 exons and 12 introns. Two transcription variants have been identified in human lung namely THXS-1 and THXS-2 [Figures 3A and 3B]. THXS-1 represents the longer protein, which contains 534 amino acids with a calculated MW of 60kDa. While the shorter THXS-2 is the C-terminal truncated form and contains 460 amino acids with a calculated MW of 52kDa. The amino acid sequence of THXS-1 is similar to those of other CYP members, and the genomic structure is similar to CYP family 3.

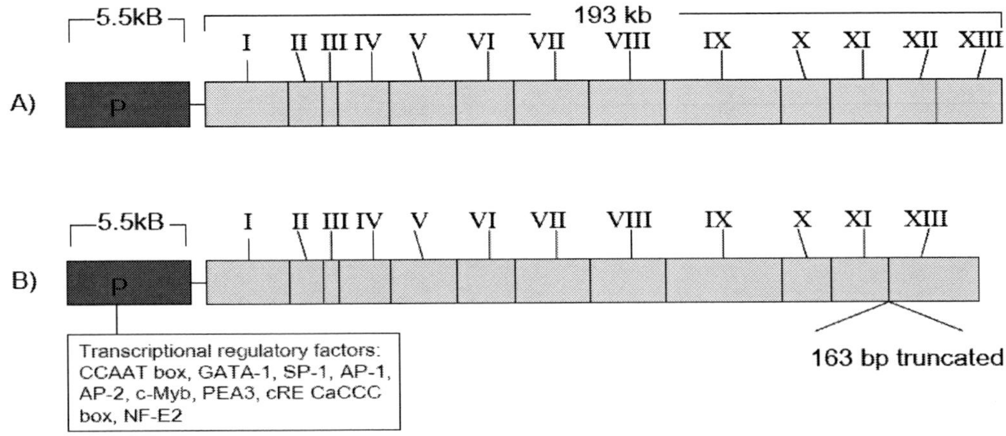

C)

Amino acid location	function
Ala^{408}, Arg^{413}, Arg^{410}, Glu^{433}	Important roles in substrate interaction and enzyme catalysis
N-terminal (around 40 amino acids), 2 segments aa 16-33 and aa 37-60	Anchoring to the endoplasmic reticular membrane
Cys^{480}	Act as 5th thiolate ligand to the heme moiety, the thromboxane synthase activity will be destroyed when replaced by serine
Amino acid 473-492 (FGAGPRSCLG)	Heme-binding domain

Figure 3. Genomic structure of human THXS. **A**) The full length of the THXS gene includes 13 exons as indicated by Roman numerals. P represents the promoter region. **B**) Splice variant of THXS which contains a full exon 12 deletion. This shorter form codes for 77 amino acids. Factors affecting the gene transcription are shown in the box under the promoter region. **C**) Tabulation of some of the important amino acids in THXS. Adapted from references [15, 29, 31, 99, 98].

Factors Affecting THXS Expression

In addition to the transcription regulatory elements found in the THXS promoter region, several factors have been identified to regulate the THXS expression. Among these factors, peroxisome proliferators-activated receptors gamma (PPARγ) is well known. PPARγis a transcriptional factor belonging to the ligand-activated nuclear receptor superfamily [32] and it is activated upon PPARγ ligand bindings [33-35]. THXS gene transcription can be regulated by PPARγ in different models. PPARγligands, 15-deoxy-$\Delta^{12,14}$-prostagladin J_2 (PGJ_2) and troglitazone (TRO) have been shown to dose-dependently inhibit the transcription of rat THXS mRNA in macrophages through the interaction of the activated PPARγ and NRF-2 binding site in THXS [36]. Moreover, the levels of TXA_2 [37, 38] and thromboxane receptor transcript expression [39, 40] can also be affected by PPARγ. In A549 lung cancer cell line NF-E2 related factor 2 (Nrf-2) increases the THXS promoter activity by 4 folds [41]. The wild type tumor suppressor protein p53 and proto-oncogenic factor ets-1 can also regulate THXS expression [42] and will be discussed in the later section.

THXS Inhibitors and Thromboxane Receptor Antagonist

Numerous THXS inhibitors and thromoboxane receptor antagonists have been identified. For examples the endoperoxide prostaglandin H_2 dervied thromboxane receptor antagonists SQ-29548, ifetroban and domitroban [43-45]. THXS inhibitors can be divided into two categories based on their chemical structures. Those with an imidzole ring include dazoxiben, dazmagrel, pimagrel, ozagrel and CS-518 [46-50] and those carried a pyridine ring are OKY1581, isobgrel and furegrelate [18-20] (Table 2). The chemical structures of some these agents are shown in [Figure 4].

Table 2. Thromboxane synthase inhibitors and thromboxane receptor antagonist

Thromboxane synthase inhibitors:
dazoxiben (UK37248), dazmagrel (UK 38485), pirmagrel (CGS 13080), OKY 1581, furegrelate,1-Benzylimidazole, NV-52
Thromboxane receptor antagonists:
ONO-3708, SQ-29548
Combined function with thromboxane synthase inhibitor and thromboxane receptor antagoinst:
BM-351, BM-567, ridogrel, picotamide

Source: adapted from references [51-54].

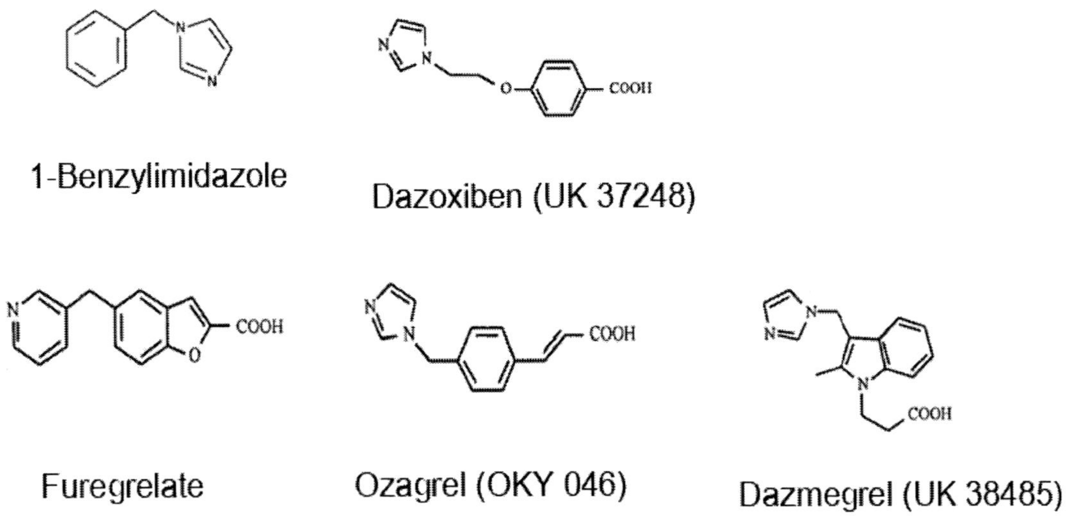

Figure 4. Chemical structures of thromboxane synthase inhibitors.

Thromboxane and Metastasis

Modulations of the thromboxane level can influence cancer metastases. There may be some communications between tumor cells and platelets and the metastatic stage may be correlated to the degree of platelet aggregation [55]. Some unidentified substances have been found to be released from platelets in order to promoter cancer cell growth. Thromboxane promotes platelet aggregation, and additionally the arachidonate product of prostaglandin endoperoxide H synthases is correlated to tumor angiogenesis. Therefore the unidentified substances are thought to be thromboxane A_2. Ketoconazole is an antagonist for both THXS and 5-Lipoxygenase (5-LOX) and it has been shown that lung metastasis from melanoma was reduced upon Ketoconazole administration [56]. TXA_2 agonist U-46619 stimulates the growth of endothelial cell migration while THXS inhibitors CI and furegrelate, and thromboxane receptor antagonist SQ29,548 significantly decrease the vascular endothelial growth factor/ basic fibroblast growth factor (VEGF/bFRF)-stimulated migrating effect [57]. An *in vivo* study has demonstrated that bFGF-induced angiogenesis in endothelial cell could be inhibited by CI. By using the mixture of COX inhibitor, THXS inhibitor and antithrombotic agent, lower pulmonary metastasis has been found in prostate tumor bearing animals [58], and a significant reduction (57%) in the metastatic liver tumor from colon cancer has been reported [59]. All these findings suggest that TXA_2 is a factor leading to cancer metastasis and that THXS inhibitors can function as a therapeutic agent to control metastatic spread.

Inhibition of Thromboxane Alters the Cell Migration, Angiogenesis and Apoptosis in Different Cancer Cells

Bladder Cancer Cells

THXS overexpresses in invasive bladder tumors [60]. When compared to the relative non-tumor tissues, THXS is highly expressed in most of the transitional cell carcinoma, squamous cell carcinoma and adenocarcioma. Moreover, 69.5% of invasive tumors exhibit elevated levels of THXS. In bladder cancer cell lines (T24 and TCC-SUP), thomboxane synthase inhibitors and thromboxane receptor antagonist reduce cell growth, migration and invasion whereas thromboxane agonist U46619 stimulates cell migration and invasion. Additionally, the level of thromboxane receptor protein is significantly elevated in tumor tissues. The finding suggests that TXA_2 pathway is involved in pathogenesis of bladder cancer.

Further study on THXS and bladder cancer has demonstrated that caspase 3, an executioner in caspase dependent apoptotic pathway was activated while level of surviving, an inhibitor for apoptosis was decreased when bladder cancer cells were treated with THXS inhibitors furegrelate and ozagrel [61], suggesting that TXA_2 may play a role in the promotion of cell survival and growth. In the presence of furegrelate and ozagrel, the number of cell death caused by paclitaxel and cisplatin was enhanced, suggesting that using THXS inhibitors increase the sensitivity of tumor cells to paclitaxel and cisplatin. Cell growth and migration are significantly decreased in th cells when THXS is silenced by THXS siRNA.

Breast Cancer

THXS level is not statistically increased in breast cancer tumor tissues compared to normal mammary tissues. However, the level of thromboxane receptor is higher in tumor tissues [62]. RNA analysis demonstrated that the thromboxane receptor expression was increased while THXS transcript level was decreased in the higher grade of breast cancer. It appears that thromboxane receptor level can be used as a prognostic marker in breast cancer because shorter disease survival is related to the higher level of thromboxane receptor. Another clinical study found that the high level of TXB_2 in breast cancer patients was associated with high lymph node metastases and large tumor sizes [24].

Colorectal Carcinoma

The level of COX-2 is upregulated in colorectal cancer [63]. The expression of THXS is found in both mRNA and protein levels in colorectal cells. In addition, THXS immunoreactivity was higher in colorectal cancer tissues than that in adjacent epithelial cells. However there was no mRNA expression level difference of thromboxane receptor between carcinoma and normal mucosa. The results suggest that THXS rather than thromboxane receptor is involved in the pathogenesis of colorectal cancer. This notion is further supported by *in vitro* studies which show that TXA_2 analogue STA_2 significantly accelerates the proliferation of colorectal cancer cells whereas THXS inhibitors OKY-1581 and Y-20811 and antisence oligonucleotide of THXS significantly inhibit it.

Glioma Cells

THXS inhibitor furegrelate can induce caspase activation, DNA fragmentation and subsequent cell death in glioma cells. The increase of DNA fragmentation by furegrelate can be inhibited by Z-VAD, a caspase inhibitor [64]. This indicates that the apoptosis caused by furegrelate is associated with a caspase-dependent pathway. Treatment of glioma cells with THXS inhibitors can produce a synergistic effect on the induction of apoptosis by camptothecin. Furegrelate can aslo block cell motility of glioma cells [65].

Lung Cancer

Several studies have been performed in the last 10 years to investigate thromboxane in lung cancer. Firstly, an *in vivo* study of 52 NSCLC samples indicated that TXB_2 level was significantly higher in tumor tissues than non-tumor tissues [66] and that the level was highly correlated with tumor stages. Secondly, thromboxane level was found to be higher in lung tissues in patients with smoking habit [66, 67]. The level of thromboxane has also been investigated in various types of lung cancers. For example, the level of TXB_2 has been found to be increased in lung squamous cell carcinoma, adenocarcinoma, small cell carcinoma, mixed cell carcinoma, bronchioloalveolar cell carcinoma, large cell undifferentiated carcinoma, bronchial carcinoid and metastatic tumors [68, 69]. The increase is particularly obvious in smokers with lung cancer. Human lung cancer carcinoma cell lines, including Calu-3, Calu-6, A549, have been demonstrated to be able to synthesize thromboxane from endogenous arachidonic acid [70], suggesting that the increase of thromboxane in lung cancer tissues may locally produced.

Our enzyme immunoassay showed that the TXB_2 level in culture medium of NCI-H23 and DMS114 were higher than that in Wi-38 and that the lowest TXB_2 level was detected in NCI-H460 [Figure 2C]. The levels of TXB_2 were approximately matched to the levels of THXS in these cell lines [Figure 2B]. These findings are reasonable as the regulation of TXA_2 level and its stable metabolite TXB_2 synthesis highly depend on the THXS activity and availability [28]. Therefore it is expected that the change in the level of TXB_2 in cells should be in proportion to the protein level of THXS. However, it is noted that the exogenous arachidonic acid is able to differentially stimulate the biosynthesis of thromboxane in lung cancer cells [70], suggesting that not only THXS but also the substrate availability for the reaction are important for the biosynthesis of thromboxane.

Taken together, the thromboxane level is generally elevated in lung cancer and such the elevation is most likely due to the increase of THXS. Cigarette smoking may be a factor that contributes to the over-production of thromboxane because it appears that the level of thromoboxane is positively correlated to the status of smoking in lung cancer patients [66, 67].

THXS and P53

Information on the relationship between thromboxane and p53 is not fully elucidated. The transcriptional factor p53 plays a critical role in cancer progression as it is identified to be involved in cell cycle control, DNA repair, cell differentiation and apoptosis by regulating multiple downstream targets such as bcl-2 family members, p21, p27 and mitogen-activated protein kinase (MARK) [71-75]. Mutation of p53 frequently occurs in different types of cancers including lung cancer [76]. p53 mutation has been found in 50% and 70% in NSCLC and SCLC respectively [77]. Our study showed that p53 protein was strongly expressed in NCI-H23 and Wi-38; slightly expressed in NCI-H460 and no p53 was detected in DMS114. NCI-H23 carries a p53 mutation at codon 246; DMS114 harbors a C-terminal truncated p53 while p53 is wild type in NCI-H460 and Wi-38 [78, 79]. The level of THXS in these cell lines may correlate with the p53 level. The highest level of THXS is observed in DMS114 that is with undetectable and mutated p53, and the low level of THXS expresses in Wi-38 and NCI-H460 both of which contain wild-type p53. THXS expression in NCI-H23 is moderate and this moderate level of THXS may also be related to p53 because the function of p53 is partially retained in NCI-H23 [78]. These findings appear to suggest that the expression of THXS is negatively associated with the level of p53 protein in lung cancer cells.

Previous study indicated that p53 is a transcriptional regulator of THXS in glioblastoma-derived cells [42]. The functional p53 is able to inhibit the expression of THXS. However, other factors also participate in the expression of THXS regulated by p53. Among these other factors, ets-1 is important. It has been demonstrated that the expression of THXS is co-regulated by both p53 and ets-1 in an antagonistic and inter-related manner, with ets-1 being a potent transcriptional activator and p53 inhibiting ets-1-dependent transcription [42]. Negative interference with ets-1 transcription requires functional p53 and is lost in cells with mutant p53 proteins. Thus wild type but not hot spot mutated p53 can negatively interfere with ets-1. ets-1 and p53 can physically associate each other *in vitro* and *in vivo* and their interaction, rather than a direct binding of p53 to the promoter of THXS, is required for the transcriptional repression of THXS by wild-type p53. Further experiments showed that the C-terminal region of p53 is essential for the p53-ets-1 association and interaction [43]. The finding is well supported by our study which shows that DMS114 cells carried a C-terminal truncated p53 express the highest level of THXS because the p53 mutation status in DMS114 cells abrogates the ability of p53 to inhibit ets-1 and thus make it being unable to control the expression of THXS. It has been noted that some of the mutated p53 are still able to interact with ets-1 [80]. This can also explain that NCI-H23 cells with a p53 point mutation can still express a reasonable level of THXS. The function of p53 is partially retained in NCI-H23 [78], though they contain a p53 mutation at codon 246. Future studies are deserved to explore in details the relationship between the pattern of p53 mutations and the level of THXS in lung cancer.

Effect of THXS Inhibitors on Lung Cancer Cell Proliferation and Apoptosis

The role of thromboxane in lung cancer cell death and apoptosis is not well defined. An *in vitro* study using human lung adenocarcinoma derived cell lines PC/9 and PC-9/CDDP reported that TXA_2 antagonist ONO-NT-126 and S-1452 significantly enhanced the apoptotic effect caused by cisplatin [81]. A similar result was obtained using another antagonist AA-2414 in the same cell lines [82]. Results from other studies support the anti-cancer features of THXS inhibitors in different types of cancers [61, 64, 65, 83].

We have used 1-Benzylimidazole (1-BI), one of the THXS inhibitors, to monitor its effect on lung cancer cell growth. Our study showed that the proliferation rate measured by MTT assay in lung cancer cells was decreased by I-BI in a time- and dose-dependent manner [Figure 5A]. By using tryphan blue exclusion assay, we demonstrated that the number of death cells in NCI-H23 and DMS114 cells was not significantly altered up to 72 hours after 1-BI treatment. On the other hand, direct cell counting demonstrated that cell death occurred obviously in NCI-H460 cells with over a ten-fold elevation at 72 hours after 1-BI treatment [Figure 5B]. Based on the above results, we believe that 1-BI triggered lung cancer cell death and the mechanism was related to the p53 status in the cells.

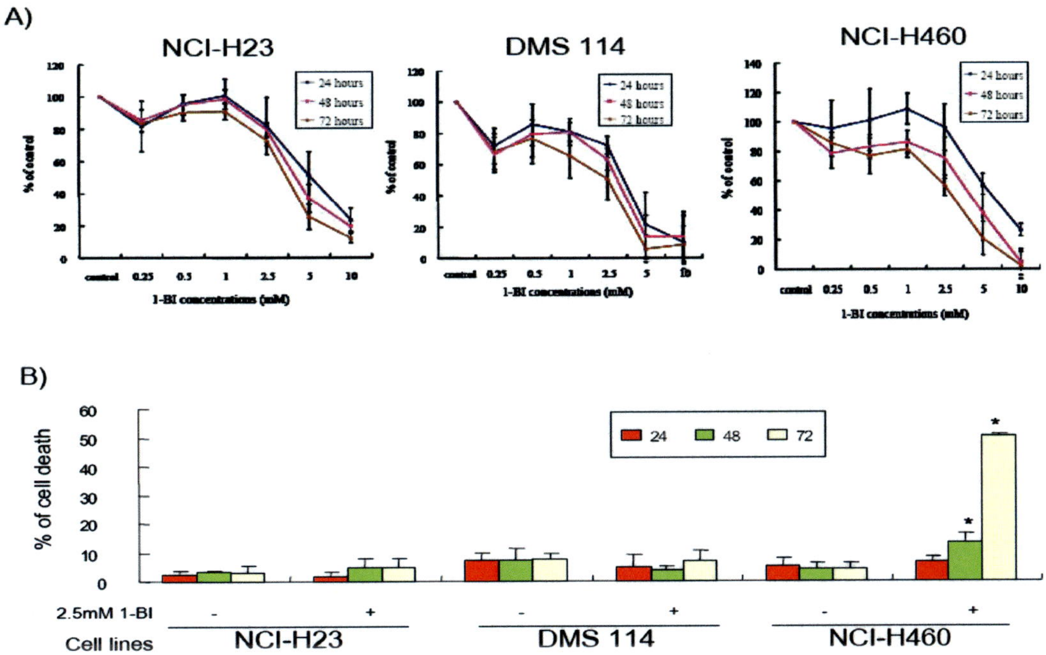

Figure 5. Effect of 1-BI on lung cancer cell proliferation and death. **A**) MTT assay measuring cell proliferation in different cells treated by 1-BI for 24, 48, 72 hours. **B**) Cell death was quantified by tryphan blue assay upon 2.5mM 1-BI treatment at 24, 48 and 72 hours. Index was calculated by (number of positive stained cells/number of negative stained cells)*100%. All of the experiments were done in triplicate. * represents significance ($p<0.05$), compared treated cells (2.5mM 1-BI) with untreated control.

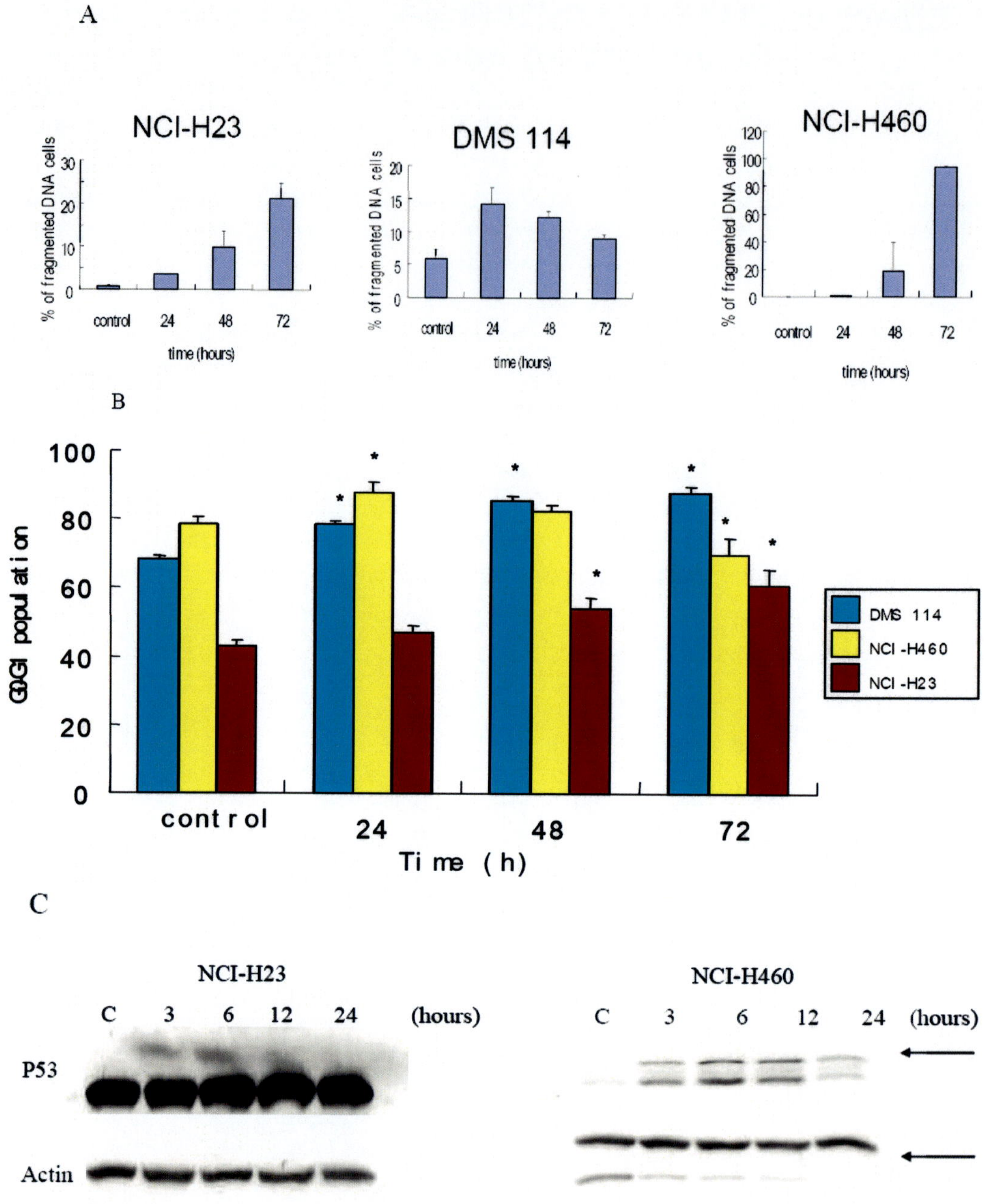

Figure 6. **A**. DNA fragmentation measured by TUNEL assay. Cells were treated by 2.5mM 1-BI, collected at 24, 48 and 72 hours and fixed by ethanol overnight and stained by APO-DIRECT™ TUNEL ASSAY kit. Fragmented DNA was detected using flow cytometry. **B**. Cell cycle analysis by PI staining. Cells were treated by 2.5mM 1-BI, collected at 24, 48 and 72 hours and fixed by ethanol overnight, followed by 20μg PI staining. G1 stage cells were detected using flow cytometry. *represents significance ($p<0.05$) between treated and untreated groups. **C**. p53 protein expression after 1-BI treatment at indicated time points. NCI-H23 and NCI-H460 were incubated with 2.5mM 1-BI and total protein was extracted. Experiment was repeated twice with similar result.

It appears that different lung cancer cells may not harbor the same molecular changes and the variation in the statue of p53 is one example. Therefore, it is likely that the mechanism responsible for THXS inhibitors-mediated cell growth inhibition may not be the same in different lung cancer cells. Both MTT and tryphan blue exclusion assay cannot define the mode of cell death. The anti-tumor function of 1-BI may be caused by different modes of molecular mechanisms such as inducing cell cycle arrest, cell senescence and apoptosis. Therefore we performed cell cycle analysis and DNA fragmentation assay to examine the 1-BI-mediated anti-tumor effect. Apoptosis is an active energy dependent process and it can be caused by intrinsic and extrinsic pathway and triggered DNA fragmentation [84]. We measured the number of fragmented DNA in the three lung cancer cells upon 1-BI treatment by using terminal dexoynucleotidyl transferease-dUTP nick end labeling (TUNEL) assay. This assay based on the binding of terminal dexoynucleotidyl transferease to the exposed 3'-OH ends of fragmentated DNA [85]. In brief the number of fragmentated DNA showed a time-dependent increase in NCI-H23 and NCI-H460, with a significant increase measured at 48 hours and 72 hours after the treatment [Figure 6A]. A significant increase in apoptosis occurred earlier (24 hours after the treatment) in DMS114.

Several chemotherapeutic drugs exert their anti-tumor effect by promoting cell cycle arrest [86, 87]. Cell cycle analysis was then performed by using PI staining [Figure 6B]. Upon 1-BI treatment, G1 phase cell cycle arrest occurred in NCI-H23 and DMS114 in a time-dependent manner, while it was peaked at 24 hours and gradually decreased at 48 and 72 hours after the treatment in NCI-H460.

According to the data from TUNEL and PI assay, it appears that lung cells with wild type p53 may be important for cell death induction by 1-BI treatment, while other mechanisms such as cell cycle arrest may be involved in the cell growth inhibition in mutated p53 cell lines. There is no doubt that different mechanisms would be triggered by chemotherapy which depends on the availabilities and status of p53 because this tumor suppressor protein possesses the ability to control both cell cycle and apoptosis in cancer cells and plays a central role in cancer treatment response [88]. Similar to lung cancer cells, glioma cells with mutated p53 responded with apoptosis while glioma cells with wild-type p53 underwent G2-M cell cycle arrest [74]. Furthermore, silencing p53 by using antisense oligonucleotides can change the response of the cells from cell cycle arrest to apoptosis.

Although p53 is essential in determining the alive or death of cancer cells and thromboxane has a role in cancer development, there are limited publications on thromboxane and p53. A previous study indicated that incubation of 10μM TXA_2 with human trophoblast triggered apoptosis characterized by increasing TUNEL positive cells and it also induced a time- and dose-dependent p53 expression [89]. However, other studies, such as TXB_2 -mediated apoptosis in rat hepatocytes [90]; apoptosis of immature thymocytes triggered by lipopolysaccharide (LPS) which is partially mediated by TXB2 [91]; the apoptotic event caused by thromboxane sythanse inhibitor furegrelate in glioma cells [65]; fail to show a correlation with p53. Our immunobloting analysis showed that the p53 level was strongly increased in NCI-H460 but not in NCI-H23 upon 1-BI treatment [Figure 6C] while no p53 expression was detected in DMS114, suggesting that the wild-type p53 but not mutated p53 in lung cancer cells responds to 1-BI treatment and p53 may be a factor that contributes to I-BI-mediated apoptotosis. In order to further prove the role of p53 in I-BI-mediated apoptosis in lung cancer cells, we transfected a wild-type p53 plasmid into NCI-H23 and DMS114 cells. After RT-PCR confirmed that p53 was overexpressed in the cells after transfection [Figure 7A], the cells were treated with I-BI and the apoptosis was determined by annexin V and PI staining. The result showed that DMS114 cells that is originally without a functional p53 and relatively resistant to I-BI treatment could well respond to I-BI after p53 transfection [Figures 7B and 7C]. The percentage of apoptotic cells induced in p53-transfected cells was much higher than those with p53 transfection. The finding provides another piece of evidence to support that p53 plays a role in the apoptosis induced by 1-BI treatment. Therefore, it is reasonable to propose that the THXS inhibitors may serve as an anti-cancer agent against lung cancer.

The relation between p53 and THXS inhibitors may mimic that between p53 and other chemotherapeutic agents used in lung cancer. For example, the chemosensitivity of NSCLC to cisplatin, etoposide and camptothecin is well dependent on the status of p53 in the tumor cells. Introduction of the wild-type p53 into mutated NSCLC cell lines (H1437, H727, H441 and H1299) enhances the cell death induced by these chemotherapeutic agents [88]. In contrast, introduction of mutated p53 into wild type p53 cells line (H460) attenuates the cell death induced by these agents. Another study also showed that the overexpression of p53 in NCI-H460 cells increased the apoptosis induced by 5-aza-2'deoxyctidine treatment [92].

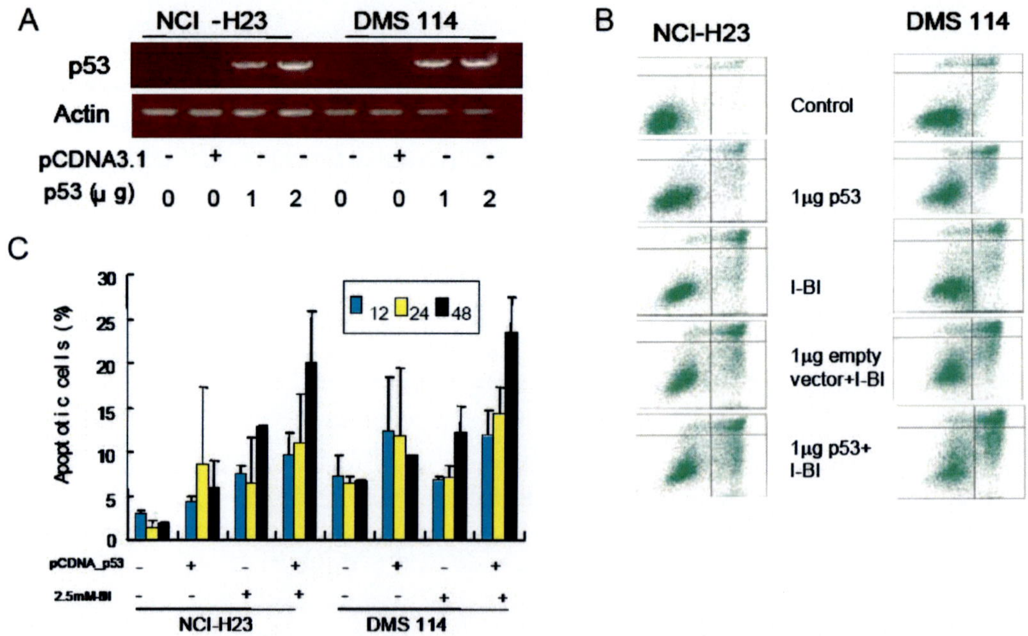

Figure 7. p53 enhances the effect of I-BI on apoptosis. **A**. The level of p53 was determined by RT-PCR in NCI-H23 and DMS 114 cells with or without p53 transfection. **B**. NCI-H23 and DMS 114 cells were transiently transfected with wild type p53 and then treated with 2.5mM 1-BI for 48 hours. Cell death was measured by Annexin V/PI staining. Cells in the lower right corner are early apoptotic cells which are positively stained with Annexin V but negatively stained with PI. **C**. Graph summarizes the number of apoptotic cells (positive for annexin V) after treatment with 2.5 mM 1-BI for 12, 24, 48 hours. Experiments were repeated twice and similar results were obtained.

All these data support that the wild-type p53 enhances the chemosensitivity [77, 93, 94], whereas the loss of p53 leads to resistance of lung cancer cells to anti-cancer therapy [95, 96]. However, it should be noted that the p53 may not be a sole molecule to determine the apoptosis of lung cancer cells induced by anti-cancer drugs. In the case of THXS inhibitors, we have observed that even in the p53 dysfunctional DMS114 cells, I-BI can still induce apoptosis in about 15% of the cells. Therefore, factors other than p53 should also participate in the cell death induced by I-BI.

Conclusion and Future Prospects

Ample evidence suggested that COX pathway and its downstream prostanoids are highly related to cancer progression and the expression of COX-2 is upregulated in several types of cancers including lung cancer [97-99]. The Non-steroidal anti-inflammatory drugs (NSAID) have been widely used in the prevention or treatment of cancers, by reducing the production of prostaglandins. Although NSAID are effective against cancers, the non-selective blocking of the prostaglandin production has been shown to generate a series of the side effects including dyspepsia and increased risk to cardiovascular diseases [97, 100]. Therefore, targeting the downstream metabolic pathway of COX such as thromboxane synthesis should significantly reduce the side effect as it will not affect other prostaglandin production.

The use of thromboxane synthesis inhibitors should also be more effective in the treatment of certain cancers as it will target the molecules that are associated with given cancers. Current experiments have shown a great potential in use of thromboxane-related inhibitors for the treatment of certain cancers including lung cancer. The thromboxane synthesis inhibitiors such as I-BI may not only suppress the growth of lung cancer but also prevent the metastasis of lung cancer. It appears that the statue of p53 in cancer cells is closely associated with the effect of thromboxane synthesis inhibitors. However how thromboxane synthesis inhibitors suppress cancer cell proliferation is yet fully elucidated. There is also lack of *in vivo* data to verify the role of thromboxane in the development of cancer and to support the anti-cancer property of thromboxane synthesis inhibitors.

Acknowledgement

The study was supported by a direct grant from CUHK (Project No. 2006.1.085).

References

[1] Harris, R. E., Beebe-Donk J, and Schuller, H. M. (2002) Chemoprevention of lung cancer by non-steroidal anti-inflammatory drugs among cigarette smokers. *Onco.l Rep.*, *9*, 693-695.

[2] Nicholson, S. A., Beasley, M. B., Brambilla, E., Hasleton, P. S., Colby, T. V., Sheppard, M. N., Falk, R. and Travis, W. D. (2002) Small cell lung carcinoma

(SCLC): a clinicopathologic study of 100 cases with surgical specimens. *Am. J. Surg. Pathol., 26*, 1184-1197.

[3] Sekido, Y., Fong, K. M. and Minna, J. D. (1998) Progress in understanding the molecular pathogenesis of human lung cancer. *Biochim. Biophys. Acta, 1378*, F21-59.

[4] Lam, T. H., Li, Z. B., Ho, S. Y., Chan, W. M., Ho, K. S., Tham, M. K., Cowling, B. J., Schooling, C. M. and Leung G. M. (2007) Smoking, quitting and mortality in an elderly cohort of 56,000 Hong Kong Chinese. *Tob. Control, 16*, 182-189.

[5] Duarte, R. L. and Paschoal, M. E. (2006) Molecular markers in lung cancer: prognostic role and relationship to smoking. *J. Bras. Pneumol., 32*, 56-65.

[6] Rom, W. N., Hay, J. G., Lee, T. C., Jiang, Y. X. and Tchou-Wong, K. M. (2000) Molecular and genetic aspects of lung cancer. *Am. J. Respir. Crit. Care Med., 161*, 1355-1367.

[7] Sanchez-Cespedes, M., Ahrendt, S. A., Piantadosi, S., Rosell, R., Monzo, M., Wu, L., Westra, W. H., Yang, S. C., Jen, J., Sidransky, D. (2001) Chromosomal alterations in lung adenocarcinoma from smokers and nonsmokers. *Cancer Res., 61*, 1309-1313.

[8] Niklinski, J., Niklinska, W., Laudanski, J., Chyczewska, E. and Chyczewski, L. (2001) Prognostic molecular markers in non-small cell lung cancer. *Lung Cancer, 34*, S53-58.

[9] Schneider, J. (2006) Tumor markers in detection of lung cancer. *Adv. Clin. Chem., 42, 1-41.*

[10] Bharti, A., Ma, P. C. and Salgia, R. (2007) Biomarker discovery in lung cancer--promises and challenges of clinical proteomics. *Mass. Spectrom. Rev., 26*, 451-466.

[11] Gharavi, N. and El-Kadi, A. O. (2004) Expression of cytochrome P450 in lung tumor. *Curr. Drug Metab., 5*, 203-210.

[12] Haurand, M. and Ullrich,V. (1985) Isolation and characterization of thromboxane synthase from human platelets as a cytochrome P-450 enzyme. *J. Biol. Chem., 260*, 15059-15067.

[13] Shen, R. F. and Tai, H. H. (1986) Monoclonal antibodies to thromboxane synthase from porcine lung. Production and application to development of a tandem immunoradiometric assay. *J. Biol. Chem., 261*, 11585-11591.

[14] Miyata, A., Yokoyama, C., Ihara, H., Bandoh, S., Takeda, O., Takahashi, E. and Tanabe, T. (1994) Characterization of the human gene (TBXAS1) encoding thromboxane synthase. *Eur. J. Biochem., 224*, 273-279.

[15] Shen, R. F. and Tai, H. H. (1986) Immunoaffinity purification and characterization of thromboxane synthase from porcine lung. *J. Biol. Chem., 261*, 11592-11599.

[16] Hamberg, M., Svensson, J. and Samuelsson, B. (1975) Thromboxanes: a new group of biologically active compounds derived from prostaglandin endoperoxides. *Proc. Natl. Acad. Sci U S A, 72*, 2994-2998.

[17] Needleman, P., Moncada, S., Bunting, S., Vane, J. R., Hamberg, M. and Samuelsson, B. (1976) Identification of an enzyme in platelet microsomes which generates thromboxane A2 from prostaglandin endoperoxides. *Nature, 261*, 558-560.

[18] Mehta, J. L., Lawson, D., Mehta, P. and Saldeen, T. (1988) Increased prostacyclin and thromboxane A2 biosynthesis in atherosclerosis. *Proc. Natl. Acad. Sci. U S A,* 85, 4511-4515.

[19] FitzGerald, G. A., Austin, S., Egan, K., Cheng, Y. and Pratico, D. (2000) Cyclo-oxygenase products and atherothrombosis. *Ann. Med., 32*, 21-26.

[20] Pratico, D., Cyrus, T., Li, H. W. and FitzGerald, G. A. (2000) Endogenous biosynthesis of thromboxane and prostacyclin in 2 distinct murine models of atherosclerosis. *Blood, 96,* 3823-3826.

[21] Whittle, B. J., Kauffman, G. L. and Moncada, S. (1981) Vasoconstriction with thromboxane A2 induces ulceration of the gastric mucosa. *Nature, 292*, 472-474.

[22] Fitzgerald, D. J., Rocki, W., Murray, R., Mayo, G. and Fitzgerald, G. A. (1990) Thromboxane A2 synthesis in pregnancy-induced hypertension. *Lancet, 335*, 751-754.

[23] Nanji, A. A. (1993) Thromboxane synthase and organ preference for metastases. *N Engl. J. Med., 329*, 138-139.

[24] Karmali, R. A., Welt, S., Thaler, H. T. and Lefevre, F. (1983) Prostaglandins in breast cancer: relationship to disease stage and hormone status. *Br. J. Cancer, 48*, 689-696.

[25] Hamberg, M. (1976) On the formation of thromboxane B2 and 12l-hydroxy-5,8,10,14-eicosatetraenoic acid (12 ho-20:4) in tissues from the guinea pig. *Biochim. Biophys. Acta, 431*, 651-654.

[26] Yokoyama, C., Miyata, A., Suzuki, K., Nishikawa, Y., Yoshimoto, T., Yamamoto, S., Nusing, R., Ullrich, V. and Tanabe, T. (1993) Expression of human thromboxane synthase using a baculovirus system. *FEBS Lett, 31*8, 91-94.

[27] Wang, L. H., Tazawa, R., Lang, A. Q. and Wu, K. K. (1994) Alternate splicing of human thromboxane synthase mRNA. *Arch. Biochem. Biophys., 315*, 273-278.

[28] Shen, R. F. and Tai, H. H. (1998) Thromboxanes: synthase and receptors. *J. Biomed. Sci., 5*, 153-172.

[29] Hunter, S. A., Burstein, S. and Sedor, C. (1984) Stimulation of prostaglandin synthesis in WI-38 human lung fibroblasts following inhibition of phospholipid acylation by p-hydroxymercuribenzoate. *Biochim. Biophys. Acta, 793*, 202-212.

[30] Ohashi, K., Ruan, K. H., Kulmacz, R. J., Wu, K. K. and Wang, L. H. (1992) Primary structure of human thromboxane synthase determined from the cDNA sequence. *J. Biol.Chem., 267*, 789-793.

[31] Tazawa, R., Green, E. D., Ohashi, K., Wu, K. K. and Wang, L. H. (1996) Characterization of the complete genomic structure of human thromboxane synthase gene and functional analysis of its promoter. *Arch. Biochem. Biophys., 334,* 349-356.

[32] Mangelsdorf, D. J. and Evans, R. M. (1995) The RXR heterodimers and orphan receptors. *Cell, 83*, 841-850.

[33] Shibata, T., Matsui, K., Nagao, K., Shinkai, H., Yonemori, F. and Wakitani, K. (1999) Pharmacological profiles of a novel oral antidiabetic agent, JTT-501, an isoxazolidinedione derivative. *Eur. J. Pharmacol., 364,* 211-219.

[34] Suh, N., Wang, Y. P., Williams, C. R., Risingsong, R., Gilmer, T., Willson, T. M. and Sporn, M. B. (1999) A new ligand for the peroxisome proliferator-activated receptor-gamma (PPAR-gamma), GW7845, inhibits rat mammary carcinogenesis. *Cancer Res., 59*, 5671-5673.

[35] Li, M. Y., Lee, T. W., Yim, A. P. C., Mok, T. S. K. and Chen. G. G. (2006) Apoptosis induced by troglitazone is both peroxisome proliferator-activated receptor-gamma- and ERK-dependent in human non-small lung cancer cells. *J. Cell. Physiol., 209*, 428-438.

[36] Ikeda, Y., Sugawara, A., Taniyama, Y., Uruno, A., Igarashi, K., Arima, S., Ito, S. and Takeuchi, K. (2000) Suppression of rat thromboxane synthase gene transcription by peroxisome proliferator-activated receptor gamma in macrophages via an interaction with NRF2. *J. Biol. Chem., 275*, 33142-33150.

[37] Yamazaki, T., Yokoo, T., Tomioka, Y., Suzuki, H., Hishinuma, T. and Mizugaki, M. (2002) Effect of peroxisome proliferator-activated receptor gamma on thromboxane A(2) and prostaglandin E(2) production in macrophage cell lines. *Prostaglandins Leukot Essent Fatty Acids, 67*, 245-251.

[38] Hishinuma, T., Yamazaki, T. and Mizugaki, M. (2000) Troglitazone has a reducing effect on thromboxane production. Prostaglandins *Other Lipid Mediat, 62*, 135-143.

[39] Sugawara, A., Uruno, A., Kudo, M., Ikeda, Y., Sato, K., Taniyama, Y., Ito, S. and Takeuchi, K. (2002) Transcription suppression of thromboxane receptor gene by peroxisome proliferator-activated receptor-gamma via an interaction with Sp1 in vascular smooth muscle cells. *J. Biol. Chem., 277*, 9676-9683.

[40] Coyle, A. T., O'Keeffe, M. B. and Kinsella, B. T. (2005) 15-deoxy Delta12,14-prostaglandin J2 suppresses transcription by promoter 3 of the human thromboxane A2 receptor gene through peroxisome proliferator-activated receptor gamma in human erythroleukemia cells. *Febs J., 272*, 4754-4773.

[41] Yaekashiwa, M. and Wang, L. H. (2003) Nrf2 regulates thromboxane synthase gene expression in human lung cells. *DNA Cell Biol., 22*, 479-487.

[42] Kim, E., Gunther, W., Yoshizato, K., Meissner, H., Zapf, S., Nusing, R. M., Yamamoto, H., Van Meir, E. G., Deppert, W. and Giese, A. (2003) Tumor suppressor p53 inhibits transcriptional activation of invasion gene thromboxane synthase mediated by the proto-oncogenic factor ets-1. *Oncogene, 22*, 7716-7727.

[43] Kaneko, N., Masuyama, J. I., Nara, H., Hirata, D., Iwamoto, M., Okazaki, H., Minota, S. and Yoshio, T. (2002) Production of thromboxane A2 and prostaglandin i2 affected by interaction of heat aggregated IgG, endothelial cells, and platelets in lupus nephritis. *J. Rheumatol., 29*, 2106-2113.

[44] Patrono, C., Ciabattoni, G., Remuzzi, G., Gotti, E., Bombardieri, S., Dimunno, O., Tartarelli, G., Cinotti, G. A., Simonetti, B. M. and Pierucci, A. (1985) Functional significance of renal prostacyclin and thromboxane A2 production in patients with systemic lupus erythematosus. *J. Clin. Invest., 76*, 1011-1018.

[45] Reilly, I. A., Roy, L. and Fitzgerald, G. A. (1986) Biosynthesis of thromboxane in patients with systemic sclerosis and Raynaud's phenomenon. *Br. Med. J. (Clin. Res. Ed.), 292*, 1037-1039.

[46] Saldeen, T. G., Saldeen, P., Nichols, W. W., Lawson, D. L., Nicolini, F. A. and Mehta, J. L. (1993) Increased production of thromboxane A2 by coronary arteries after thrombolysis. *Am. Heart J., 125*, 277-284.

[47] Saldeen, P., Nilsson, I. M. and Saldeen, T. (1983) Increased synthesis of thromboxane B2 and 6-keto-PGF1 alpha in hand veins from patients with deep venous thrombosis. *Thromb. Res., 32*, 461-467.

[48] Fitzgerald, D. J., Wright, F. and FitzGerald, G. A. (1989) Increased thromboxane biosynthesis during coronary thrombolysis. Evidence that platelet activation and

thromboxane A2 modulate the response to tissue-type plasminogen activator in vivo. *Circ. Res., 65*, 83-94.

[49] Jankov, R. P., Belcastro, R., Ovcina, E., Lee, J., Massaeli, H., Lye, S. J. and Tanswell, A. K. (2002) Thromboxane A(2) receptors mediate pulmonary hypertension in 60% oxygen-exposed newborn rats by a cyclooxygenase-independent mechanism. *Am. J. Respir. Crit. Care Med., 166*, 208-214.

[50] Dogne, J. M., de Leval, X., Benoit, P., Rolin, S., Pirotte, B. and Masereel, B. (2002) Therapeutic potential of thromboxane inhibitors in asthma. *Expert Opin. Investig Drugs, 11*, 275-281.

[51] Dogne, J. M., Hanson, J., de Leval, X., Pratico, D., Pace-Asciak, C. R., Drion, P., Pirotte, B. and Ruan, K. H. (2006) From the design to the clinical application of thromboxane modulators. *Curr. Pharm. Des., 12*, 903-923.

[52] Johnson, R. A., Nidy, E. G., Aiken, J. W., Crittenden, N. J. and Gorman, R. R. (1986) Thromboxane A2 synthase inhibitors. 5-(3-Pyridylmethyl)benzofuran-2-carboxylic acids. *J. Med. Chem., 29*, 1461-1468.

[53] Pettigrew, L. C., Grotta, J. C., Rhoades, H. M. and Wu, K. K. (1989) Effect of thromboxane synthase inhibition on eicosanoid levels and blood flow in ischemic rat brain. *Stroke, 20*, 627-632.

[54] Howes, L. G., James, M. J., Florin,T. and , Walker, C. (2007) Nv-52: a novel thromboxane synthase inhibitor for the treatment of inflammatory bowel disease. *Expert. Opin. Investig. Drugs, 16*, 1255-1266.

[55] Poggi, A., Vicenzi, E., Cioce, V. and Wasteson, A. (1988) Platelet contribution to cancer cell growth and migration: the role of platelet growth factors. *Haemostasis, 18*, 18-28.

[56] Nardone, P. A., Slotman, G. J. and Vezeridis, M. P. (1988) Ketoconazole: a thromboxane synthetase and 5-lipoxygenase inhibitor with antimetastatic activity in B16-F10 melanoma. *J. Surg. Res., 44*, 425-429.

[57] Nie, D., Lamberti, M., Zacharek, A., Li, L., Szekeres, K., Tang, K. Q., Chen, Y. C. and Honn, K. V. (2000) Thromboxane A(2) regulation of endothelial cell migration, angiogenesis, and tumor metastasis. *Biochem. Biophys. Res. Commun., 267*, 245-251.

[58] Drago, J. R. and Al-Mondhiry, H. A. (1984) The effect of prostaglandin modulators on prostate tumor growth and metastasis. *Anticancer Res., 4*, 391-394.

[59] Yokoyama, I., Hayashi, S., Kobayashi, T., Negita, M., Yasutomi, M., Uchida, K. and Takagi, H. (1995) Prevention of experimental hepatic metastasis with thromboxane synthase inhibitor. *Res. Exp. Med. (Berl), 195*, 209-215.

[60] Moussa, O., Yordy, J. S., Abol-Enein, H., Sinha, D., Bissada, N. K., Halushka, P. V., Ghoneim, M. A. and Watson, D. K. (2005) Prognostic and functional significance of thromboxane synthase gene overexpression in invasive bladder cancer. *Cancer Res., 65*, 11581-11587.

[61] Moussa, O., Riker, J. M., Klein, J., Fraig, M., Halushka, P. V. and Watson, D. K. (2007) Inhibition of thromboxane synthase activity modulates bladder cancer cell responses to chemotherapeutic agents. *Oncogene*, (in press).

[62] Watkins, G., Douglas-Jones, A., Mansel, R. E. and Jiang, W. G. (2005) Expression of thromboxane synthase, TBXAS1 and the thromboxane A2 receptor, TBXA2R, in human breast cancer. *Int. Semin. Surg. Oncol., 2*, 23.

[63] Sakai, H., Suzuki, T., Takahashi, Y., Ukai, M., Tauchi, K., Fujii, T., Horikawa, N., Minamimura, T., Tabuchi, Y., Morii, M., Tsukada, K. and Takeguchi, N. (2006) Upregulation of thromboxane synthase in human colorectal carcinoma and the cancer cell proliferation by thromboxane A2. *FEBS Lett, 580*, 3368-3374.

[64] Yoshizato, K., Zapf, S., Westphal, M., Berens, M. E. and Giese, A. (2002) Thromboxane synthase inhibitors induce apoptosis in migration-arrested glioma cells. *Neurosurgery, 50*, 343-354.

[65] Kurzel, F., Hagel, C., Zapf, S., Meissner, H., Westphal, M. and Giese, A. (2002) Cyclo-oxygenase inhibitors and thromboxane synthase inhibitors differentially regulate migration arrest, growth inhibition and apoptosis in human glioma cells. *Acta Neurochir (Wien), 144*, 71-87.

[66] Chen, G. G., Lee, T. W., Yip, J. H. Y., Xu, H., Lee, I. K. Y., Mok, T. S. K., Warner, T. D.and Yim, A. P. C. (2006) Increased thromboxane B(2) levels are associated with lipid peroxidation and Bcl-2 expression in human lung carcinoma. *Cancer Lett, 234*, 193-198.

[67] Tanizawa, H., Iwanaga, T. and Tai, H. H. (1987) Increase in thromboxane B2 and decrease in prostaglandin E2 and 6-keto-prostaglandin F1 alpha release into rat bronchoalveolar fluid as a consequence of cigarette smoking. *Prostaglandins Leukot Med, 28*, 195-201.

[68] McLemore, T. L., Hubbard, W. C., Litterst, C. L., Liu, M. C., Miller, S., McMahon, N. A., Eggleston, J. C. and Boyd, M. R. (1988) Profiles of prostaglandin biosynthesis in normal lung and tumor tissue from lung cancer patients. *Cancer Res., 48*, 3140-3147.

[69] Prisco, D., Paniccia, R., Coppo, M., Filippini, M., Francalanci, I., Brunelli, T., Comeglio, P. and Abbate, R. (1995) Platelet activation and platelet lipid composition in pulmonary cancer. *Prostaglandins Leukot Essent Fatty Acids, 53*, 65-68.

[70] Hubbard, W. C., Alley, M. C., McLemore, T. L. and Boyd, M. R. (1988) Evidence for thromboxane biosynthesis in established cell lines derived from human lung adenocarcinomas. *Cancer Res., 48*, 2674-2647.

[71] Wu, G. S. (2004) The functional interactions between the p53 and MAPK signaling pathways. *Cancer Biol. Ther., 3*, 156-61.

[72] Collot-Teixeira, S., Bass, J., Denis, F. and Ranger-Rogez, S. (2004) Human tumor suppressor p53 and DNA viruses. *Rev. Med. Virol., 14*, 301-319.

[73] Brown, L., Boswell, S., Raj, L. and Lee, S. W. (2007) Transcriptional targets of p53 that regulate cellular proliferation. *Crit. Rev. Eukaryot. Gene Expr., 17*, 73-85.

[74] Datta, K., Shah, P., Srivastava, T., Mathur, S. G., Chattopadhyay, P. and Sinha, S. (2004) Sensitizing glioma cells to cisplatin by abrogating the p53 response with antisense oligonucleotides. *Cancer Gene Ther., 11*, 525-531.

[75] Jiang, P., Du, W. and Wu, M. (2007) p53 and Bad: remote strangers become close friends. *Cell Res., 17*, 283-285.

[76] Levine, A. J. (1997) p53, the cellular gatekeeper for growth and division. *Cell, 88*, 323-331.

[77] Viktorsson, K., De Petris, L. and Lewensohn, R. (2005) The role of p53 in treatment responses of lung cancer. *Biochem. Biophys. Re.s Commun., 331*, 868-880.
[78] Pellizzaro, C., Coradini, D., Daniotti, A., Abolafio, G. and Daidone, M. G. (2001) Modulation of cell cycle-related protein expression by sodium butyrate in human non-small cell lung cancer cell lines. *Int. J. Cancer., 91*, 654-657.
[79] David-Pfeuty, T., Chakrani, F., Ory, K. and NouvianDooghe, Y. (1996) Cell cycle-dependent regulation of nuclear p53 traffic occurs in one subclass of human tumor cells and in untransformed cells. *Cell Growth Differ, 7*, 1211-1225.
[80] Sampath, J., Sun, D. X., Kidd, V. J., Grenet, J., Gandhi, A., Shapiro, L. H., Wang, Q. J., Zambetti, G. P. and Schuetz, J. D. (2001) Mutant p53 cooperates with ETS and selectively up-regulates human MDR1 not MRP1. *J. Biol. Chem., 276*, 39359-39367.
[81] Fujimura, M., Kasahara, K., Shirasaki, H., Heki, U., Iwasa, K., Ueda, A. and , Matsuda, T. (1999) Up-regulation of ICH-1L protein by thromboxane A2 antagonists enhances cisplatin-induced apoptosis in non-small-cell lung-cancer cell lines. *J. Cancer. Res. Clin. Oncol., 125*, 389-394.
[82] Kasahara, K., Fujimura, M., Bando, T., Shibata, K., Shirasaki, H. and Matsuda, T. (1996) Modulation of sensitivity to cis-diamminedichloroplatinum (II) by thromboxane A2 receptor antagonists in non-small-cell lung cancer cell lines. *Br. J. Cancer., 74*, 1553-1558.
[83] Nie, D., Che, M. X., Zacharek, A., Qiao, Y., Li, L., Li, X. L., Lamberti, M., Tang, K. Q., Cai, Y. L., Guo, Y. D., Grignon, D. and Honn, K. V. (2004) Differential expression of thromboxane synthase in prostate carcinoma: role in tumor cell motility. *Am. J. Pathol., 164*, 429-439.
[84] Vermeulen, K., Van Bockstaele, D. R. and Berneman, Z. N. (2005) Apoptosis: mechanisms and relevance in cancer. *Ann. Hematol., 84*, 6276-6239.
[85] Huerta, S., Goulet, E. J., Huerta-Yepez, S. and Livingston, E. H. (2007) Screening and detection of apoptosis. *J. Surg. Res., 139*, 143-156.
[86] Lin, M. S., Chen, W. C., Bai, X. and Wang, Y. D. (2007) Activation of peroxisome proliferator-activated receptor gamma inhibits cell growth via apoptosis and arrest of the cell cycle in human colorectal cancer. *J. Dig. Dis., 8*, 82-88.
[87] Kim, J. H., Kim, H. P., Jung, C. H., Hong, M. H., Hong, M. C., Bae, H. S., Lee, S. D., Park, S. Y., Park, J. H. and Ko, S. G. (2006) Inhibition of cell cycle progression via p27Kip1 upregulation and apoptosis induction by an ethanol extract of Rhus verniciflua Stokes in AGS gastric cancer cells. *Int J Mol Med, 18*, 201-208.
[88] Lai, S. L., Perng, R. P. and Hwang, J. (2000) p53 gene status modulates the chemosensitivity of non-small cell lung cancer cells. *J. Biomed. Sci., 7*, 64-70.
[89] Yusuf, K., Smith, S. D., Levy, R., Schaiff, W. T., Wyatt, S. M., Sadovsky, Y. and Nelson, D. M. (2001) Thromboxane A(2) limits differentiation and enhances apoptosis of cultured human trophoblasts. *Pediatr .Res., 50*, 203-209.
[90] Hrushka, N. H., Korniichuk, H. M., Makohon, N. V. and Aleksieieva, I. M. (2006) Effect of thromboxane B2 on morphological features of apoptosis in cultured rat hepatocytes exposed to hepatotropic substances. *Fiziol. Zh, 52*, 34-40.

[91] Rocha, P. N., Plumb, T. J., Robinson, L. A., Spurney, R., Pisetsky, D., Koller, B. H. and Coffman, T. M. (2005) Role of thromboxane A2 in the induction of apoptosis of immature thymocytes by lipopolysaccharide. *Clin. Diagn. Lab. Immunol., 12*, 896-903.

[92] Gomyo, Y., Sasaki, J., Branch, C., Roth, J. A., Mukhopadhyay, T. (2004) 5-aza-2'-deoxycytidine upregulates caspase-9 expression cooperating with p53-induced apoptosis in human lung cancer cells. *Oncogene, 23*, 6779-6787.

[93] Clarke, A. R., Purdie, C. A., Harrison, D. J., Morris, R. G., Bird, C. C., Hooper, M. L. and Wyllie, A. H. (1993) Thymocyte apoptosis induced by p53-dependent and independent pathways. *Nature, 362*, 849-852.

[94] Lowe, S. W., Ruley, H. E., Jacks, T. and Housman, D. E. (1993) p53-dependent apoptosis modulates the cytotoxicity of anticancer agents. *Cell, 74*, 957-967.

[95] El-Deiry, W. S., (2003) The role of p53 in chemosensitivity and radiosensitivity. *Oncogene, 22*, 7486-7495.

[96] Vogelstein, B., D. Lane, and Levine, A. J. (2000)Surfing the p53 network. *Nature, 408*, 307-310.

[97] Harris, R. E., (2007) Cyclooxygenase-2 (cox-2) and the inflammogenesis of cancer. *Subcell Biochem., 42*, 93-126.

[98] Mann, J. R., Backlund, M. G. and DuBois, R. N. (2005) Mechanisms of disease: Inflammatory mediators and cancer prevention. *Nat. Clin. Pract. Oncol., 2*, 202-210.

[99] Gasparini, G., Longo, R., Sarmiento, R. and Morabito, A. (2003) Inhibitors of cyclo-oxygenase 2: a new class of anticancer agents? *Lancet Oncol., 4*, 605-615.

[100] Mardini, I. A. and FitzGerald, G. A. (2001) Selective inhibitors of cyclooxygenase-2: a growing class of anti-inflammatory drugs. *Mol. Interv., 1*, 30-38.

In: Oncogene Proteins: New Research
Editors: A. H. Malloy and E.C. Carson, pp. 217-231
ISBN 978-1-60456-290-3

Chapter 8

Current Controversies in Bcl-2-Regulated Apoptosis

Erinna F. Lee and W. Douglas Fairlie
The Walter and Eliza Hall Institute of Medical Research,
1G Royal Parade, Parkville, Victoria, 3050, Australia

Abstract

Programmed cell death (apoptosis) is essential for the development and survival of all metazoans. Apoptosis regulated by the Bcl-2 family of proteins involves a complex interplay between members that promote cell survival and those that promote cell death. However, despite extensive investigation, the molecular mechanisms underlying the control of apoptosis signalling via the Bcl-2 pathway remain controversial. Two major models have emerged: "indirect" and "direct" activation. These models differ significantly, both in terms of the role of the pro-survival proteins, and how the essential mediators of cell death, Bax and Bak, become activated following receipt of death stimuli. In this chapter, these opposing models are described in detail and the validity of each discussed. How these models impact upon strategies for the development of novel anti-cancer therapeutics will also be explored.

Introduction

The genetically programmed elimination of cells (apoptosis) occurs during the development of all metazoans. This is best exemplified in the landmark studies of the nematode worm *Caenorhabditis elegans* where precisely 131 of the 1090 somatic cells that are generated to form an adult hermaphrodite die (Horvitz, 1999). In mature organisms, apoptosis is essential for the removal of cells which are no longer needed, that are damaged (such as occurs upon exposure to ionising radiation), or that may prove dangerous (for example, cells that are infected with viruses, or which are auto-reactive). Critically, defective apoptosis signalling can have dramatic consequences: too much cell death can lead to degenerative diseases whilst too little cell death can result in diseases such as cancer.

Two main pathways lead to apoptosis with both converging on the activation of the same cellular demolitionists – the caspases. The first is referred to as the "extrinsic" pathway and signals through "death receptors". The ligand/receptor system responsible for managing apoptosis via this pathway is the tumour-necrosis factor (TNF) cytokine/receptor family (Bodmer et al., 2002). Probably the best-studied ligands within the family are TNF, FAS ligand, and TNF-related apoptosis-inducing ligands (TRAIL), which bind to their respective receptors: TNF receptor 1 (TNFR1), FAS and TRAIL receptor 1 (TRAILR1) or 2 (TRAILR2). Binding of the ligands to their receptors leads to the hierarchical assembly of the death-inducing signalling complex (DISC) (Ashkenazi, 2002; Kischkel et al., 1995; Peter and Krammer, 2003). Receptors containing cytosolic death domains cluster in membranes upon ligand binding, recruiting adaptor proteins such as the FAS-associated proteins with death domain (FADD). In turn, FADD binds to and activates caspase-8, initiating a proteolytic cascade that results in the disassembly and death of the cells.

The second pathway to apoptosis is commonly referred to as the "intrinsic" pathway which is activated in response to cellular stresses such as cytokine withdrawal, DNA damage or viral infection. This pathway is regulated by a finely orchestrated process governed largely by interactions between the opposing pro-survival and pro-apoptotic factions of the Bcl-2 family of proteins. Whilst both pathways have been extensively studied, there still remain a number of questions about the underlying molecular mechanisms that lead to apoptosis. Understanding these mechanisms may be critical for the development of drugs to treat diseases where apoptosis has gone awry. In this chapter we will focus on the intrinsic pathway controlled by the Bcl-2 family of proteins. Here there is still significant controversy regarding how many of the key players within the pathway are functioning, although two major models have emerged: the "indirect" and "direct" activation models (for recent commentaries see: Adams and Cory, 2007a, b; Galonek and Hardwick, 2006; Green, 2006, 2007; Letai, 2006; Youle, 2007). Each of these models will be described in detail and we will discuss how our understanding of the molecular mechanisms involved can impact upon the development and application of drugs that target the key players of the Bcl-2-regulated apoptotic pathway.

The Key Players

The *bcl-2* gene was the first molecularly defined component of the apoptotic machinery, and was identified at the site of the t(14:18) chromosomal translocation characteristic of follicular lymphoma (Tsujimoto et al., 1984). As a consequence of this translocation event, the *bcl-2* gene is brought under the control of the immunoglobulin heavy chain enhancer, resulting in high level constitutive expression of Bcl-2 in B-cells (Graninger et al., 1987). The strong association of this translocation with neoplastic B-cells suggested that *bcl-2* might be a candidate oncogene. However, unlike its oncogenic predecessors such as *ras* or *v-abl* (Cook et al., 1985; McCoy et al., 1983), Bcl-2 was shown to promote cellular survival rather than proliferation, hence establishing that these were independent processes under separate genetic control (Vaux et al., 1988).

In the nearly 20 years since the discovery of the function of Bcl-2 (Vaux et al., 1988), the number of proteins involved in this pathway has grown significantly with almost twenty members that have been well characterised. These proteins are all related to each other by the presence of 1-4 regions of sequence homology referred to as Bcl-2 homology domains (BH domains), and fall into opposing pro-survival and pro-death factions (Cory et al., 2003). The pro-survival group includes Bcl-2 itself, as well as Bcl-x_L, Bcl-w, Mcl-1 and A1, and all can promote cell survival in response to various death stimuli if over-expressed in cells. However, due to the partial redundancy in the function of each pro-survival protein, and their complex expression patterns, inactivation of individual genes in mice results in diverse phenotypes (Hamasaki et al., 1998; Motoyama et al., 1995; Nakayama et al., 1994; Print et al., 1998; Rinkenberger et al., 2000; Veis et al., 1993).

The pro-death faction is divided into two further classes: the BH3-only proteins and the Bax/Bak molecules. The BH3-only proteins (which includes Bim, Bad, Puma, Noxa, Bmf, Bik, Bid, Hrk) are largely unrelated in sequence to either Bcl-2, or indeed each other, except for the signature BH3 domain (Huang and Strasser, 2000). This short conserved sequence motif facilitates interactions with other members of the family and is essential for their killing function. Detailed structural and structure-function studies have shown that in the complex structures that have been solved to date, the BH3 domain forms an amphipathic helix which binds into a large hydrophobic cleft induced upon interaction with the pro-survival molecules (Czabotar et al., 2007; Liu et al., 2003; Liu et al., 2006; Petros et al., 2000; Sattler et al., 1997; Yan et al., 2004). Very recently, it has become apparent that all BH3-only proteins are not equal in their abilities to engage pro-survival molecules (Certo et al., 2006; Chen et al., 2005; Kim et al., 2006; Kuwana et al., 2005). Whilst some are promiscuous (such as Bim and Puma) and can bind to all with high affinity, others such as Bad, Noxa and Bmf are more selective and bind to limited subsets. For example, Bad only binds Bcl-2, Bcl-x_L and Bcl-w with high affinity, whilst Noxa has the complementary binding profile and only engages Mcl-1 and A1 tightly (Certo et al., 2006; Chen et al., 2005). This selectivity has important consequences for how (and if) cell death proceeds, and will be discussed further below.

The BH3-only proteins have been referred to as the sentinels of cellular welfare as they sense and respond to a variety of death stimuli (Cory et al., 2003). The different BH3-only proteins perform distinct duties as some appear to monitor particular subcellular compartments for stress or damage, and some are expressed only in certain cell types (Cory et al., 2003). For example, *noxa* and *puma* are targets of the tumour suppressor, p53 (Oda et al., 2000; Yu et al., 2001), and have been shown to be important for apoptosis induced by genotoxic damage, and in the case of Puma, for death induced by drugs such as etoposide (Villunger et al., 2003). Bad mediates cell death following glucose or growth factor deprivation (Danial et al., 2003; Ranger et al., 2003) and Bim is up-regulated in response to calcium flux amongst other stimuli (Cante-Barrett et al., 2006). These proteins form the apical step in the Bcl-2 pathway and depending on your viewpoint, trigger the apoptotic cascade by either engaging the pro-survival proteins or the Bax/Bak molecules (or both).

The Bax/Bak proteins are essential mediators of apoptosis. Although *bax*$^{-/-}$ mice (Knudson et al., 1995) or *bak*$^{-/-}$ mice (Lindsten et al., 2000) display limited phenotypic abnormalities, the majority of mice lacking both these proteins die perinatally, with less than 10% surviving into adulthood (Lindsten et al., 2000). The mice that did survive displayed multiple developmental defects, including persistence of interdigital webs and an accumulation of excess cells within both the central nervous and haematopoietic systems (Lindsten et al., 2000). Furthermore, cells doubly-deficient for both Bax and Bak are extremely resistant to multiple apoptotic stimuli (such as staurosporine treatment, ultra-violet irradiation, etoposide and growth factor deprivation). Interestingly, these proteins are structurally most closely related to the pro-survival molecules as both possess BH1-3 domains and appear to fold into a similar helical-bundle type structure (Moldoveanu et al., 2006; Suzuki et al., 2000).

Within the cell Bax and Bak differ in their sub-cellular localization: Bak is primarily located on the mitochondria whilst Bax is mostly cytosolic (Hsu et al., 1997). Both molecules however appear to mediate cell death in a similar fashion by causing the permeabilisation of the mitochondrial outer membrane which results in the release of components of the intermembrane space, particularly cytochrome *c*, into the cytosol. Binding of cytochrome *c* to the adapter molecule Apaf-1 enables the assembly of the "apoptosome" which includes caspase-9 (Bao and Shi, 2007). This in turn results in activation of caspase-9 and sets off the caspase cascade that eventually leads to the destruction of the cell following cleavage of critical intracellular components (Creagh and Martin, 2001; Shi, 2004). How Bax and Bak permeabilise the mitochondrial membrane is still poorly understood, but has been shown to involve a conformational change(s) and oligomerisation of these molecules (Antonsson et al., 2001; Griffiths et al., 1999; Hsu et al., 1997; Nechushtan et al., 2001) to possibly form pores similar to those observed with bacterial toxins that possess some structural similarities. Activation of Bax differs from Bak, though, as it must also translocate from the cytosol to the mitochondria (Wolter et al., 1997).

The manner by which Bax and Bak are activated and/or held in check is perhaps the most controversial aspect of the molecular mechanisms underlying apoptotic regulation by the Bcl-2 pathway. Genetic evidence suggests that the BH3-only proteins exert their pro-apoptotic function upstream of Bax and Bak as cells deficient in both these proteins remain resistant to the BH3-only proteins (Cheng et al., 2001; Zong et al., 2001). However, how the BH3-only proteins cause Bax and Bak to become activated remains questionable. The two major models that have emerged, referred to as "direct" or "indirect" activation, confer different roles for both the pro-survival and BH3-only proteins, as discussed below.

The "Direct" Activation Model

This model appears to have been gradually refined over the last couple of years, but at its heart suggests that certain members of the BH3-only family ("activators") can directly engage Bax and Bak (Certo et al., 2006; Cheng et al., 2001; Del Gaizo Moore et al., 2007; Deng et al., 2007; Kim et al., 2006; Kuwana et al., 2005; Letai et al., 2002) (Figure 1). This binding event, which involves the BH3 domain on the BH3-only molecules, results in the conformational change that enables Bax and Bak to oligomerise, and subsequently permeabilise the mitochondrial outer membrane. The proteins that are best characterised as activator BH3-only proteins are Bim and tBid (the caspase-cleaved form of Bid) (Kuwana et al., 2005; Letai et al., 2002), though Puma has also been implicated (Cartron et al., 2004; Kim et al., 2006). The role of the pro-survival proteins in this model is a passive one – they act as "sinks" to sequester the activators and prevent them from engaging Bax and Bak. Hence, here survival is the default state of the cell.

How does the death signal get transmitted to Bax and Bak? The proposed model is that this task is performed by a second class of BH3-only proteins termed "sensitisers" (or "derepressors") that are unable to bind directly to Bax and Bak but are able to bind to the pro-survival proteins with high affinity (Certo et al., 2006; Cheng et al., 2001; Kuwana et al., 2005; Letai et al., 2002). Therefore, upon reception of a death stimulus, these sensitiser BH3-only proteins (which includes Bad, Noxa, Bmf and Hrk) are upregulated, bind to the pro-survival proteins and out-compete the activators. The activators are then free to engage Bax and/or Bak which results in their activation.

The "Indirect" Activation Model

This model differs from the "direct" activation model in several aspects (Figure 1). Firstly, the pro-survival proteins play an important and active role in directly engaging and antagonizing the pro-apoptotic activity of Bax and Bak, and this interaction is likely mediated by the BH3 domain of the pro-apoptotic molecules (Sattler et al., 1997; Willis et al., 2005). This implies that, in contrast to the "direct" activation model, death is the default state of the cell (Willis et al., 2005; Willis et al., 2007). Secondly, in this model the role of the BH3-only proteins is to bind to the pro-survival proteins and displace Bax and/or Bak, freeing them to oligomerise and permeabilise the mitochondria. Hence, there is no requirement for the BH3-only molecules to interact with Bax and/or Bak directly.

This model is perhaps best understood for Bak-mediated killing (Willis et al., 2005). It is now well established that Bak is restrained by both (and only) Mcl-1 and Bcl-x_L. These interactions can be demonstrated in co-immunopecipitation studies, and are consistent with affinity measurements of the Bak BH3 domain for pro-survival proteins; it only binds to Bcl-x_L and Mcl-1 with high affinity (Willis et al., 2005). For Bak-mediated apoptosis to occur both of these pro-survivals proteins must be neutralised. Hence, BH3-only proteins such as Bad or Noxa that can only engage one or the other of Bcl-x_L or Mcl-1 are ineffective killers, whilst those that can engage both (such as Bim or Puma) are potent killers (Chen et al., 2005; Willis et al., 2005). Similarly, the combination of two selective ligands with complementary binding profiles is also sufficient for Bak-mediated cell death (Chen et al., 2005; Willis et al., 2005).

The situation seems to be very similar for Bax-mediated cell death though neutralisation of a wider range of pro-survival proteins, including Bcl-2 (and possibly others), together with Mcl-1 and Bcl-x_L appears to be required (Willis et al., 2007). Hence, BH3-only proteins that are promiscuous binders (Bim and Puma) are again more potent than their selective-binding counterparts because of their ability to neutralize the full complement of pro-survival proteins, and not because they are able to directly activate the multi-domain pro-apoptotic proteins.

An important issue that still requires further investigation in this model is the signal required for Bax translocation to the mitochondria. One possibility is that whilst Bax is predominantly cytosolic, a small population of molecules is associated with the mitochondria of healthy cells, possibly in a "primed" conformation that is restrained by the pro-survival proteins (Adams and Cory, 2007b; Willis et al., 2007). This would explain, for example, why Bcl-2, which is predominantly located on the mitochondria, is still an important brake on Bax.

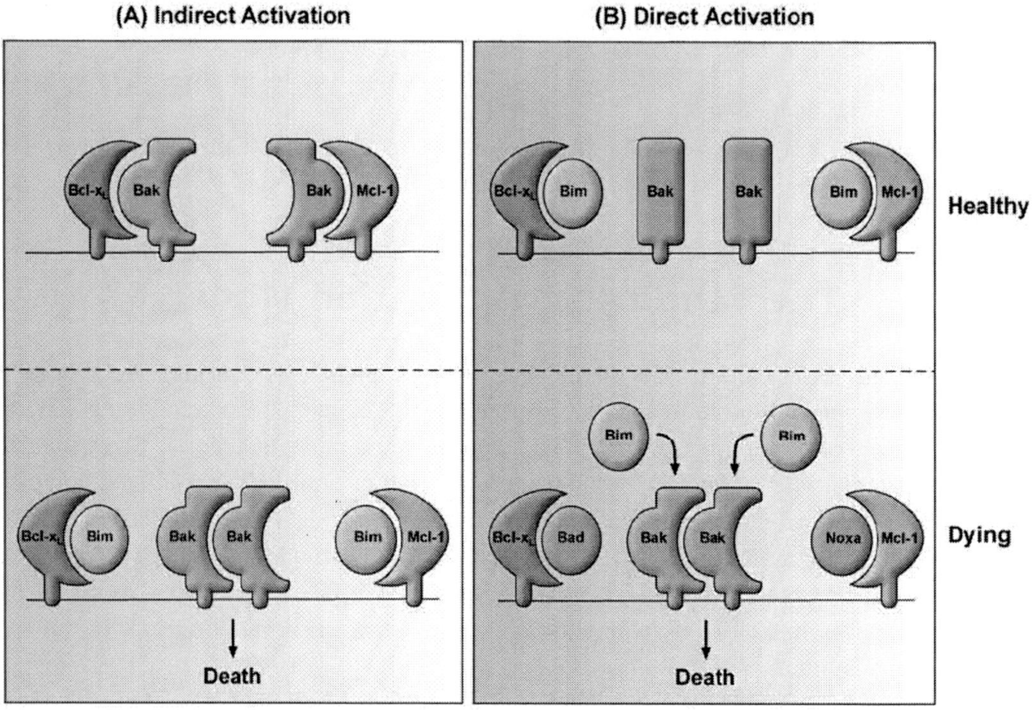

Figure 1. Models of apoptosis. (A) According to the *indirect activation* model, Bax/Bak are held in check by the pro-survival proteins in a healthy cell (upper panel). Upon receipt of a death stimulus (lower panel), the BH3-only proteins indirectly activate Bax/Bak by displacing them from their pro-survival guardians, hence committing a cell to death. (B) The *direct activation* model proposes the division of BH3-only proteins into two classes: "activators" (yellow) and "sensitisers" (orange). The activator BH3-only proteins, including Bim, tBid, and perhaps Puma, are capable of binding directly to, and activating Bax/Bak. In contrast, the sensitiser BH3-only proteins, comprising the remaining BH3-only proteins, only bind the pro-survival proteins but not Bax/Bak. In a healthy cell (upper panel), the activator BH3-only proteins are restrained away from Bax/Bak through their sequestration by the pro-survival proteins. Upon receipt of a death stimulus (lower panel), the sensitiser BH3-only proteins are activated and bind the pro-survival proteins, displacing and freeing the activator BH3-only proteins to activate Bax/Bak, hence initiating apoptosis.

Hence, the release of the primed Bax from the restraint of the pro-survival molecules by BH3-only proteins may, at least, provide an initial signal for the translocation of additional Bax molecules from the cytosol to the mitochondria, although this is yet to be established. A similar mechanism involving a primed form has also been proposed for Bak-mediated apoptosis (Adams and Cory, 2007b; Willis et al., 2007).

Direct or Indirect?

Is it possible to establish which of the proposed models is correct from the available data? This is a difficult proposition as there is strong evidence for both, and data from many experiments can be interpreted according to either. For example, over-expression of the Bad/Noxa combination results in potent cell death of mouse embryonic fibroblasts (Chen et al., 2005). As no direct activator has been introduced, it might be concluded that direct activation is not occurring. However, this can be countered by the possibility that these BH3-only proteins are acting by releasing the sequestered activator BH3-only molecules from the pro-survival proteins (Certo et al., 2006; Kim et al., 2006; Kuwana et al., 2005). This scenario has recently been extended somewhat further, using genetic studies, to enable a slightly more definitive conclusion, to be reached. Huang and colleagues generated mice lacking both of the best characterised of the putative direct activators, Bim and Bid (Willis et al., 2007). The direct activation model would predict that these mice should be severely compromised, similar to the *bax*$^{-/-}$*bak*$^{-/-}$ mice, because in this model these proteins are proposed to be essential for apoptosis to occur. Instead, these mice were born at the expected Mendelian frequency, appeared normal and remained healthy for over 12 months (Willis et al., 2007). Furthermore, *bim*$^{-/-}$*bid*$^{-/-}$ immortalized mouse embryonic fibroblasts remained sensitive to apoptotic stimuli, such as DNA damage as well as over-expression of sensitiser BH3-only proteins. Whilst Puma has also been suggested as an activator BH3-only protein (Cartron et al., 2004; Kim et al., 2006), this notion similarly has been challenged (Certo et al., 2006; Kuwana et al., 2005; Ming et al., 2006). Regardless, decreased Puma expression using RNA interference in *bim*$^{-/-}$*bid*$^{-/-}$ mouse embryonic fibroblasts still did not affect the sensitivity of these cells to apoptotic stimuli (Willis et al., 2007). Interestingly, in a separate study, *bim*$^{-/-}$ cells in which Puma had been knocked-down were somewhat more, though not completely, resistant to apoptosis induced by over-expression of the Bad/Noxa combination compared to wild-type cells (Kim et al., 2006). Whilst combined these results suggest that direct activation is not essential, they are still not entirely conclusive and probably await the generation of mice triply deficient in Bim, Bid and Puma. Even then however, proponents of the direct activation model have suggested that alternative yet-to-be discovered activators may exist (Deng et al., 2007; Letai, 2006). This argument is difficult to challenge because, of course, it could be true, however it also essentially prevents a definitive conclusion ever being drawn from genetic studies of this sort.

Biochemical approaches to this question have been similarly inconclusive with a range of arguments and counter-arguments put forward. For example, one stumbling block that confounds the direct activation model is the difficulty in detecting stable complexes of Bim and/or tBid with either Bax or Bak. In the case of Bax, such complexes can only be observed when non-ionic detergents are present (Willis et al., 2007). However, these detergents also induce conformational changes in Bax (Hsu and Youle, 1998), which may not be representative of the form of Bax normally present within the cell. Moreover, for Bim, this interaction is only observed with the least abundantly expressed isoform, namely Bim_S, even in the presence of detergents. Nevertheless a counter-argument to the difficulty in detecting such complexes is that Bim and tBid function via a so-called "hit-and-run" or "kiss-and-run" catalytic-type mechanism where a transient interaction is sufficient to induce the necessary conformational change(s) required for mitochondrial translocation and self-assembly of Bax oligomers, thereby precluding the necessity for formation of a stable complex. Interestingly Walensky *et al.* recently generated a helix-stabilized form of the Bid BH3 domain (Walensky et al., 2004; Walensky et al., 2006) that allowed its affinity for full-length Bax protein to be measured in the absence of detergents (Walensky et al., 2006). The authors suggest that these stabilized BH3 domain variants more accurately mimic the helical structure of the BH3 domains in the context of their parent full-length proteins, though this does not explain why affinity measurements have yet to be reported for tBid or full-length Bim binding to Bax.

One line of evidence that is consistent with the idea that direct activation can occur (regardless of whether it is essential) comes from a model system using protein-free liposomes (Kuwana et al., 2005). In these studies, incubation of Bax with tBid or peptides corresponding to the BH3 domains of Bim or Bid, but not from other BH3-only proteins, induced Bax to become associated with, and permeabilize liposomes. Importantly, the association of Bax with the liposome was inhibited by addition of Bcl-x_L, though it is not clear whether this is due to sequestration of the activator BH3-only protein or direct inhibition of Bax (Kuwana et al., 2005). It is also not clear how accurately this artificial system replicates the cellular environment. Even more ambiguous results have been observed with tBid variants. Mutants of tBid that have lost the ability to engage pro-survival proteins but retained Bax binding can still cause apoptosis (Kim et al., 2006). However a tBid mutant that retains binding to pro-survival proteins but not to Bax (even in the presence of detergent) can also cause cell death (Willis et al., 2007).

An important point to consider is that many non-Bcl-2 family members have also been implicated in regulating Bax and Bak. In particular, a minor isoform of the voltage-dependent anion channel-2 (VDAC2) has been proposed to play a role in keeping the pro-apoptotic activity of Bak in check on the mitochondrial outer membrane (Cheng et al., 2003; Kim et al., 2006). In the case of Bax, proteins such as Ku70, 14-3-3 isoforms σ, θ, ε, ζ, and Hsp70-dj1 and Hsp70-dj2, are thought to be responsible for maintaining it in the cytosol under resting conditions, hence preventing its translocation to the mitochondria (Lucken-Ardjomande and Martinou, 2005).

Implications for Bcl-2-Targeted Therapeutics

Is it necessary to resolve which of the models is correct for the development of novel therapies targeting the Bcl-2 pathway? The answer is probably not, given that a highly promising small-molecule targeting pro-survival proteins, ABT-263 (derived from ABT-737), has already entered Phase I clinical trials as an anti-cancer therapeutic. This compound is a small molecule BH3 mimetic that binds to Bcl-2, Bcl-w and Bcl-x_L with high affinity (Oltersdorf et al., 2005), and high-resolution crystallographic analysis has demonstrated that it targets the hydrophobic groove on Bcl-x_L that normally engages BH3 domains (Lee et al., 2007). A detailed understanding of the apical events in the apoptotic pathway (i.e. how BH3-only proteins engage pro-survival proteins), which was revealed primarily by structural studies, was essential for the discovery of ABT-737, the precursor molecule to ABT-263.

Interestingly, this initial binding event is one point that is agreed between both models (though it is the consequences that differ).

Nevertheless, determining which model is correct could have important consequences for identifying the best target proteins for such BH3-mimetics. The direct model predicts that over-expressed pro-survival proteins, such as Bcl-2 in many lymphomas, may act as sinks that can "load" cells with activator BH3-only proteins such as Bim (Certo et al., 2006; Del Gaizo Moore et al., 2007; Deng et al., 2007). Hence, such cells will be more sensitive to "sensitiser"-type BH3 mimetic drugs targeting Bcl-2. This possibility has been studied extensively by the Letai group, who have termed such cells "primed for death" (Certo et al., 2006). Indeed, this group has shown that quantitation of Bcl-2:Bim complex levels can predict sensitivity to ABT-737 (Deng et al., 2007). This model would also suggest that molecules that can directly target and hence activate Bax and/or Bak, like the activator BH3-only proteins, should also be able to induce apoptosis in cells. However, a potential problem with such molecules is that they may be highly toxic as most cell types express Bax and/or Bak.

Proponents of the indirect model have taken a different view. As both Bcl-x_L and Mcl-1 are the only restraints on Bak in this model, BH3 mimetics that target one or the other in cells that are deficient or express low levels of the complementary pro-survival protein should succumb to the drug. There is now a significant body of data supporting this hypothesis. In particular, ABT-737 which targets Bcl-2, Bcl-x_L and Bcl-w but not Mcl-1 does appear to be most effective as a single agent against cell-lines, including cancers, that possess low or undetectable levels of Mcl-1 (Konopleva et al., 2006; Lin et al., 2006; van Delft et al., 2006). Moreover, cells that are insensitive to ABT-737 can be rendered sensitive by using strategies that reduce Mcl-1 (Chauhan et al., 2007; Chen et al., 2007; Konopleva et al., 2006; Lin et al., 2006; Trudel et al., 2007; van Delft et al., 2006). This is particularly important as it suggests a number of strategies for combination therapies to broaden the range of tumours against which ABT-263 will be effective. Indeed a number of small-molecule drugs such as etoposide, Seliciclib and BAY43-900 can reduce Mcl-1 levels in cells and have been shown to synergise with ABT-737 (Chauhan et al., 2007; Chen et al., 2007; Konopleva et al., 2006; Lin et al., 2006; Trudel et al., 2007; van Delft et al., 2006).

Conclusion

Despite extensive study, a number of unanswered questions about the molecular mechanisms underlying the apoptotic Bcl-2 pathway still remain to be definitively resolved. Clearly both of the major models have strengths. One possibility is that aspects of both models may be correct. For example, it is feasible that pro-survival proteins do function to restrain Bax and Bak, however, these multi-domain pro-apoptotic proteins may also be directly activated by activator BH3-only proteins at some point during the apoptotic cascade. It has been proposed, for example, that the direct activation of Bax/Bak by activator BH3-only proteins may serve to amplify the apoptotic response (Fletcher and Huang, 2006). This view is consistent with the fact that Bid remains in an inactive form until cleaved by caspases to generate tBid, (which is the form that has been shown to bind Bax) downstream of mitochondrial damage. This is also supported by the recent finding that Bax translocation (and cytochrome *c* release) is impaired in *caspase-3/caspase-7* double knock-out cells (Lakhani et al., 2006). It is not clear, however, whether Bim, which does not require processing by caspases to become activated, also acts in a similar manner to amplify the apoptotic signal. Questions like precisely how the pro-survival proteins restrain Bax and Bak, and what is the nature of the Bax/Bak oligomeric state also still persist.

Despite the above controversies and questions, our current understanding of how the Bcl-2 pathway is regulated has advanced enormously since the function of Bcl-2 was determined nearly 20 years ago. As a consequence, promising therapeutics have now emerged to enable Bcl-2 proteins, which are key mediators of oncogenesis, to be targeted. Future studies aimed at unravelling some of the points of contention are likely to enhance the development of the next generation of such drugs.

References

Adams, J.M., and Cory, S. (2007a). Bcl-2-regulated apoptosis: mechanism and therapeutic potential. *Curr .Opin. Immunol. 19*, 488-496.

Adams, J.M., and Cory, S. (2007b). The Bcl-2 apoptotic switch in cancer development and therapy. *Oncogene 26*, 1324-1337.

Antonsson, B., Montessuit, S., Sanchez, B., and Martinou, J.C. (2001). Bax is present as a high molecular weight oligomer/complex in the mitochondrial membrane of apoptotic cells. *J. Biol. Chem. 276*, 11615-11623.

Ashkenazi, A. (2002). Targeting death and decoy receptors of the tumour-necrosis factor superfamily. *Nat. Rev. Cancer 2*, 420-430.

Bao, Q., and Shi, Y. (2007). Apoptosome: a platform for the activation of initiator caspases. *Cell Death Differ. 14*, 56-65.

Bodmer, J.L., Schneider, P., and Tschopp, J. (2002). The molecular architecture of the TNF superfamily. *Trends Biochem. Sci. 27*, 19-26.

Cante-Barrett, K., Gallo, E.M., Winslow, M.M., and Crabtree, G.R. (2006). Thymocyte negative selection is mediated by protein kinase C- and Ca2+-dependent transcriptional induction of bim. *J. Immunol. 176*, 2299-2306.

Cartron, P.F., Gallenne, T., Bougras, G., Gautier, F., Manero, F., Vusio, P., Meflah, K., Vallette, F.M., and Juin, P. (2004). The first alpha helix of Bax plays a necessary role in its ligand-induced activation by the BH3-only proteins Bid and PUMA. *Mol. Cell 16*, 807-818.

Certo, M., Del Gaizo Moore, V., Nishino, M., Wei, G., Korsmeyer, S., Armstrong, S.A., and Letai, A. (2006). Mitochondria primed by death signals determine cellular addiction to antiapoptotic BCL-2 family members. *Cancer Cell 9*, 351-365.

Chauhan, D., Velankar, M., Brahmandam, M., Hideshima, T., Podar, K., Richardson, P., Schlossman, R., Ghobrial, I., Raje, N., Munshi, N., and Anderson, K.C. (2007). A novel Bcl-2/Bcl-X(L)/Bcl-w inhibitor ABT-737 as therapy in multiple myeloma. *Oncogene 26*, 2374-2380.

Chen, L., Willis, S.N., Wei, A., Smith, B.J., Fletcher, J.I., Hinds, M.G., Colman, P.M., Day, C.L., Adams, J.M., and Huang, D.C. (2005). Differential targeting of prosurvival Bcl-2 proteins by their BH3-only ligands allows complementary apoptotic function. *Mol. Cell* 17, 393-403.

Chen, S., Dai, Y., Harada, H., Dent, P., and Grant, S. (2007). Mcl-1 down-regulation potentiates ABT-737 lethality by cooperatively inducing Bak activation and Bax translocation. *Cancer Res. 67*, 782-791.

Cheng, E.H., Sheiko, T.V., Fisher, J.K., Craigen, W.J., and Korsmeyer, S.J. (2003). VDAC2 inhibits BAK activation and mitochondrial apoptosis. *Science 301*, 513-517.

Cheng, E.H., Wei, M.C., Weiler, S., Flavell, R.A., Mak, T.W., Lindsten, T., and Korsmeyer, S.J. (2001). BCL-2, BCL-X(L) sequester BH3 domain-only molecules preventing BAX- and BAK-mediated mitochondrial apoptosis. *Mol. Cell 8*, 705-711.

Cook, W.D., Metcalf, D., Nicola, N.A., Burgess, A.W., and Walker, F. (1985). Malignant transformation of a growth factor-dependent myeloid cell line by Abelson virus without evidence of an autocrine mechanism. *Cell 41*, 677-683.

Cory, S., Huang, D.C., and Adams, J.M. (2003). The Bcl-2 family: roles in cell survival and oncogenesis. *Oncogene 22*, 8590-8607.

Creagh, E.M., and Martin, S.J. (2001). Caspases: cellular demolition experts. *Biochem. Soc. Trans. 29*, 696-702.

Czabotar, P.E., Lee, E.F., van Delft, M.F., Day, C.L., Smith, B.J., Huang, D.C., Fairlie, W.D., Hinds, M.G., and Colman, P.M. (2007). Structural insights into the degradation of Mcl-1 induced by BH3 domains. *Proc. Natl. Acad. Sci. U S A 104*, 6217-6222.

Danial, N.N., Gramm, C.F., Scorrano, L., Zhang, C.Y., Krauss, S., Ranger, A.M., Datta, S.R., Greenberg, M.E., Licklider, L.J., Lowell, B.B., *et al.* (2003). BAD and glucokinase reside in a mitochondrial complex that integrates glycolysis and apoptosis. *Nature 424*, 952-956.

Del Gaizo Moore, V., Brown, J.R., Certo, M., Love, T.M., Novina, C.D., and Letai, A. (2007). Chronic lymphocytic leukemia requires BCL2 to sequester prodeath BIM, explaining sensitivity to BCL2 antagonist ABT-737. *J. Clin. Invest. 117*, 112-121.

Deng, J., Carlson, N., Takeyama, K., Dal Cin, P., Shipp, M., and Letai, A. (2007). BH3 profiling identifies three distinct classes of apoptotic blocks to predict response to ABT-737 and conventional chemotherapeutic agents. *Cancer Cell 12*, 171-185.

Fletcher, J.I., and Huang, D.C. (2006). BH3-only proteins: orchestrating cell death. *Cell Death Differ. 13*, 1268-1271.

Galonek, H.L., and Hardwick, J.M. (2006). Upgrading the BCL-2 network. *Nat. Cell Biol. 8*, 1317-1319.

Graninger, W.B., Seto, M., Boutain, B., Goldman, P., and Korsmeyer, S.J. (1987). Expression of Bcl-2 and Bcl-2-Ig fusion transcripts in normal and neoplastic cells. *J. Clin .Invest. 80*, 1512-1515.

Green, D.R. (2006). At the gates of death. *Cancer Cell 9*, 328-330.

Green, D.R. (2007). Life, death, BH3 profiles, and the salmon mousse. *Cancer Cell 12*, 97-99.

Griffiths, G.J., Dubrez, L., Morgan, C.P., Jones, N.A., Whitehouse, J., Corfe, B.M., Dive, C., and Hickman, J.A. (1999). Cell damage-induced conformational changes of the pro-apoptotic protein Bak in vivo precede the onset of apoptosis. *J. Cell Biol. 144*, 903-914.

Hamasaki, A., Sendo, F., Nakayama, K., Ishida, N., Negishi, I., Nakayama, K., and Hatakeyama, S. (1998). Accelerated neutrophil apoptosis in mice lacking A1-a, a subtype of the bcl-2-related A1 gene. *J. Exp. Med. 188*, 1985-1992.

Horvitz, H.R. (1999). Genetic control of programmed cell death in the nematode *Caenorhabditis elegans*. *Cancer Research 59*, 1701s-1706s.

Hsu, Y.T., Wolter, K.G., and Youle, R.J. (1997). Cytosol-to-membrane redistribution of Bax and Bcl-X(L) during apoptosis. *Proc. Natl .Acad. Sci. U S A 94*, 3668-3672.

Hsu, Y.T., and Youle, R.J. (1998). Bax in murine thymus is a soluble monomeric protein that displays differential detergent-induced conformations. *J. Biol. Chem. 273*, 10777-10783.

Huang, D.C., and Strasser, A. (2000). BH3-Only proteins-essential initiators of apoptotic cell death. *Cell 103*, 839-842.

Kim, H., Rafiuddin-Shah, M., Tu, H.C., Jeffers, J.R., Zambetti, G.P., Hsieh, J.J., and Cheng, E.H. (2006). Hierarchical regulation of mitochondrion-dependent apoptosis by BCL-2 subfamilies. *Nat. Cell Biol. 8*, 1348-1358.

Kischkel, F.C., Hellbardt, S., Behrmann, I., Germer, M., Pawlita, M., Krammer, P.H., and Peter, M.E. (1995). Cytotoxicity-dependent APO-1 (Fas/CD95)-associated proteins form a death-inducing signaling complex (DISC) with the receptor. *Embo. J. 14*, 5579-5588.

Knudson, C.M., Tung, K.S., Tourtellotte, W.G., Brown, G.A., and Korsmeyer, S.J. (1995). Bax-deficient mice with lymphoid hyperplasia and male germ cell death. *Science 270*, 96-99.

Konopleva, M., Contractor, R., Tsao, T., Samudio, I., Ruvolo, P.P., Kitada, S., Deng, X., Zhai, D., Shi, Y.X., Sneed, T., *et al.* (2006). Mechanisms of apoptosis sensitivity and resistance to the BH3 mimetic ABT-737 in acute myeloid leukemia. *Cancer Cell 10*, 375-388.

Kuwana, T., Bouchier-Hayes, L., Chipuk, J.E., Bonzon, C., Sullivan, B.A., Green, D.R., and Newmeyer, D.D. (2005). BH3 domains of BH3-only proteins differentially regulate Bax-mediated mitochondrial membrane permeabilization both directly and indirectly. *Mol. Cell 17*, 525-535.

Lakhani, S.A., Masud, A., Kuida, K., Porter, G.A., Jr., Booth, C.J., Mehal, W.Z., Inayat, I., and Flavell, R.A. (2006). Caspases 3 and 7: key mediators of mitochondrial events of apoptosis. *Science 311*, 847-851.

Lee, E.F., Czabotar, P.E., Smith, B.J., Deshayes, K., Zobel, K., Colman, P.M., and Fairlie, W.D. (2007). Crystal structure of ABT-737 complexed with Bcl-x(L): implications for selectivity of antagonists of the Bcl-2 family. *Cell Death Differ. 14*, 1711-1713.

Letai, A. (2006). Restoring cancer's death sentence. *Cancer Cell 10*, 343-345.

Letai, A., Bassik, M.C., Walensky, L.D., Sorcinelli, M.D., Weiler, S., and Korsmeyer, S.J. (2002). Distinct BH3 domains either sensitize or activate mitochondrial apoptosis, serving as prototype cancer therapeutics. *Cancer Cell 2*, 183-192.

Lin, X., Morgan-Lappe, S., Huang, X., Li, L., Zakula, D.M., Vernetti, L.A., Fesik, S.W., and Shen, Y. (2006). 'Seed' analysis of off-target siRNAs reveals an essential role of Mcl-1 in resistance to the small-molecule Bcl-2/Bcl-X(L) inhibitor ABT-737. *Oncogene 26*, 3972-3979.

Lindsten, T., Ross, A.J., King, A., Zong, W.X., Rathmell, J.C., Shiels, H.A., Ulrich, E., Waymire, K.G., Mahar, P., Frauwirth, K., *et al.* (2000). The combined functions of proapoptotic Bcl-2 family members bak and bax are essential for normal development of multiple tissues. *Mol .Cell 6*, 1389-1399.

Liu, X., Dai, S., Zhu, Y., Marrack, P., and Kappler, J.W. (2003). The structure of a Bcl-xL/Bim fragment complex: implications for Bim function. *Immunity 19*, 341-352.

Liu, X., Zhu, Y., Dai, S., White, J., Peyerl, F., Kappler, J.W., and Marrack, P. (2006). Bcl-xl does not have to bind Bax to protect T cells from death. *J. Exp. Med. 203*, 2953-2961.

Lucken-Ardjomande, S., and Martinou, J.C. (2005). Newcomers in the process of mitochondrial permeabilization. *J Cell Sci. 118*, 473-483.

McCoy, M.S., Toole, J.J., Cunningham, J.M., Chang, E.H., Lowy, D.R., and Weinberg, R.A. (1983). Characterization of a human colon/lung carcinoma oncogene. *Nature 302*, 79-81.

Ming, L., Wang, P., Bank, A., Yu, J., and Zhang, L. (2006). PUMA Dissociates Bax and Bcl-XL to Induce Apoptosis in Colon Cancer Cells. *J. Biol. Chem. 281*, 16034-16042.

Moldoveanu, T., Liu, Q., Tocilj, A., Watson, M., Shore, G., and Gehring, K. (2006). The X-ray structure of a BAK homodimer reveals an inhibitory zinc binding site. *Mol. Cell 24*, 677-688.

Motoyama, N., Wang, F., Roth, K.A., Sawa, H., Nakayama, K., Nakayama, K., Negishi, I., Senju, S., Zhang, Q., Fujii, S., and et al. (1995). Massive cell death of immature hematopoietic cells and neurons in Bcl-x-deficient mice. *Science 267*, 1506-1510.

Nakayama, K., Nakayama, K., Negishi, I., Kuida, K., Sawa, H., and Loh, D.Y. (1994). Targeted disruption of Bcl-2 alpha beta in mice: occurrence of gray hair, polycystic kidney disease, and lymphocytopenia. *Proc. Natl .Acad .Sci. U S A 91*, 3700-3704.

Nechushtan, A., Smith, C.L., Lamensdorf, I., Yoon, S.H., and Youle, R.J. (2001). Bax and Bak coalesce into novel mitochondria-associated clusters during apoptosis. *J. Cell Biol. 153*, 1265-1276.

Oda, E., Ohki, R., Murasawa, H., Nemoto, J., Shibue, T., Yamashita, T., Tokino, T., Taniguchi, T., and Tanaka, N. (2000). Noxa, a BH3-only member of the Bcl-2 family and candidate mediator of p53-induced apoptosis. *Science 288*, 1053-1058.

Oltersdorf, T., Elmore, S.W., Shoemaker, A.R., Armstrong, R.C., Augeri, D.J., Belli, B.A., Bruncko, M., Deckwerth, T.L., Dinges, J., Hajduk, P.J., *et al.* (2005). An inhibitor of Bcl-2 family proteins induces regression of solid tumours. *Nature 435*, 677-681.

Peter, M.E., and Krammer, P.H. (2003). The CD95(APO-1/Fas) DISC and beyond. *Cell Death Differ 10*, 26-35.

Petros, A.M., Nettseheim, D.G., Wang, Y., Olejniczak, E.T., Meadows, R.P., Mack, J., Swift, K., Matayoshi, E.D., Zhang, H., Thompson, C.B., and Fesik, S.W. (2000). Rationale for Bcl-x_L/Bad peptide complex formation from structure, mutagenesis, and biophysical studies. *Protein Science 9*, 2528-2534.

Print, C.G., Loveland, K.L., Gibson, L., Meehan, T., Stylianou, A., Wreford, N., de Kretser, D., Metcalf, D., Kontgen, F., Adams, J.M., and Cory, S. (1998). Apoptosis regulator bcl-w is essential for spermatogenesis but appears otherwise redundant. *Proc. Natl .Acad. Sci*

U S A 95, 12424-12431.

Ranger, A.M., Zha, J., Harada, H., Datta, S.R., Danial, N.N., Gilmore, A.P., Kutok, J.L., Le Beau, M.M., Greenberg, M.E., and Korsmeyer, S.J. (2003). Bad-deficient mice develop diffuse large B cell lymphoma. *Proc. Natl. Acad. Sci. U S A 100*, 9324-9329.

Rinkenberger, J.L., Horning, S., Klocke, B., Roth, K., and Korsmeyer, S.J. (2000). Mcl-1 deficiency results in peri-implantation embryonic lethality. *Genes Dev. 14*, 23-27.

Sattler, M., Liang, H., Nettesheim, D., Meadows, R.P., Harlan, J.E., Eberstadt, M., Yoon, H.S., Shuker, S.B., Chang, B.S., Minn, A.J., *et al.* (1997). Structure of Bcl-x_L-Bak peptide complex: recognition between regulators of apoptosis. *Science 275*, 983-986.

Shi, Y. (2004). Caspase activation, inhibition, and reactivation: a mechanistic view. *Protein Sci 13*, 1979-1987.

Suzuki, M., Youle, R.J., and Tjandra, N. (2000). Structure of Bax: coregulation of dimer formation and intracellular localization. *Cell 103*, 645-654.

Trudel, S., Stewart, A.K., Li, Z., Shu, Y., Liang, S.B., Trieu, Y., Reece, D., Paterson, J., Wang, D., and Wen, X.Y. (2007). The Bcl-2 Family Protein Inhibitor, ABT-737, Has Substantial Antimyeloma Activity and Shows Synergistic Effect with Dexamethasone and Melphalan. *Clin. Cancer Res. 13*, 621-629.

Tsujimoto, Y., Finger, L.R., Yunis, J., Nowell, P.C., and Croce, C.M. (1984). Cloning of the chromosome breakpoint of neoplastic B cells with the t(14;18) chromosome translocation. *Science 226*, 1097-1099.

van Delft, M.F., Wei, A.H., Mason, K.D., Vandenberg, C.J., Chen, L., Czabotar, P.E., Willis, S.N., Scott, C.L., Day, C.L., Cory, S., *et al.* (2006). The BH3 mimetic ABT-737 targets selective Bcl-2 proteins and efficiently induces apoptosis via Bak/Bax if Mcl-1 is neutralized. *Cancer Cell. 10*, 389-399.

Vaux, D.L., Cory, S., and Adams, J.M. (1988). Bcl-2 gene promotes haemopoietic cell survival and cooperates with c-myc to immortalize pre-B cells. *Nature 335*, 440-442.

Veis, D.J., Sorenson, C.M., Shutter, J.R., and Korsmeyer, S.J. (1993). Bcl-2-deficient mice demonstrate fulminant lymphoid apoptosis, polycystic kidneys, and hypopigmented hair. *Cell 75*, 229-240.

Villunger, A., Michalak, E.M., Coultas, L., Mullauer, F., Bock, G., Ausserlechner, M.J., Adams, J.M., and Strasser, A. (2003). p53- and drug-induced apoptotic responses mediated by BH3-only proteins *puma* and *noxa*. *Science 302*, 1036-1038.

Walensky, L.D., Kung, A.L., Escher, I., Malia, T.J., Barbuto, S., Wright, R.D., Wagner, G., Verdine, G.L., and Korsmeyer, S.J. (2004). Activation of apoptosis in vivo by a hydrocarbon-stapled BH3 helix. Science *305*, 1466-1470.

Walensky, L.D., Pitter, K., Morash, J., Oh, K.J., Barbuto, S., Fisher, J., Smith, E., Verdine, G.L., and Korsmeyer, S.J. (2006). A stapled BID BH3 helix directly binds and activates BAX. *Mol. Cell 24*, 199-210.

Willis, S.N., Chen, L., Dewson, G., Wei, A., Naik, E., Fletcher, J.I., Adams, J.M., and Huang, D.C. (2005). Proapoptotic Bak is sequestered by Mcl-1 and Bcl-xL, but not Bcl-2, until displaced by BH3-only proteins. *Genes Dev. 19*, 1294-1305.

Willis, S.N., Fletcher, J.I., Kaufmann, T., van Delft, M.F., Chen, L., Czabotar, P.E., Ierino, H., Lee, E.F., Fairlie, W.D., Bouillet, P., *et al.* (2007). Apoptosis initiated when BH3 ligands engage multiple Bcl-2 homologs, not Bax or Bak. *Science 315*, 856-859.

Wolter, K.G., Hsu, Y.T., Smith, C.L., Nechushtan, A., Xi, X.G., and Youle, R.J. (1997). Movement of Bax from the cytosol to mitochondria during apoptosis. *J. Cell Biol. 139*, 1281-1292.

Yan, N., Gu, L., Kokel, D., Chai, J., Li, W., Han, A., Chen, L., Xue, D., and Shi, Y. (2004). Structural, biochemical, and functional analyses of CED-9 recognition by the proapoptotic proteins EGL-1 and CED-4. *Mol .Cell 15*, 999-1006.

Youle, R.J. (2007). Cell biology. Cellular demolition and the rules of engagement. *Science 315*, 776-777.

Yu, J., Zhang, L., Hwang, P.M., Kinzler, K.W., and Vogelstein, B. (2001). PUMA induces the rapid apoptosis of colorectal cancer cells. *Mol. Cell 7*, 673-682.

Zong, W.X., Lindsten, T., Ross, A.J., MacGregor, G.R., and Thompson, C.B. (2001). BH3-only proteins that bind pro-survival Bcl-2 family members fail to induce apoptosis in the absence of Bax and Bak. *Genes Dev. 15*, 1481-1486.

In: Oncogene Proteins: New Research
Editors: A. H. Malloy and E.C. Carson, pp.233-244
ISBN 978-1-60456-290-3

Chapter 9

Oncogene Proteins as Tumor Markers

Viroj Wiwanitkit
Department of Laboratory Medicine, Faculty of Medicine, Chulalongkorn University, Bangkok Thailand 10330

Abstract

Tumor markersis a group of important laboratory diagnostic tools. In this review, details on the general concepts of tumor markers their usefullness and limitations are collected and presented. Basically, a specific tumor marker is a fusion protein associated with a malignant process in which an oncogene is translocated. Oncogene proteins can serve as tumor markers. Here, important oncogene proteins as serum and tissue tumor markers, will be presented and discussed. In addition, the clinical application of oncogene proteins as tumor markers for important tumors in medicine will be mentioned.

Introduction to Tumor Markers

Cancer is an important pathological condition. The diagnosis of cancer is hard. Without laboratory investigation, cancer is usually detected by chance due to the history taking and physical examination. The mass is the common clinical manifestation. For cancer clinical pathology, the tool for diagnosis of cancer is focused for a long time. Tumor marker is a group of laboratory investigations. Molecules occurring in the human body, associated with cancer are tumor markers. Tumor marker can be used for diagnosis or management of disease. An ideal marker is a "blood test" that is positive only when the patient has the disease. In clinical pathology, tumor markers have a very long history. The first tumor marker is the Bence Jones protein, a urine protein that was first described by Bence Jones in 1847 [1]. This protein is the tumor marker for multiple myeloma, a cancer of the plasma cell. The other classical tumor markers include acid phosphatase in advanced prostate cancer which was discovered by Gutman and Gutman in 1938 [2]. Oncofetal antigens and prostate specific antigen are examples of new tumor markers discovered in late twentieth century. Five groups of serum tumor markers are available at present [3]. The enzyme and hormone markers are not specific since these two groups can be secreted from the normal cells. Of five groups, tumor antigens are the most specific. Two subgroups of tumor antigens, oncofetal and non-oncofetal proteins are mentioned [3].

Basically, protein synthesis starts within the nucleus. DNA translates via mRNA into protein in cytosol and secretes out to the membrane and is transported through cell membranes into the external environment. These proteins can then be detected in serum. Any abnormal proteins due to the carcinogenesis process can be a candidate for the tumor marker test. Monoclonal antibodies can be developed and used to detect these proteins on the cancer cell surface or in the serum. The five main advantages of tumor markers includes screening, diagnosis, monitoring therapy, monitoring remission and following-up (Table 2).

Table 1. Examples of tumor markers

Types	Example	Diagnosis usefulness
Enzymes	Acid Phosphatase	Leukemia, cholanigiocarcinoma
	Neuron specific enolase	Neurological malignancy
Hormones	HCG	Choriocarcinoma
	ADH	Lung cancer
	Calcitonin	Thyroid C cell cancer
	Parathormone	Parathyroid cancer
Tumor antigens	CEA	Colon cancer
	CA 125	Ovarian cancer
	CA 19-9	Gastrointestinal cancer
	PSA	Prostate cancer
	AFP	Liver cancer
Immunoglobulins	Immunoglobulin	Plasmacytoma
Other	Bence Jones Protein	Multiple myeloma

Table 2. Examples of advantage of some tumor markers

Advantages	Example	Diagnosis usefulness
1. screening	PSA	Prostate cancer
2. diagnosis	Bence Jones Protein	Multiple myeloma
3. monitoring therapy	AFP	Liver cancer
4. monitoring remission	Calcitonin	Thyroid C cell cancer
5. following-up	HCG	Choriocarcinoma

Some Problems on the Present Tumor Marker Testing

1. Lack of Good Test

Screening and diagnosis are both focused properties of good tumor markers. For screening, high sensitivity is needed. While for diagnosis, high specificity is needed. Any test with these two combined properties are difficult to find. Of several tumor markers, PSA is the best marker in clinical pathology [4].

Most of the present tumor markers lack specificity due to the high false negatives. Benign diseases positive CA 125 [5] or CEA [6] are common. Some physiological underlying can affect the analytical result such as smokers having a raised CEA [7]. Also, cancer heterogeneity usually exists for many organs [8]. In addition, many patients die with, not from, cancers.

2. How can We Manage the Tumor Marker Test?

Due to the preventive medicine principle, early diagnosis is good preventive action. Screening is a widely used tumor marker. However, there are many problems such as, "Is a negative reassuring?" and "What does a positive indicate?". Management of anxiety due to the test result is a necessary point [9 - 10]. Sometimes, the increased level of tumor markers can be seen in benign diseases. On the other hand, some increase can be detected in the late stage, unresectable mass. These facts are the flaws of tumor marker screening.

For following up, a decrease after therapy usually means response, however, "normal" does not mean remission [8]. The query in palliative therapy is treating the symptom or the tumor marker level. Using the laboratory results accompanied with the signs and symptoms is suggested. However, it should be noted that not all symptoms are cancer progression. At the end of therapy, the physician should beware of misleading. A normal level does not mean cured, does not mean complete remission, may not even mean improvement. This can create anxiety, false reassurance and false hope [8].

3. Decision Making Due to the Tumor Marker Investigation Result

Decision making due to the tumor marker investigation result is sometimes very hard since tumor marker results sometimes seems aberrant. Some common aberrant results include increased to two markers. Indeed, the possible clues for this phenomenon are 1) there are two existing growing tumors within the patient, which is rare but possible, 2) there is something wrong in the investigation and 3) there is possibly asecond tumor on top of the primary lesion. In addition, it should be noted that the tumor marker sometimes increases in the improper time such as increase in the benign disease due to the interference of external variation especially for cigarette smoking, which can bring to the over therapeutic options.

On the other hand, it sometimes can be seen that the tumor marker just rise when the disease is very advanced and untreatable, resection is not possible. These facts bring the conclusion to the physicians that they should rely on the tumor marker accompanied with the patients' signs and symptoms in making a decision.

4. Communication Gaps between the Patients and Physician

"Does the patient understand?" is the main question when a physician in charge ordering the tumor marker testing. Explanation is necessary. Creating of anxiety have to be managed. For the oncologist, interpretation of the test result should be carefully performed. Sometimes, multiple markers elevated and sometimes marker falling, mass increasing can be seen. In addition, the oncologist should be accessible to the family. Physicians who knows the patient best, must answer the questions "why?" or "what next"? from the family. The special technique to tell the bad news to the patients and his/her family must be used.

5. Availability of the Tumor Marker Testing

The tumor marker testing is considered a special test in laboratory test in medicine. The availability of the tumor marker testing is a concern. It cannot be denied that there is no equilibrium in health service assessment in different settings. Rural and remote setting usually lack for many medical facilities including the laboratory investigation. Tumor marker testing is usually lack in those distance settings. However, the good referring system and good clinical skill in decision making for request the test in the limited resource setting may be the clue for resolving of the problem of this disequilibrium.

Oncogene Proteins as Tumor Marker Candidates

Oncogenesis is a rather complicated process. As previously mentioned, the underlying genetic abnormalities accompanied with the external environmental insults are the main contributors to cancer development. Recent advances brought the identification of cellular genes which are involved in the starting and progression of tumorigenesis. Several types of cancers have been shown to be derived from cells that have extensively mutated both alleles of one or both of these genes, resulting in a loss-of-function mutation [11 – 13]. The proto-oncogenes, which normally participate in the regulation of cell proliferation and differentiation, can become oncogenes through alterations in the regulation of their expression and/or their coding sequences [11]. The proto-oncogenes can be grouped into different categories based on their protein products, i.e. protein kinases, growth factors, growth factor receptors, and DNA binding proteins. The analysis of the molecular mechanisms governing multistep carcinogenesis became experimentally approachable since the identification and characterization in tumor cells of altered or activated versions of cellular oncogenes that normally control cell growth and differentiation [12]. The activating mutations confer new properties to the oncogene products and should therefore be considered as gain of function mutations [12]. The anti-oncogenes or tumor suppressor genes or recessive oncogenes are normally implicated in a negative regulation of cellular proliferation [11]. The suppression or reversion of the malignant phenotype by the introduction of a normal chromosome into a tumor cell line has lent support to the idea that a family of cellular genes are coding for factors capable to interact with the cell-growth control machinery [12]. These genes seem to reconstitute the normal control of cell growth even in the presence of an activated oncogene [12]. Anderson and Spandidos noted for evidences for the existence of tumor suppressor genes based on the following points: a) recessive cancer genes in higher and lower eukaryotes have been detected, b) studies on cell hybrids have implicated genes which suppress various stages in the malignant conversion of normal cells, c) the isolation from virally induced transformants of flat, non-tumorigenic revertants in which expression of the transforming gene is not down-regulated suggests the presence of genes which suppress the effects of transformation. and d) blocks to differentiation can be bypassed by inducing compounds or differentiation factors [14]. The contribution of oncogene to the tumorigenic phenotype is dominant or positive regulation since they act also in the presence of the homologous wild-type allele while the loss of then tumor suppressor activity contributes to tumorigenesis in a recessive manner since the tumor phenotype appears when both alleles of an onco-suppressor gene are inactivated and the mutations affecting their normal functions belong to the type "loss of function" [11 - 12]. Balancing between oncogene and tumor suppressor gene lead to no tumorogenesis in normal condition. Common to most tumors, several regulatory circuits are altered during multistage tumor progression, most importantly, the control of proliferation, the balance between cell survival and programmed cell death (apoptosis), the communication with neighbouring cells and the extracellular matrix, the induction of tumor neovascularization and, finally, tumor cell migration, invasion and metastatic dissemination [15]. De-regulation of each of these processes represents a rate-limiting step for tumor development and, hence, has to be achieved by tumor cells in a highly selective manner during tumor progression [15].Apoptosis is a genetically controlled response by which eukaryotic cells undergo programmed cell death [16]. This phenomenon plays a major role in developmental pathways, provides a homeostatic balance of cell populations, and is deregulated in many diseases including cancer [16]. Elucidating the mechanisms that link cell cycle control with apoptosis will be of key importance in understanding tumor progression and designing new models of effective tumor therapy [16]. One may reasonably hope that defining a complete profile of alterations of both oncogene and tumor suppressor types of essential genes in the tumor or in the clinically healthy individual will be invaluable to assess prognosis or genetic predisposition, respectively [17].

At present, the main actions of proto-oncogenes during cellular proliferation are outlined and oncogenes are classified to their specific roles [18]. The relationships between proto-, viral-, and cellular oncogenes are analyzed and the several steps of the evolution of a p-onc towards a c-onc are well described. The oncogenes' products, their weight and cellular position are also presented [18]. The onco-suppressor genes, oncogenes' counterparts are also described, their localization being deduced from analysis of frequent breakpoints or mutation site/s in cancers [18]. Basically, in cancer cases, overaction of oncogene than suppressor genes can be expected. These genes are part of the normal complement of cells, and become altered in their structure or expression, during the development of the neoplastic phenotype [19]. Oncogene protein is an important product of oncogene that has many clinical implications.

Besides aiding in tumor diagnosis, the detection of such tumor-associated genes and their products allows the identification of individuals with an inherited predisposition to neoplastic growths, and the overexpression of many of these oncogene products has been shown to be a potential marker of tumor behavior and a predictor of treatment outcome and response [20]. The ability to utilize DNA and RNA probes for nucleic acid hybridization and polymerase chain reaction procedures in cell and tissue preparations of solid tumors and lymphoid proliferations expands and complements the information provided by immunohistochemical techniques [20].

Basically, a specific tumor marker is a fusion protein associated with a malignant process in which an oncogene is translocated. As previously mentioned, the main function of an oncogene is to control the growth of cells to make excessive proteins which further make the cells grow out of control and can result in cancer. Several oncogene proteins are mentioned in oncology. Those oncogene proteins can be served as tumor markers. Second generation tests could easily include PCR on mRNA, and/or in situ hybridization, both of which could be performed using blood samples [21]. Both methods would provide a faster means of testing a large number of cells, however, the methodologies must be improved through automation and computer-aided image analysis, respectively, in order to become useful routine tests [21]. Here, important oncogene proteins as serum and tissue tumor markers will be presented and discussed. In addition, the clinical application of oncogene proteins as tumor markers for important tumors in medicine will be mentioned.

1. Ras Oncogene Protein

The ras oncogene protein or P21 is an important oncogene protein. Point mutations in the cellular c- ras gene can also result in a mutant p21 protein that can transform human and other mammalian cells. Oncogene protein p21has been directly implicated in human tumors and might accounti for as much as 15-20% of all human tumors. One of the most commonly found transforming *ras* oncogenes in human tumours has a valine codon replacing the glycine codon at position 12 of the normal c-Ha-*ras* gene [22]. Tong et al found that one of the major differences between *ras* oncogene protein and normal protein is that the loop of the transforming ras protein that binds the β-phosphate of the guanine nucleotide is enlarged [22]. They also proposed that such a change in the "catalytic site' conformation could explain the reduced GTPase activity of the mutant,which keeps the protein in the GTP bound 'signal on9 state for a prolonged period of time [22]. The ras oncogene protein is known to occur frequently in human tumors, including those of occupational concern such as lung cancer [23]. Knowledge of their mechanisms of action may lead to new opportunities for preventing occupational cancer [23]. Mutations of the K-ras gene is frequent genetic changes of oncogenes in lung cancer [24]. Niklinska et al said that K-RAS oncogene exerts a crucial impact on cell cycle regulation and appears to be of major clinical significance for lung cancer evaluation [25]. Point mutations of the K-ras gene, which are found in 10 to 30% of lung adenocarcinomas, are regarded as being an early event during the carcinogenesis [26]. Autonomous vigorous motility of neoplastic cells, as well as growth and survival advantages, are considered to be necessary for cancer development and progression [26]. However, Ramakrishna et al reported that lung tumors did not have more total K-ras p21 or K-ras p21 GTP than normal lung tissue, nor were high levels of these proteins found in tumors with mutant K-ras [27]. They noted that normal K-ras p21 activity was associated with growth arrest of lung type II cells, and that the exact contribution of mutated K-ras p21 to tumor development remained to be discovered [27]. Okudela et al concluded that suggest that the K-ras-Akt pathway might facilitate the motility of neoplastic cells during the early period of carcinogenesis in lung adenocarcinomas, and might contribute to their non-invasive expansion along the alveolar septa, rather than invasion or metastasis [26].

In addition to lung cancer, the ras oncogene protein is also widely studied for its possibility in using as the tumor marker in pancreas and colorectal tumor marker. For pancreas cancer, genetic analysis of pancreatic juice is a promising aid for the accurate and early diagnosis of pancreatic cancer [28]. K- ras mutation is frequently observed in pancreatic cancer; however, it is not specific for carcinoma because pancreatic adenoma and pancreatitis also show this mutation [28]. Wenker et al also found that K-ras mutations lacked specificity to discriminate malignant pancreatic disease from chronic inflammation in tissue and stool [29]. For colorectal cancer, point mutations of the K-RAS gene at codon 12 are found in about 40% of cases with colorectal cancer, however, the diagnostic implications of the detection of these mutations and their clinical utility are still unclear [30]. Salbe et al noted that K-ras mutations could be found in circulating DNA extracted from serum samples of patients with colorectal cancer and showed that there was a correspondence between serum and tissue K-ras patterns [30]. It is noted that ras-p21 protein could be used as effective biomarkers for colorectal carcinogenesis in human chemopreventive trials [31].

De Biasi et al found that more ras p21 was observed in breast carcinomas compared with their respective normal counterparts [32]. It has been suggested that the immunocytochemical demonstration of the p21 ras oncogene product is a useful marker of malignancy in breast disease [33 – 34]. Rundle et al noted that ras p21 may be useful in the early detection of breast tumors and in post-surgical follow-up of patients, giving patients and physicians new tools for managing breast cancer [33]. However, Candlish et al proposed that the presence of ras p21 protein as demonstrated by the antibody technique was not a useful marker of malignancy or of proliferating epithelium but was rather a normal feature of certain cell types [34]. More ras p21 was also demonstrated in the majority of tumors of the stomach, colon and bladder compared with their respective adjacent normal tissues [32].

2. Fes Oncogene Protein

The human c-fes protooncogene encodes a protein-tyrosine kinase (c-fes) distinct from c-Src, c-Abl and other nonreceptor tyrosine kinases [35]. Although originally identified as the cellular homolog of several transforming retroviral oncoproteins, fes was later found to exhibit strong expression in myeloid hematopoietic cells and to play a direct role in their differentiation [35 - 36]. While it contains a tyrosine kinase and SH2 domain, there is no SH3 domain or carboxy terminal regulatory phosphotyrosine such as found in the Src family of kinases [36]. Fes has a unique N-terminal domain of over 400 amino acids of unknown function [36]. It has been implicated in signaling by a variety of hematopoietic growth factors, and is predominantly a nuclear protein [36].

The fes oncogene protein is also in the topic of occupational medicine at present, especially those who are at risk for leukemia [23].

3. Neu Oncogene Protein

The neu (c-erbB-2) oncogene and its p185 protein is mentioned as an important oncogene protein. The neu oncogene encodes a transmembrane phosphoglycoprotein [37 - 39]. This molecule appears to be a growth factor receptor in the family of tyrosine kinase growth factor receptors [37 -39]. The neu gene encodes a novel growth factor receptor but, in contrast to met, its activation involves a single T:A→A:T point mutation in the region of the neu gene encoding the receptor transmembrane domain [37]. Makar et al suggested that the neu oncogene was an independent prognostic factor, which might predict the development of distant metastasis [38]. Preclinical studies have suggested that HER-2/neu overexpression enhances the metastatic potential of breast cancer cells [40]. Amplification and/or overexpression of c-erbB-2 in breast adenocarcinomas occurs frequently and its occurrence implies a more advanced malignancy [39]. This functional tumor marker is readily identified by appropriate DNA and antibody probes [39]. Breast carcinomas express high levels of ErbB receptors and their ligands, and their overexpression has been associated with a more aggressive clinical behavior [41]. Amplification of the HER-2/neu gene resulting in overexpression of the p185 occurs in approximately 25% of early stage breast cancers [42]. HER-2/neu has been established as an important independent prognostic factor in early stage breast cancer [42]. Overexpression /amplification of the HER-2/neu has been associated with a worse outcome in patients with breast cancer [42 – 45]. New data also suggest that HER-2/neu may be useful not only as a prognostic factor but also as a predictive marker for projecting response to chemotherapeutics, antiestrogens, and therapeutic anti-HER-2/neu monoclonal antibodies [42 - 44]. At present, monoclonal antibodies directed against HER-2/neu have been developed and used in clinical practice [40]. Clinical activity with one of these antibodies, trastuzumab, a humanized anti-ErbB2 MAb, has been documented in patients with breast cancer in a series of clinical trials and has recently been approved for the therapy of patients with metastatic ErbB2 overexpressing breast cancer [41]. Herceptin, a "humanized" murine monoclonal antibody directed against the extracellular domain of the HER-2/neu protein, is being used to treat breast cancer that overexpresses HER-2/neu [45]. The status of Her-2/neu in the tumor has become a critical factor in the management strategy of a breast cancer patient [45]. Therapeutic strategies that target other molecules involved in breast cancer development and progression are on the same parallel way [46]. In addition, in the breast cancer case, HER2/neu is a focus as a target for vaccine development [47]. It is crucial that pathologists become aware of these advances and assume a pivotal role in the development and application of assays to evaluate these new molecular targets [46].

4. myc Oncogene Protein

The myc proteins are nuclear proteins that exert their biological functions at least in part through the transcriptional regulation of large sets of target genes. Recent microarray analyses show that several percent of all genes may be directly regulated by myc [48]. The myc transcription factor functions as a downstream effector of most mitogenic signals [49]. Myc is synthesized rapidly in response to extracellular mitogenic signals, and blocking myc induction abolishes or at least severely attenuates any mitogenic response [49]. Furthermore, ectopic myc expression can often bypass the requirement for extracellular signals for entry into S phase [49]. It can be concluded that myc regulates to some degree every major process in the cell. Proliferation, growth, differentiation, apoptosis, and metabolism are all under Myc control and these processes feed back to adjust the level of c-myc expression [50]. Although myc is regulated at every level from RNA synthesis to protein degradation, c-myc transcription is particularly responsive to multiple diverse physiological and pathological signals [50]. However, it has been difficult to gain a precise understanding of how myc drives cancer because myc acts rather weakly at many of its target loci, and it has been reported to regulate as many as 10% to 15% of all cellular genes [51].

MYCN is a member of the myc family of oncogenes that encode nuclear proteins serving as transcription factors. This myc oncogene protein is mentioned for its tumor marker role for some cancers especially for neurological cancer. The correlation of N-myc gene amplification with poor prognosis in neuroblastoma was one of the first examples of prognostic data supplied by a c-oncogene [52].

In neuroblastoma, amplified MYCN is a strong prognostic indicator of poor prognosis [53]. Identification of amplified MYCN in neuroblastomas has marked the clinical debut of oncogenes, and MYCN status is now being widely used as a standard marker for neuroblastoma stratification [53 - 54].

5. P53

The tumor suppressor p53 is a transcription factor that is frequently inactivated in human tumors, therefore, restoring its function has been considered an attractive approach to restrain cancer [55]. Typically, p53-dependent growth arrest, senescence and apoptosis of tumor cells have been attributed to transcriptional activity of nuclear p53 [55]. Hence, p53 is not a pure oncogene protein tumor marker and will not be discussed in this chapter.

References

[1] Stone MJ. Henry Bence Jones and his protein. *J. Med. Biogr.* 1998 Feb;6(1):53-7.

[2] Sproul EE. Acid phosphatase and prostate cancer: historical overview. *Prostate.* 1980;1(4):411-3.

[3] Ochi Y, Okabe H, Inui T, Yamashiro K. Tumor marker--present and future. *Rinsho Byori.* 1997 Sep;45(9):875-83.

[4] Frydenberg M, Wijesinha S. Diagnosing prostate cancer - what GPs need to know. *Aust. Fam. Physician.* 2007 May;36(5):345-7.

[5] Devine PL, McGuckin MA, Ward BG. Circulating mucins as tumor markers in ovarian cancer (review). *Anticancer Res.* 1992 May-Jun;12(3):709-17.

[6] Fletcher RH. Carcinoembryonic antigen. Ann Intern Med. 1986 Jan;104(1):66-73.

[7] Ishii M. Limitation of clinical usefulness of tumor marker. *Gan. To Kagaku Ryoho.* 1995 Aug;22(9):1139-45.

[8] Motoo Y, Watanabe H, Sawabu N. Sensitivity and specificity of tumor markers in cancer diagnosis. *Nippon Rinsho.* 1996 Jun;54(6):1587-91.

[9] Evans R, Edwards AG, Elwyn G, Watson E, Grol R, Brett J, Austoker J. "It's a maybe test": men's experiences of prostate specific antigen testing in primary care. *Br. J. Gen. Pract.* 2007 Apr;57(537):303-10.

[10] Marteau TM, Kidd J, Michie S, Cook R, Johnston M, Shaw RW. Anxiety, knowledge and satisfaction in women receiving false positive results on routine prenatal screening: a randomized controlled trial. *J. Psychosom. Obstet. Gynaecol.* 1993 Sep;14(3):185-96.

[11] Monier R. Oncogenes and anti-oncogenes in tumorigenesis. *Reprod. Nutr. Dev.* 1990;30(3):445-54.

[12] Della Porta G, Radice P, Pierotti MA. Onco-suppressor genes in human cancer. *Tumori.* 1989 Aug 31;75(4):329-36.

[13] Levine AJ. Tumor suppressor genes. Bioessays. 1990 Feb;12(2):60-6.

[14] Anderson ML, Spandidos DA. Onco-suppressor genes and their involvement in cancer (review). *Anticancer Res.* 1988 Sep-Oct;8(5A):873-9.

[15] Compagni A, Christofori G. Recent advances in research on multistage tumorigenesis.Br *J. Cancer.* 2000 Jul;83(1):1-5.

[16] Fotedar R, Diederich L, Fotedar A. Apoptosis and the cell cycle. *Prog. Cell Cycle Res.* 1996;2:147-63.

[17] Jeanteur P, Theillet C, Pujol H. Oncogenes, anti-oncogenes and their alterations in human tumors. *Rev. Med. Interne.* 1990 May-Jun;11(3):216-20.

[18] Turcanu V. An overview of oncogenesis. *Rev. Med. Chir. Soc. Med. Nat. Iasi.* 1992;96 Suppl:13-8.

[19] Renan MJ. Cancer genes: current status, future prospects, and applications in radiotherapy/oncology. *Radiother .Oncol.* 1990 Nov;19(3):197-218.

[20] Leong AS, Robbins P, Spagnolo DV. Tumor genes and their proteins in cytologic and surgical specimens: relevance and detection systems. *Diagn. Cytopathol.* 1995 Dec;13(5):411-22.

[21] McKenzie SJ Diagnostic utility of oncogenes and their products in human cancer. *Biochim. Biophys. Acta.* 1991 Dec 10;1072(2-3):193-214.

[22] Tong LA, de Vos AM, Milburn MV, Jancarik J, Noguchi S, Nishimura S, Miura K, Ohtsuka E, Kim SH. Structural differences between a *ras* oncogene protein and the normal protein. *Nature.* 1989 Jan 5;337(6202):90-3.

[23] Brandt-Rauf PW. Oncogenes and oncoproteins in occupational carcinogenesis. *Scand. J. Work Environ. Health.* 1992;18 Suppl 1:27-30.

[24] Mitsudomi T, Takahashi T. Genetic abnormalities in lung cancer and their prognostic implications. *Gan To Kagaku Ryoho.* 1996 Jul;23(8):990-6.

[25] Niklinska W, Chyczewski L, Niklinski J. New molecular approaches to lung cancer: biological and clinical implications of P53, P16 and K-RAS studies. *Folia Histochem. Cytobiol.* 2001;39(2):99-103.

[26] Okudela K, Hayashi H, Ito T, Yazawa T, Suzuki T, Nakane Y, Sato H, Ishi H, KeQin X, Masuda A, Takahashi T, Kitamura H. K-ras gene mutation enhances motility of immortalized airway cells and lung adenocarcinoma cells via Akt activation: possible contribution to non-invasive expansion of lung adenocarcinoma. *Am. J. Pathol.* 2004 Jan;164(1):91-100.

[27] Ramakrishna G, Sithanandam G, Cheng RY, Fornwald LW, Smith GT, Diwan BA, Anderson LM. K-ras p21 expression and activity in lung and lung tumors. *Exp. Lung Res.* 2000 Dec;26(8):659-71.

[28] Mizumoto K, Tanaka M. Genetic diagnosis of pancreatic cancer. *J. Hepatobiliary Pancreat Surg.* 2002;9(1):39-44.

[29] Wenger FA, Zieren J, Peter FJ, Jacobi CA, Muller JM. K-ras mutations in tissue and stool samples from patients with pancreatic cancer and chronic pancreatitis. *Langenbecks Arch. Surg.* 1999 Apr;384(2):181-6.

[30] Salbe C, Trevisiol C, Ferruzzi E, Mancuso T, Nascimbeni R, Di Fabio F, Salerni B, Dittadi R. Molecular detection of codon 12 K-RAS mutations in circulating DNA from serum of colorectal cancer patients. *Int .J. Biol .Markers.* 2000 Oct-Dec;15(4):300-7.

[31] Jia X, Han C. Biomarkers in the studies on chemoprevention of colorectal cancer. Wei Sheng Yan Jiu. 2000 Mar 30;29(2):109-11.

[32] De Biasi F, Del Sal G, Hand PH. Evidence of enhancement of the ras oncogene protein product (p21) in a spectrum of human tumors. *Int. J. Cancer.* 1989 Mar 15;43(3):431-5.

[33] Rundle A, Tang D, Brandt-Rauf P, Zhou J, Kelly A, Schnabel F, Perera FP. Association between the ras p21 oncoprotein in blood samples and breast cancer. *Cancer Lett.* 2002 Nov 8;185(1):71-8.

[34] Candlish W, Kerr IB, Simpson HW. Immunocytochemical demonstration and significance of p21 ras family oncogene product in benign and malignant breast disease. *J. Pathol.* 1986 Nov;150(3):163-7.

[35] Smithgall TE, Rogers JA, Peters KL, Li J, Briggs SD, Lionberger JM, Cheng H, Shibata A, Scholtz B, Schreiner S, Dunham N. The c-Fes family of protein-tyrosine kinases. *Crit. Rev. Oncog.* 1998;9(1):43-62.

[36] Yates KE, Gasson JC. Role of c-Fes in normal and neoplastic hematopoiesis. *Stem .Cells.* 1996 Jan;14(1):117-23.

[37] Cooper CS. The role of non-ras transforming genes in chemical carcinogenesis. *Environ. Health Perspect.* 1991 Jun;93:33-40.

[38] Makar AP, Desmedt EJ, De Potter CR, Vanderheyden JS, Schatteman EA. Neu (C-erbB-2) oncogene in breast cancer and its possible association with the risk of distant metastases. A retrospective study and review of literature. *Acta Oncol.* 1990;29(7):931-4.

[39] Maguire HC Jr, Greene MI. Neu (c-erbB-2), a tumor marker in carcinoma of the female breast. *Pathobiology.* 1990;58(6):297-303.

[40] Sahin AA. Biologic and clinical significance of HER-2/neu (cerbB-2) in breast cancer. *Adv. Anat. Pathol.* 2000 May;7(3):158-66.

[41] Albanell J, Baselga J. The ErbB receptors as targets for breast cancer therapy. *J. Mammary Gland Biol .Neoplasia.* 1999 Oct;4(4):337-51.

[42] Pegram MD, Pauletti G, Slamon DJ. HER-2/neu as a predictive marker of response to breast cancer therapy. *Breast Cancer Res. Treat.* 1998;52(1-3):65-77.

[43] Stearns V, Yamauchi H, Hayes DF. Circulating tumor markers in breast cancer: accepted utilities and novel prospects. Breast Cancer Res Treat. 1998;52(1-3):239-59.

[44] Wisecarver JL. HER-2/neu testing comes of age. *Am. J. Clin. Pathol.* 1999 Mar;111(3):299-301.

[45] Kaptain S, Tan LK, Chen B. Her-2/neu and breast cancer. *Diagn. Mol .Pathol.* 2001 Sep;10(3):139-52.

[46] Schnitt SJ. Breast cancer in the 21st century: neu opportunities and neu challenges. *Mod. Pathol.* 2001 Mar;14(3):213-8.

[47] Wiwanitkit V. Predicted B-cell epitopes of HER-2 oncoprotein by a bioinformatics method: a clue for breast cancer vaccine development. *Asian Pac. J. Cancer Prev.* 2007 Jan-Mar;8(1):137-8.

[48] Kleine-Kohlbrecher D, Adhikary S, Eilers M. Mechanisms of transcriptional repression by Myc. Curr. *Top Microbiol. Immunol.* 2006;302:51-62.

[49] Cole MD, Nikiforov MA. Transcriptional activation by the Myc oncoprotein. *Curr. Top. Microbiol. Immunol.* 2006;302:33-50.

[50] Liu J, Levens D. Making myc. Curr Top Microbiol Immunol. 2006;302:1-32.

[51] Knoepfler PS. Myc goes global: new tricks for an old oncogene. *Cancer Res.* 2007 Jun 1;67(11):5061-3.

[52] Duffy MJ. Cellular oncogenes and suppressor genes as prognostic markers in cancer. *Clin. Biochem.* 1993 Dec;26(6):439-47.

[53] Schwab M. MYCN in neuronal tumours. Cancer Lett. 2004 Feb 20;204(2):179-87.

[54] Phillips WS, Stafford PW, Duvol-Arnold B, Ghosh BC. Neuroblastoma and the clinical significance of N-myc oncogene amplification. *Surg. Gynecol. Obstet.* 1991 Jan;172(1):73-80.

[55] Fuster JJ, Sanz-Gonzalez SM, Moll UM, Andres V. Classic and novel roles of p53: prospects for anticancer therapy. *Trends Mol. Med.* 2007 May;13(5):192-9.

In: Oncogene Proteins: New Research
Editors: A. H. Malloy and E.C. Carson, pp. 245-272
ISBN 978-1-60456-290-3

Chapter 10

Apoptosis Induced by PPARγ Ligands in Lung Cancer

Ming-Yue Li[1], Tony S.K. Mok[2], and George G. Chen*

[1]Department of Surgery, [2]Department of Clinical Oncology,
The Chinese University of Hong Kong, Shatin, N.T., Hong Kong, China

Abstract

Lung cancer is the world's leading cause of cancer death. Currently there is no accepted adjuvant or palliative treatment modalities that have been conclusively shown to prolong survival in lung cancer. In order to improve outcomes with this disease, search for novel treatments is critical. Peroxisome proliferator activated receptors gamma (PPARγ) is a transcriptional factor belonging to the ligand-activated nuclear receptor superfamily. PPARγ is highly expressed in adipose tissue and has a dominant regulatory role in adipocyte differentiation. In addition to adipogenic and anti-inflammatory effects, PPARγ is expressed in multiple types of tissues, including the breast, colon, lung, ovary, and placenta in humans. The activation of PPARγ plays an inhibitory role in cell proliferation and growth and this property makes the activation of PPARγ by its natural and synthetic ligands an attractive tool in cancer treatment and prevention. A number of PPARγ ligands have been identified, such as natural prostaglandins and the synthetic anti-diabetic thiazolidinediones (TZDs). The discovery of PPARγ agonists enables the recognition of the mechanisms involved in PPARγ multiple effects such as inhibition of tumor cell growth. The importance of this transcription factor and its ligands in physiology and pathophysiology has instigated much research in this field. In this article, current situations in lung cancer therapy, structural features of PPARγ, mechanisms of PPARγ gene transcription and recent development in the discovery of their biological functions on growth inhibition and apoptosis of lung tumors are reviewed. Prospects for future research leading to new therapies for lung cancer are also discussed.

* Correspondence: George G Chen; Department of Surgery, Prince of Wales Hospital, The Chinese University of Hong Kong, Shatin, N.T., Hong Kong. Tel: 852 2632 3934, Fax: 852 2645 0605, email:gchen@cuhk.edu.hk.

1. Introduction

Lung cancer is the world's major cause of cancer death [1]. It is classified into two major types: small cell lung cancer (SCLC) and non-small cell lung cancer (NSCLC). SCLC accounts for about 25% of bronchogenic carcinomas and NSCLC constitutes almost all of the remainder. As a group, the NSCLC are subdivided into squamous, adenocarcinoma and large cell carcinoma phenotypes [2-4]. Although it is still unclear whether SCLC and NSCLC are derived from the same or different stem cells, there has been much progress made towards understanding the molecular etiology of lung cancer, particularly involving mutations in oncogenes and tumor suppressor genes (recessive oncogenes), including *myc* family members, K-*ras*, *TP53*, *RB*, *p16-CDKN2*, and candidate tumor suppressor gene regions on chromosomes 3p, 5q22 (*APC-MCC* region), 8p, 11p, and others [4, 5]. Activating mutations in Ras are associated with NSCLC, with a mutation at codon 12 of the Ki-Ras gene observed in approximately 30% of adenocarcinomas, and just under 10% of other NSCLC types. These mutations appear to be virtually absent from SCLC [6]. Overexpression of the c-myc gene is also frequently observed in NSCLC, but appears to be more prevalent in SCLC [7]. Alterations in tumour suppressor genes have also been reported. Mutations in p53 have been detected in 90% of SCLC and 50% of NSCLC [7]. Recent advances in cancer biology have led to the identification of signaling proteins that may be rational therapeutic targets.

The average 5-year survival rate for localized and metastatic lung cancer is 48% and 2.5% respectively [8, 9]. With only 15% of all patients are diagnosed at an early, localized stage because lung cancer begins to grow silently without any symptoms until the cancer has developed into an advanced stage [9]. For the treatment of non-small cell lung cancer (NSCLC), cisplatin- based chemotherapy, when compared with best supportive care alone, the differences are fairly small with an increased median survival of 6 weeks and an improvement in the 1-year survival rate of only 10% in metastatic disease. For small-cell lung cancer (SCLC), chemotherapy is rarely curative, except in limited-stage patients, and those patients are candidates for radiotherapy and chemotherapy combinations. However, multiple strategies designed to improve systemic treatments have not yielded significant breakthroughs [10, 11]. There is also currently no accepted adjuvant or palliative treatment modalities that have been conclusively shown to prolong survival in lung cancer [8]. Therefore, translational research to improve outcomes with this disease is critical [12]. New methods are needed for early detection, treatment, and prevention of this disease.

The epidemic of lung cancer is directly attributable to cigarette and therefore major efforts are needed to prevent its initiation and to aid with cessation of smoking [13]. Smoking is the single most extensively documented risk factor for all histological types of lung cancer. The major cause of lung cancer is tobacco smoke which contains thousands of substances including carcinogens, co-carcinogens, and tumor promoters. [14]. The three major classes of carcinogens in tobacco smoke are the polycyclic hydrocarbons (such as benzo[a]pyrene), nitrosamines, and aromatic amines. It is estimated that tobacco smoke is responsible for 85-90% of all cases of lung cancer [15]. Among the numerous toxic and carcinogenic agents contained in tobacco products, the nicotine-derived 4-(*N*-methyl-*N*-nitrosoamino)-1-(3-pyridyl)-1-butanone (NNK) is the most potent carcinogen in laboratory animals (16) and the total dose of NNK experienced by a smoker in a lifetime of smoking is remarkably close to the lowest dose shown to induce lung cancer in rats [17]. Therefore, NNK is considered to be a major contributor to lung carcinogenesis in smokers [18, 19].

Peroxisome proliferator-activated receptors (PPARs) are members of the nuclear hormone receptors superfamily of ligand-activated transcriptional factors that include receptors for steroids, thyroid hormone, vitamin D and retinoid acid [20]. PPAR binds to peroxisome proliferator responsive element (PPRE) as a heterodimer with the retinoid X receptor (RXR) [21]. To date, three PPAR subtypes have been isolated and are termed as PPARα, PPARβ (also called δ, NUC-1 or FAAR), and PPARγ [20, 22]. PPARα is a key regulator of lipid metabolism. Specifically, activation of PPARα results in proliferation of peroxisomes and the induction of gene involved in β-oxidation of fatty acids [23]. PPARβ is ubiquitously expressed and has been implicated in the repression of PPARα and γ transcriptional activity [24] as well as in wound healing [25]. PPARγ is highly expressed in adipose tissues and has a dominant regulatory role in adipocyte differentiation [26, 27]. In addition to adipogenic and anti-inflammatory effects, PPARγ is expressed in multiple types of tissues, including the breast, colon, lung, ovary, and placenta in humans [28-30]. The activation of PPARγ plays an inhibitory role in cell proliferation and growth and this property makes the activation of PPARγ by its natural and synthetic ligands an attractive tool in cancer treatment and prevention [30-32]. Lung cancer cells are known to express PPARγ but its activation is believed being defected [33]. However, application of PPARγ synthetic ligands can activate the PPARγ pathway and lead to differentiation, growth arrest and apoptosis of lung cancer cells, suggesting that activation of PPARγ and its ligands may be useful in slowing or arresting the proliferation of dedifferentiated lung tumor cells.

2. Peroxisome Proliferator Activated Receptor gamma (PPAR) γ

PPARγ was discovered in Xenopus [34] and mammals [35] based on its similarity to PPARα, a member of the nuclear receptor superfamily that is activated by natural fatty acids and the fibrate class of hypolipidaemic drugs [36]. PPARγ possesses similar structural and functional features of the nuclear receptor superfamily. Principally, four functional domains have been identified, named A/B, C, D and E/F (23) (Figure 1). The N-terminal A/B domain contains a ligand-independent activation function 1 (AF-1) [37] responsible for the phosphorylation of PPARγ. The DNA binding domain (DBD) or C domain promotes the binding of PPARγ to the peroxisome proliferator response element (PPRE) in the promoter region of target genes [38]. The D site is a docking domain for cofactors. The E domain or ligand-binding domain (LBD) is responsible for ligand specificity and activation of PPAR binding to the PPRE, which increases the expression of targeted genes [39]. Recruitment of PPAR co-factors to assist the gene transcription processes is carried out by the ligand-dependent activation function 2 (AF-2), which is located in the E/F domain. The E/F domain also exhibits a region involved in the dimerization with a partner nuclear receptor, the retinoid X receptor (RXR) [39].

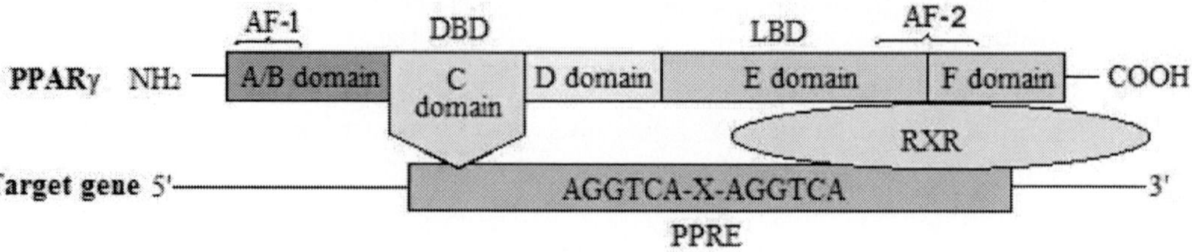

Figure 1. PPARγ structure and activation. AF-1: ligand-independent transcriptional activating function-1, responsible for the phosphorylation of PPARγ; DBD: DNA binding domain, promotes the binding of PPARγ to the peroxisome proliferator response element (PPRE) in the promoter region of target genes; D domain, a docking domain for cofactors. LBD: ligand-binding domain, responsible for ligand specificity and activation of PPAR binding to the PPRE; AF-2: ligand-dependent activating function-2, responsible for recruitment of PPAR co-factors to assist the gene transcription processes; RXR: retinoid X receptor; PPRE: peroxisome proliferator response element.

The PPARγ gene contains three promoters that yield three isoforms, namely, PPARγ1, PPARγ2 and PPARγ3 [40]. PPARγ1 and γ3 RNA transcripts translate into the identical PPARγ1 protein and PPARγ2 protein contains an additional 30 amino acids at its N-terminus. PPARγ expression is tissue-dependent. PPARγ1 is found in a broad range of tissues, whereas PPARγ2 is restricted to adipose tissues. PPARγ3 is abundant in macrophages, the large intestine and white adipose tissues [40, 41].

3. PPARγ Ligands

A number of PPARγ ligands have been identified, including certain polyunsaturated fatty acids, natural prostaglandins, and the synthetic anti-diabetic thiazolidinediones (TZDs). Polyunsaturated fatty acids such as linoleic acid and arachidonic acid are endogenous ligands of PPARγ. Prostaglandin (PG)-related compounds such as 15-deoxy-delta12-14-PGJ2 (15d-PGJ2) have been identified as beneficial PPARγ activation. Oxidized lipids such as 9-HODE, 13-HODE and 15-HETE have also been found to be effective activators of PPARγ [42, 43]. TZDs such as troglitazone(TGZ), ciglitazone, pioglitazone, rosiglitazone are the most notable synthetic compounds with PPARγ activation properties but they show inhibitory activity towards PPARα or PPARβ [44]. Non-TZDs such as isoxazolidinedione JTT-501 [45] and the tyrosine-based GW7845 [46] also act on PPARγ activation. The structures of these ligands are described in Figure 2. In addition to synthetic ligands, some natural potent PPAR-γ agonists from several medicinal plants such as Saurufuran A from *Saururus chinensis* (*Saururaceae*)[47], flavonoids such as chrysin and kampferol [48] and phenolic compounds from *Glycyrrhiza uralensis* (*Fabaceae*) [49] have also been found. Novel PPARγ partial agonists [50] and antagonists [51] have been recently identified. Although these compounds have less clinical significance, they may be useful in understanding PPARγ physiology and the identification of new ligands.

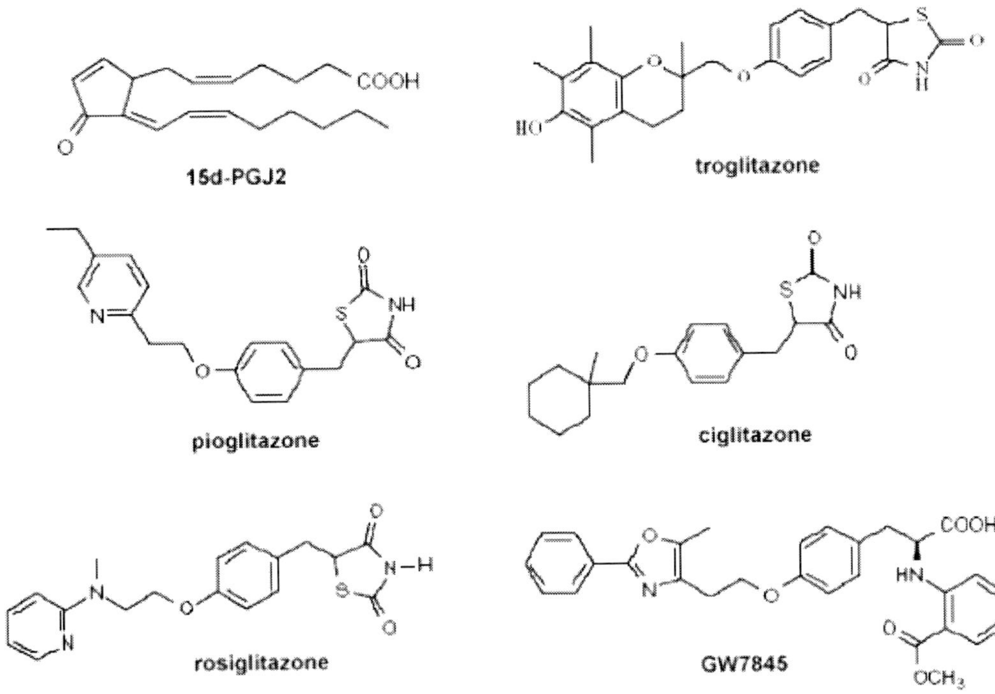

Figure 2. Chemical structure of commonly used PPARγ lignads.

4. Ligands-Dependent Activation of PPARγ

The activation of PPARγ by natural and synthetic ligands leads to a cascade molecular event: PPARγ heterodimerization with retinoid X receptor (RXR), binding of this heterodimer to PPRE, dissociation of corepressors and recruitment of transcriptional coactivators, and finally the regulation of the target gene expression (Figure 3). The LBD domain facilitates the heterodimerisation of PPARγ with the RXR and the resultant heterodimer subsequently binds to PPRE with the recruitment of cofactors [21]. PPRE consists of direct repeat (DR)-1 elements of two hexanucleotides with the AGGTCA sequence separated by a single nucleotide spacer [52, 53] and the DR-1 pattern is specific for PPAR–RXR heterodimer.

PPAR coactivators or corepressors mediate the ability of nuclear receptors to initiate or suppress the transcription process. They interact with nuclear receptors in a ligand-dependent manner [53]. Without ligand stimulation, a heterodimerised nuclear receptor associates with co-repressors containing histone deacetylase activity. The deacetylated state of histone inhibits transcription [54]. Upon ligand binding, coactivators such as steroid receptor co-activator (SRC)-1 and the PPAR binding protein (PBP) with histone acetylase activity [55, 56] initiate a cascade event, which induces the PPAR target gene transcription.

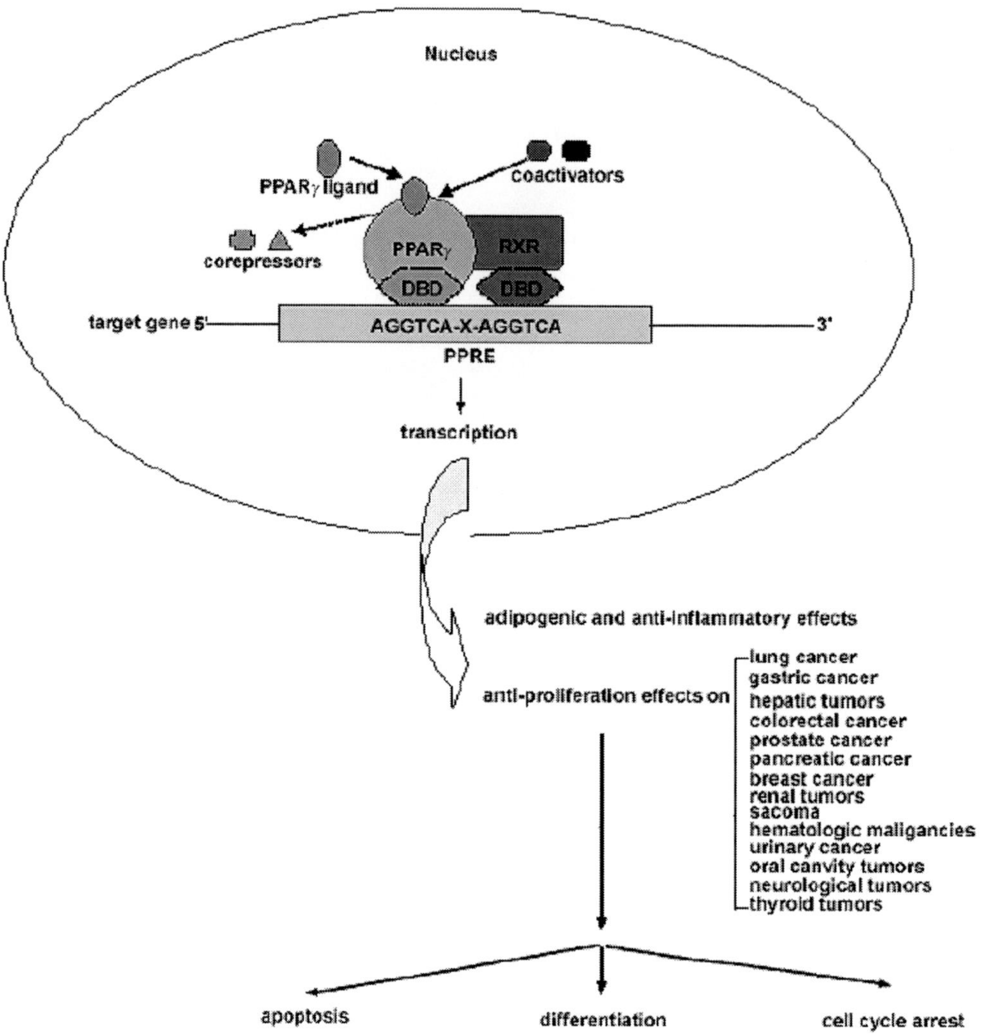

Figure 3. Schematic diagram of PPARγ signaling. PPARγ functions as heterodimer with its obligate partner, RXRa. The dimmer interacts with co-regulators, either co-activators or co-repressors. Binding of specific PPARγ agonists leads to coactivator recruitment and corepressor release before PPARγ/RXRa binds to the PPRE that is present in the promoters of target genes and activates their transcription. Activation of PPARγ is implicated in the expression of genes involved in control of various physiological and pathophysiological processes.

Distinct ligands selectively activate PPARγ, and such a selection is dependent on the types of cells. For example, rosglitizone, ciglitizone and PGJ_2 are all potent agonists of PPARγ transactivation in lung adenocarcinoma cell lines. However, these same ligands have no effect on PPARγ transactivation in squamous cell or large cell carcinomas of the lung [57]. Our results showed that PPARγ ligand-TGZ up-regulated PPARγ expression and induced PPARγ transactivity not only in human lung normal cells but also NSCLC, such as NCI-H23 cells [58].

5. PPARγ Activity Tuned by its Phosphorylation Status in a Ligand-Dependent or Independent Manner

In addition to their ligand-mediated activation, PPARγ, as a member of nuclear receptors whose activity is finely tuned by their phosphorylation [59]. *In vitro* and *in vivo* experiments have revealed that the activation of PPARγ is achieved via a phosphorylated process, which is regulated by all three of the well known mammalian MAPKs: extracellular signal-regulated kinase (ERK), p38, and c-Jun N-terminal kinase (JNK) [27, 60]. For example, stimulation of rat liver epithelial cells (GN4 cells) with PPARγ agonists initiates two independent kinase-driven signaling cascades, which result in the activated ERK and p38. The activation of ERK and p38 can phosphorylate PPARγ to regulate PPARγ activity [61]. However, information about the relationship between MAPKs and the activation of PPARγ in human lung cancer is limited. The activity of PPARγ is regulated by its phosphorylation status in a ligand-dependent or -independent manner, according to the isoform and/or the cellular context and it has been further indicated that modification of PPARγ by phosphorylation may be linked with positive or negative regulation of the PPARγ transcriptional activity, which is discussed below.

5.1. Phosphorylation Linked with Positive Regulation of the PPARγ Transcriptional Activity

In Opossum kidney cells, the PPARγ agonists 15d-PGJ_2, ciglitazone, and TGZ are found to significantly stimulate transcriptional activation and phosphorylation of PPARγ. ERK can positively regulate PPARγ activation [62], suggesting that ERK is involved in the ligands-dependent activation of PPARγ in Opossum kidney cells probably via the phosphorylation of PPARγ. Indeed, it has been demonstrated that constitutively active ERK MAPK-kinase (MKK1) can stimulate PPARγ phosphorylation in CHO cells [63]. In the same type of the cell, insulin-stimulated PPARγ activation is mediated by MAPK/ERK and PPARγ transactivation is much higher when the cell is treated by both insulin and TZD. Thus, insulin can increase the ligand-independent transcriptional activity of PPARγ and enhance the TZD-induced PPARγ transactivating function [63]. However, ERK activity may be not required for the activation of PPARγ in some other types of cells. For example, in opossum kidney proximal tubular cells, PPARγ activity can be induced by C-peptide or insulin via a ligand-independent pathway that is unrelated to ERK but PI3-kinase [64].

Information on the relationship between PPARγ and ERK in human lung cancer cells is limited. Our recent studies have found that TGZ, one of the PPARγ ligands, up-regulated PPARγ expression and stimulated PPARγ transcriptional activity associated with a sustained increase of ERK1/2 activation in human lung cancer cells [58, 65]. Considering that ERK1/2 can mediate a series of reactions leading to cell cycle arrest, differentiation and apoptosis [66], our result suggests that ERK1/2 may function at the downstream of PPARγ and play a central role in the inhibitory pathway of PPARγ. Interestingly, not only can PPARγ talks to ERK1/2, but also ERK1/2 to PPARγ. Further experiments from our laboratory have showed that PPARγ can co-localize with activated ERK in the nucleus [65, 66], suggesting a potential interaction between them and that the activation of ERK may stimulate the phosphorylation of PPARγ. Inhibition of ERK1/2 by U0126, a specific inhibitor of ERK1/2, led to a significant decrease in the level of PPARγ expression and reduction of its accumulation in nuclear [65, 66]. The result indicates that the activation of PPARγ is ERK-dependent and there was a positive crosstalk between ERK and PPARγ in human NSCLC. A non-ERK-dependent activation of PPARγ is not found in NSCLC (non-published data). Therefore, the above results may suggest that there is an auto-regulatory feedback mechanism in PPARγ pathway of lung cancer cells to amplify the effect of ERK1/2 on cell growth arrest and apoptosis [65, 66].

5.2. Phosphorylation Linked with Negative Regulation of the PPARγ Transcriptional Activity

Indeed, epidermal growth factor (EGF) and platelet-derived growth factor (PDGF) decrease the ligand-dependent transcriptional activity of PPARγ though they increase PPARγ phosphorylation through MAPK signaling in adipocyte cell lines [27, 67]; Substitution of mouse PPARγ serine 112 by alanine leads to a loss of PDGF-mediated repression of PPARγ activity [67]; suggesting that PDGF is able to stimulate phosphorylation of PPARγ which counteracts the ligand-dependent activation of PPARγ. Therefore, the regulation of PPARγ activation by PDGF is dependent on the intact of PPARγ serine 112. In human, Serine 84 of PPARγ1 is phosphorylated by ERK2- and JNK-MAPK which inhibits both PPARγ ligand-dependent and -independent transactivating function [68]. These results may suggest that the location of phosphorylated serine may play a vital role in determining the activity of PPARγ. The treatment of macrophages with tumor growth factor (TGF) β1 and β2 inhibits PPARγ ligand-induced expression of CD36 through phosphorylation of PPARγ via an activation of the ERK-MAPK pathway [69]. It is unknown whether such a negative regulation of the PPARγ transcriptional activity by phosphorylation is present in human lung cells. However, according to our result, PPARγ transcriptional activity in human lung cancer cells is enhanced by its phosphorylation.

6. PPARγ and its Ligands in Cancer Treatment

Since the discovery at previous decade, the increasing evidnce has demonstrated that PPARγ is an important modulator in the regulation of complex pathways not only in physiological conditions but also in pathological conditions such as cancer (Figure 3). TGZ and 15d-PGJ2 treatment inhibited the growth of human lung cancer cells through the induction of apoptosis [58, 70]. Human gastric cancer cell line MKN45 can express PPARγ mRNA and protein. TGZ or pioglitazone treatment inhibits the growth of MKN45 cells via the induction of apoptosis [71]. PPARγ is constitutively expressed in various hepatoma cell lines including HLF, HuH-7, HAK-1B and HAK-5. TGZ has been found to induce cell cycle arrest on the G0/G1 phase in all these hepatoma cell lines [72]. Human colon carcinoma cell lines express PPARγ, and ligand activation of this receptor caused most of them to undergo a differentiative response and reverse their malignant phenotype [73]. Human prostate cancer cell lines PC-3 cultured with TGZ showed dramatic morphological changes by light and electron microscopy, suggesting that the cells became less malignant, although cell cycle or differentiation related molecules were not affected [74]. TZDs treatment resulted in inhibition of both cellular and clonogenic growth and G1 cell cycle arrest in various human pancreatic cancer cell lines (Capan-1, AsPC-1, Capan-2, HPAF-II, BxPC-3, PANC-1 and MIA PaCa-2) [75, 76]. PPARγ was expressed at significant levels in human primary and metastatic breast adenocarcinomas. Ligand activation in cultured breast cancer cells caused a more differentiated, less malignant state, a reduction of growth rate and clonogenic capacity of these cells. These data suggest that the activated PPARγ pathway is able to induce terminal differentiation of malignant breast epithelial cells and thus might provide a novel, nontoxic therapy for human breast cancer [30]. PPARγ is expressed in human renal cell carcinoma (RCC) specimens and cell lines. Synthetic PPARγ ligands as TZD compounds and the endogeneous PPARγ ligand as 15d-PGJ2 inhibited the growth of RCC cells [77]. PPARγ is expressed at high levels in each of the major histologic types of human liposarcoma. Additionally, primary liposarcoma cells can be induced to undergo terminal differentiation by treatment with the PPARγ ligand pioglitazone [78]. TGZ or pioglitazone treatment induces cell cycle arrest and apoptosis of the promyelocytic leukemia cell line HL60 [79, 80]. A non neoplastic urothelial cell line (1T-1), a low-grade (RT4) carcinoma cell line, and two high-grade (T24 and 253J) ones can express PPARγ mRNA and protein. Different PPARγ ligands as 15d-PGJ2, TGZ and pioglitazone suppress the growth of nonneoplastic and neoplastic urothelial cells in a dose dependent manner [81]. The antineoplastic effect of the NSAID sulindac on human oral squamous cell carcinoma (SCCa) cells has been examined. Sulindac treatment results in a significant growth inhibitory effect on SCCa cells, mediated at least in part through induction of apoptosis. Treatment with antisense PPARγ oligonucleotides abolished sulindac's growth inhibitory effect. These data suggest that PPARγ may be, at least partially, responsible for the antiproliferative action of sulindac [82, 83]. Positive PPARγ expression has been noted in 95% of human glioma cases of different pathological stages examined. Additionally, two cell lines either derived from malignant glioma (SK-MG-1) or neuroblastoma (NB-1) are PPARγ positive. TGZ treatment induced growth inhibition in both cell lines caused by a PPARγ-mediated induction of apoptosis [84]. PPARγ immunoreactivity is noted in neoplastic than in adjacent normal thyroid tissue in papillary thyroid carcinoma cases. The PPARγ ligands, 15d-PGJ2, TGZ and BRL49653 induce apoptosis in PPARγ expressing thyroid carcinoma cell lines (BHP2-7, 7-13, 10-3 and 18-21). Additionally, TGZ has been shown to significantly inhibit tumor growth and prevent distant metastasis of BHP18-21 tumors in nude mice *in* vivo [85].

7. PPARγ and its Ligands in Lung Cancer

Recent evidence demonstrates that PPARγ is overexpressed in many different tumor types. The expression of PPARγ is often increased in human primary tumors, compared to corresponding normal tissues by both immunochemical staining and Western blotting [31,

86]. PPARγ is known to express in both human small cell lung cancer (SCLC) cells and non-small cell lung cancer(NSCLC) cells [31, 58, 70]. Activation of PPARγ in lung cancer cell lines indicate that PPARγ can be selectively modulated [57]. PPARγ ligands such as TGZ and PGJ_2 and rosiglitazone inhibit the growth of human lung cancer cells through ligand-induced differentiation and growth arrest and induction of apoptosis [31, 58, 70, 87-91]. These results raise the possibility that PPARγ and its ligands may play role in the pathogenesis and progression of human lung cancer and that PPARγ agonists may serve as anti-tumor agents against lung cancer.

7.1. PPARγ and its Ligands Effect on Lung Cancer Cell Differentiation

Cancer cells have been shown to dedifferentiate prior to acquiring proliferative and antiapoptotic phenotype [92]. Therefore, the induction and maintenance of a differentiated state has been an important strategy in the search for cancer therapeutics. Gelsolin, an actin regulatory protein, is expressed at low levels in most cancer cell lines and is up-regulated during *in vitro* differentiation induced by agents such as phorbol esters and histone deacetylase inhibitors in leukemic and epithelial cell lines [93, 94]. The expression of gelsolin is decreased in lung cancer tissues compared with histologically normal surrounding lung tissues [95]. Mad, a member of the myc family of interacting proteins that has been shown to be up-regulated during leukemic and keratinocyte differentiation [96]. Expression of the cyclin-dependent kinase inhibitor p21 (Wafl) is increased during *in vitro* differentiation in multiple cell types [97, 98]. Lung lineage-specific markers *MUC1* and SP-A, are both specific for the type II pneumocyte in the alveolar epithelium [93]. The expression of SP-A is decreased during lung cell differentiation [31, 99]. Treatments with PPARγ ligands, ciglitizone and 15d-PGJ_2 result in a prominent induction of gelsolin, mad and p21 expression and markedly down-regulate *MUC1* and SP-A espression in lung cancer cell lines. Thus, PPARγ ligands induce differentiation and reversal of the transformed phenotype in NSCLC (99). We have recently found that the expression of gelsolin is increased whereas the level of SP-A is decreased by TGZ treatment in NCI-H23 lung cancer cells, suggesting that a more mature and slower growing phenotype develops in TGZ-treated lung cancer cells [65].

In NSCLC cell lines (H2122) transfected with wild-type PPARγ, stable overexpression of PPARγ promotes a more differentiated and epithelial- polarized like phenotype in three-dimensional (3-D) tissue culture, although no significant effect on cell proliferation or apoptosis under standard tissue culture conditions. Furthermore, orthotopic implantation of H2122-PPARγ cells into the lungs of nude rats inhibited tumor growth and metastasis *in vivo* and formed polarized spheroid structures similar to those observed with normal lung epithelial cells and prolonged survival. Thus, PPARγ may inhibit the invasive metastasis of lung tumor cells and activate the pathway of promoting a more differentiated epithelial phenotype [87].

In lung tumor tissues, the activated form of MMP-2 has been shown to be significantly associated with tumor spread [100]. Ciglitizone treatment resulted in a marked decrease of activated form of MMP-2 as well as the decreased total amount of MMP-2 secreted by the cells, which suggests decreased malignant potential of lung cancer cells [99]. Thus, ciglitizone may inhibit cancer cell growth via inhibit MMP-2.

TGZ or pioglitazone treatment significantly reduces primary tumor growth in lung A549 tumor-bearing SCID mice and these PPARγ ligands inhibit the growth of cells *in vitro* and *in vivo* in part by promoting tumor cell differentiation [31]. ELR+CXC chemokines, the CXC chemokine family consists of a three–amino-acid motif of glutamic acid–leucine–arginine (ELR motif) residues, including CXCL8, CXCL5, and CXCL1, mediate both *in vivo* and *in vitro* angiogenic activities by directly stimulating proliferation and migration of endothelial cells (101, 102). A549 cells constitutively secrete significant amounts of CXCL8, CXCL5, and CXCL1, and increase the expression of chemokines when stimulation with TNFα. Treatment of A549 cells with the PPARγ ligand-TGZ blocks constitutive and TNFα induced expression of CXCL8, CXCL5, and CXCL1 in a dose-dependent manner with no significant effects on proliferation or apoptosis. Importantly, TGZ treatment also abolished the TNFα–induced endothelial cell chemotaxis in the A549 conditioned media. And the modulation of NF-κB activity is involved in PPARγ ligand-induced inhibition of ELR+CXC chemokine expression [103].

Previous studies have demonstrated that extracellular matrixes, including Fn, affect cell growth and differentiation [104, 105]. Although the exact consequence of Fn expression in lung carcinoma cells remains unclear, it is reported that Fn stimulates the proliferation and survival of NSCLC [106]. Thus both host cell- and tumor cell-derived Fn may serve as a mitogen/autocrine factor for lung carcinoma cells and, as such, represents a potential target for developing anti-cancer therapies. Treatment of H1838 lung carcinoma cells with the PPARγ ligands 15d-PGJ2, BRL49653, or troglitazone significantly decreased Fn mRNA and protein levels compared with nontreated control cells, suggesting a novel role for PPARγ ligands in the development of therapeutic interventions that are able to prevent or inhibit lung carcinoma progression [107].

Since the induction of differentiation is a relatively nontoxic and therapeutic approach, PPARγ ligands such as TGZ and PGJ_2 may be candidates for a novel approach to the treatment of human lung cancer.

7.2. Effect of PPARγ and its Ligands on Lung Cancer Cell Growth Arrest

Our study has shown that TGZ, a PPARγ ligand, causes a decrease in cellular growth of lung cells as evidence of an increase in the portion of cells in the G0-G1 phase of the cell cycle with a corresponding decrease in cells in S and G2-M phases [58]. Similar to our result, treatment of lung adenocarcinoma cells (A549) with TGZ induced a dose-dependent inhibition of cell growth due to the inhibition of cell proliferation since the cell cycle profile reveals an arrest at G_0/G_1 with the concomitant downregulation of G_0/G_1 cyclins D and E [31]. Similar to TGZ, nonsteroidal anti-inflammatory drugs (NSAIDs) mediate cyclooxygenase-independent inhibition of lung cancer cell growth through PPARγ activation [88]. Other studies also provided evidence that PPARγ ligands induce growth arrest in lung cancer cells [90, 99].

In many tumors, the overexpression of the fibronectin integrin receptor $\alpha 5\beta 1$ is associated with a more malignant phenotype. $\alpha 5\beta 1$ is generally not found in normal lung tissue, but it is expressed in a considerable number of lung carcinomas [108]. The expression of $\alpha 5\beta 1$ in NSCLC lung carcinomas correlates with tumor progression, and its overexpression is associated with decreased survival [108]. Recent study showed that PPARγ ligands decrease mRNA and protein expression of the $\alpha 5$ subunit of the $\alpha 5\beta 1$ heterodimer in NSCLC associated with reduced NSCLC adhesion to fibronectin [109]. Therefore, more malignant phenotype of lung cancer cells can still respond to PPARγ ligand treatment. In addition to $\alpha 5\beta 1$, the inhibitory action of PPARγ may also be linked to prostaglandin E_2 (PGE2), which is reported to be a key modulator of tumor development [110]. PGE2 has also been implicated in the pathogenesis of NSCLC carcinomas [111, 112]. Human NSCLC cell lines (H1838 and H2106) express PGE2 receptors, and the inhibition of cell growth by PPARγ ligands was associated with a significant decrease in EP2 (PGE2 receptor) mRNA and protein levels. PPARγ ligands may function in a PPARγ-dependent or independent pathway to inhibit cell growth for the inhibitory effects of rosiglitazone and ciglitazone can be reversed by a specific PPARγ antagonist GW9662, whereas the inhibitory effects of PGJ2 can not be reversed by GW9662, suggesting the involvement of both the PPARγ-dependent and PPARγ-independent mechanism in PPARγ ligands-mediated inhibition of human lung carcinoma cell growth through decrease the expression of EP2 [89].

7.3. Effect of PPARγ and its Ligands on Lung Cancer Cell Apoptosis

The reduction of mitochondria membrane potential ($\Delta\psi m$) is a well recognized marker for early apoptosis [113]. Bcl-2 family members including Bcl-2 and Bcl-w are functional at the mitochondria to regulate apoptosis [114]. Caspase 9 is specifically associated with the mitochondrial-mediated apoptosis (115). And caspase 3 not only is an executive caspase but also has an amplification role in the apoptosis [116]. We have recently demonstrated that TGZ, a PPARγ ligand, induced apoptosis in human lung cancer cells via a mitochondrial pathway, reflecting by reduction of $\Delta\psi m$, down-regulation of Bcl-2 and Bcl-w expressions, release of cytochrome c into cytosol, and activation of caspase 9 and 3 [65].

Experiments have identified that TGZ can stimulate the expression of the growth arrest and DNA-damage inducible (GADD) 153 gene in NSCLC cells in a PPARγ-dependent manner. The findings collectively indicate that activation of PPARγ by TGZ can cause growth inhibition and apoptosis of NSCLC cells and that GADD153 might be a candidate factor implicated in these processes [90].

PPARγ ligands PGJ2 and ciglitazone inhibit the growth and induce the apoptosis of several human lung carcinoma cell lines. Treatment of lung carcinoma cells with the PPARγ ligands PGJ2, ciglitazone, TGZ, and GW1929 elevated p21 at both mRNA and protein levels and reduced cyclin D1 mRNA levels. p21 antisense oligonucleotides inhibit PPARγ ligand-induced p21 protein expression and significantly prevents the PPARγ ligand-mediated inhibition of lung carcinoma cell growth. Further experiments demonstrate that PPARγ ligands can increase the nuclear binding activities of Sp1 and NF-IL6 (C/EBP), two transcription factors with regulatory elements in the promoter region of the p21 gene. Thus,

PPARγ ligands inhibit human lung carcinoma cell growth and induce apoptosis by stimulating the cyclin-dependent kinase inhibitor p21 and by reducing cyclin D1 gene expression [117]. However, how the change in p21 and cyclin D1 induced by PPARγ is linked to mitochondria-medicated apoptotic pathway is not clearly known in lung cancer cells.

RARß plays a critical role in mediating the growth-inhibitory effects of retinoids in various cancer cells. The aberrant expression or loss of RARß in a variety of cancer cell lines suggests that decreased RARß expression may contribute to retinoid resistance [118, 119], implying that RARß may act as a tumor suppressor. Regulation of RARß gene expression is principally mediated by the retinoic acid response element (ßRARE) in its promoter [120]. The PPARγ ligand ciglitazone and the RXR ligand SR11237 cooperatively inhibit the growth and induce apoptosis of lung cancer cells, Calu-6 cells [121]. Ciglitazone strongly induces RARß expression in human lung cancer cell lines when used together with RXR ligand-SR11237. It is demonstrated that PPARγ can bind to the ßRARE and ciglitazone together with SR11237 strongly induced ßRARE activity to form a PPARγ/RXR heterodimer. These results show that ciglitazone can induce lung cancer apoptosis in a PPARγ dependent manner that is involved in a novel RARß-mediated mechanism [121].

15(S)-HETE has been demonstrated to induce apoptosis of A549 lung cancer cells by binding to PPARγ (122). Others also verified that PPARγ ligands such as TGZ and PGJ2 can inhibit the growth of human lung cancer cells through the induction of apoptosis (70, 99).

7.4. Activation of Mitogen-Activated Protein Kinases (Mapks) - Extracellular Signal-Regulated Kinase (ERK) Signal Pathway in Lung Cancer Cells Treated with PPARγ Ligands

Traditionally, ERK is activated in response to growth stimuli and it functions as a survival signal to promote cell proliferation and growth [123]. However, recent studies have generated increasing evidence to challenge the concept that ERK functions solely as a survival factor. For example, activation of ERK is required for quercetin-induced apoptosis in A549 lung cancer cells [124]. Cisplatin-induced apoptosis in human malignant testicular germ cells depends on ERK activation [125]. Nuclear localization of activated ERK may impede S-phase entry and induce apoptosis in pulmonary epithelial and mesothelial cells [126]. Activation of ERK is specifically required for retinoid-induced apoptosis of breast cancer cells [127]. Down-regulation of ERK suppresses apoptosis caused by death-associated protein kinase (128). ERK signaling pathway plays an active role in mediating H_2O_2-induced apoptosis of human glioma cells (129). Several molecules are targeted by ERK to inhibit proliferation and promote apoptosis. They include Bcl-2, caspase 3, cyclin D1 and death-associated protein kinase [125, 126, 128, 130].

Our recent study has found that U0126, a specific inhibitor of ERK1/2, blocks ERK1/2 activation and counteracts the effect of TGZ, a PPARγ ligand, on the expression of gelsolin and SP-A by decreasing the level of the former and increasing the level of the latter. Therefore, ERK activation appears to play a stimulating role on TGZ-induced NCI-H23 differentiation. In agreement with the above findings, PPARγ ligand GW1929 activates the

ERK, and prevents fibronectin-induced cell growth by inhibiting α5 expression in NSCLC through ERK-related signals. The suppressive effect of the GW1929 on α5 gene expression can be reversed by GW9662, an antagonist of PPARγ, suggesting the involvement of PPARγ-dependent mechanisms [109]. There are also studies demonstrating that the inhibition of PGE2 receptor expression by PPARγ ligand treatment is associated with phosphorylation of ERK, and that the effect of PPARγ ligands is inhibited by PD98095, an inhibitor of the MEK-1/ERK pathway [89]. The inhibitory effects of the ligands can be reversed by a specific PPARγ antagonist GW9662, suggesting the involvement of PPARγ-dependent mechanisms. Studies also reveal that the effect of PPARγ and ERK on the PGE2 can mediate the proliferative effect of PGE_2 on lung carcinoma cells [89]. All these results demonstrate that ERK plays a positive role in the regulation of lung cancer cell differentiation, growth and apoptosis by PPARγ and its ligands.

Transcriptional changes induced by PPARγ agonists can be dissociated into distinct MAPK- and PPARγ-dependent pathways, suggesting that MAPKs mediate some of the cellular effects of PPARγ ligands. Thus, kinase activation by PPARγ agonists appears to play an important role in the action of these compounds. Sustained activation of ERK1/2 is often associated with cellular differentiation and/or growth arrest in a cell- and tissue-specific manner [131]. According to the result from our laboratory, ERK activation plays a critical role during the inhibition of lung cancer cell growth by PPARγ and its ligands. We have demonstrated that TGZ, a PPARγ ligand, induces cell growth arrest and differentiation and finally evokes apoptosis in human lung cancer cells. TGZ up-regulates PPARγ expression and activates ERK in the cytosol. More importantly, it induces PPARγ and ERK1/2 accumulation in the nucleus, where the co-localization of both proteins was found. Both PPARγ siRNA (small interfering RNA) and U0126, a specific inhibitor of ERK1/2, are able to block the effect of TGZ, suggesting that apoptosis induced by TGZ is PPARγ- and ERK1/2-dependent. Inhibition of ERK1/2 by U0126 also leads to a significant decrease in the level of PPARγ, indicating a positive cross-talk between PPARγ and ERK1/2, which may form an auto-regulatory feedback mechanism to amplify the effect of ERK1/2 on cell growth arrest and apoptosis [65].

PPARγ, as a member of nuclear receptors, are phosphoproteins and phsphorylation has been shown to modulate its activity as described above. It is assumed that a physical contact between kinases and their targets is a necessary step in the phosphorylation of target molecules [132]. Therefore, the co-localization of PPARγ and pERK in the nucleus should create a suitable environment for the phosphorylation to occur. An auto-regulatory and positive feedback loop between PPARγ and ERK in this ligand-induced apoptotic pathway should be deemed to be important in the view of PPARγ ligands as an anti-cancer agent since they can greatly enhance the killing of tumor cells without increasing the concentration of anti-cancer agents used.

Taken together, PPARγ activation by its ligands inhibits lung cancer cells growth through increased cell differentiation, cell-cycle arrest and induction of apoptosis. The activation of ERK by PPARγ ligands plays a positive role in promoting the inhibitory effect of PPARγ on lung cancer cell growth. The interaction between PPARγ and ERK may create an auto-regulatory and positive feedback loop to enhance the effect of the degree of growth inhibition occurred in lung cancer cells [66].

7.5. Troglitazone (TGZ), a Ligand of PPARγ, Inhibits the Proliferation of Non-Small Lung Cancer Induced by 4-(N-Methyl-N-Nitrosamino)-1-(3-Pyridyl)-1-Butanone (NNK)

The epidemic of lung cancer is directly attributable to cigarette. However, it is still not completely known the molecular pathway of cigarette smoking in the pathogenesis of lung cancer. Among the carcinogenoic chemicals of cigarette smoking, 4-(*N*-Methyl-*N*-nitrosamino)-1-(3-pyridyl)-1-butanone (NNK) is the most potent, which induces lung cancer in all animal species tested (16). Unlike PPARγ ligands, NNK can promote cell proliferation and growth. It is interesting to know whether PPARγ ligands can inhibit the growth-promoting function of NNK. Our unpublished data demonstrated that NNK stimulated NCI-H23 lung cancer cells proliferation, induced the DNA binding activity of nuclear factor-κB (NF-κB), down-regulated Bad expression, and up-regulated PPARγ protein expressions accompanied with the increase of PPARγ transcriptional activity by TGZ (Figure 4). Inhibition of NF-κB nuclear translocation leads to the suppression of NNK-mediated Bad expression, indicating that NNK may regulate Bad expression through the activation of NF-κB. TGZ significantly inhibits cell proliferation induced by NNK. Though TGZ did not affect nuclear factor-κB (NF-κB) activity, it up-regulated Bad expression. Taken together, TGZ can efficiently inhibit the proliferation of lung cancer cells induced by NNK via Bad- and PPARγ- related pathways, which may not be directly relevant to the activity of NF-κB.

7.6. Effect of Combination of PPARγ Ligands with other Chemicals on Lung Cancer Cell

It is reported that inhibitors of the arachidonic acid pathway and PPAR ligands have superadditive effects on lung cancer growth inhibition [133]. A murine lung alveolar carcinoma cell model (Line 1) treated with TGZ leads to the transcriptional induction of lipogenic enzymes, increases glucose consumption to support lipogenesis and shifts cellular energy metabolism towards β-oxidative respiration. As such, TGZ can sensitize Line 1 tumor cells to metabolic interference by trimetazidine, an inhibitor of β-oxidative respiration. Associated with the cancer chemo-preventive activity of PPARγ agonists, inhibition of β-oxidative respiration may be potentially a therapeutic method [134]. Gefitinib exhibits antitumor activity in patient with non-small cell lung cancer. However, only 10-20% of patients exhibit clinical response to this drug. PPARγ ligand-rosiglitazone potentiated gefitinib's anti-proliferative effects by increased of PTEN expression in A549 cell lines [135]. A dramatic synergism between lovastatin and troglitazone is found in anticancer at clinically achievable concentrations. This synergism has been shown in far majority of lung cell line accompanied by synergistic modulation of the transcription factor E2F-1 and p27(Kip1), which are reported to mediate the anticancer activities of lovastatin and troglitazone, respectively [136]. The dramatic synergism may provide significant clinical benefits for patients suffering from cancers.

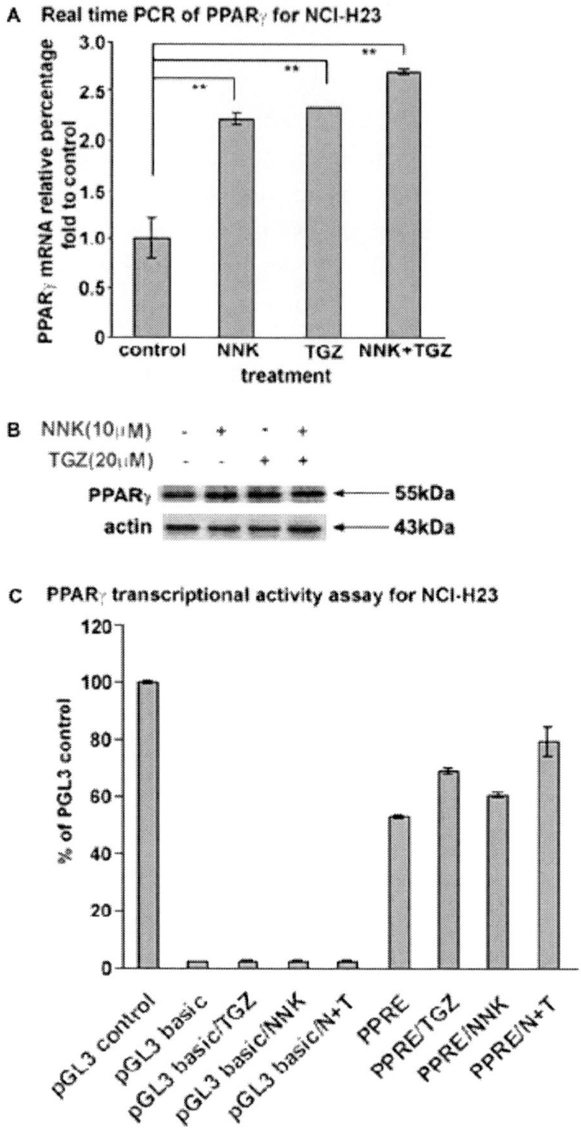

Figure 4. TGZ induces PPARγ transcription activity to counteract NNK founction in lung cancer cells. NCI-H23 cells were treated with NNK 10 μM or TGZ 20 μM or NNK 10 μM plus TGZ (20 μM) for 24h. Cells without the treatment were set up as a control. After the incubation, the cells were harvested for different investigations as shown in A and B. (A) Real time PCR analyzed PPARγ mRNA level in NCI-H23 cells. PPARγ mRNA values (mean± SD) are expressed as fold induction compared with Control (n=3). **, p<0.01 between treated and control cells. (B) PPARγ protein was measured by Western blot. Equal loading was confirmed by probing with antibodies against actin in total protein. (C) PPARγ activity was analyzed in lung cells using a luciferase-based reporter-gene transactivation assay. Cells were transiently transfected with either PPRE-tk-Luc or tk-Luc PGL3 basic plasmid, or PGL3 control plasmid as the positive control of transfection efficiency and luciferase activity. After 24 h treatment, the luciferase activity was assayed using a luciferase assay kit. The data were represented as mean ± SD for three replicate determinations for each group. Experiments were repeated two times and similar results were obtained.

7.7. PPARγ Ligands May Play Role in PPARγ Independent Manner

Several studies have demonstrated that PPARγ ligands can act through PPARγ-dependent and -independent signals [90, 137] such as PGJ_2 has been shown to act through PPARγ-independent mechanisms in several cell systems and this is likely due to the activation of unrelated transcription factors [89, 138]. PGJ_2 may have a receptor other than PPARγ and this receptor may act through cAMP or another mechanism [139]. Both PGJ_2 and TZDs have anti-inflammatory effects that are independent of PPARγ [137]. These results emphasize that caution needs to be exercised in ascribing functions to PPARγ on the basis of the action of its ligands.

8. Significance and Future Prospects

Studies have clearly demonstrated that PPARγ is a regulator of the expression of many genes relevant to carcinogenesis [140] and that the activation of PPARγ plays an important role in the regulation of cell proliferation, growth and apoptosisin in lung cancer. The observation that PPARγ expression is elevated in several human malignancies but tumor cells can still grow at a fast speed raises the question if adequate levels of natural ligands are also present. Our lab unpublished data show that natural ligands of PPARγ are lower than the normal level. *In vitro* and *in vivo* studies have substantiated the importance of PPARγ ligands in mediating cell cycle related events. PPARγ ligands through different mechanisms can induce growth arrest, apoptosis and differentiation in a variety of transformed cells. For further clinical studies, in vivo study is needed to delineate the optimal dosage, the duration of therapy and possibly the efficacy of PPARγ ligands alone or in combination with other agents in the fight against lung cancer. At the fundamental level, much more work is needed in order to discover or synthesize novel agents and to further understand the molecular basis of known PPARγ ligands action. It becomes so far evident that these PPARγ ligands can be used as promising anticancer drugs for cancer chemotherapy and chemoprevention in the future.

The activation of PPARγ by natural and synthetic ligands leads to a cascade molecular event and finally the regulation of the target gene expression (Figure 3). In most studies, lung cancer cells were treated only by PPARγ ligand which leading to the cell growth inhibition. To further the study, it is possible to combine PPARγ ligands with its partner-RXR for treatment of cancer cells.

Overall, potential targets in preventive and therapeutic strategies to reduce the risk of cancer and to inhibit cancer growth involve agonists of endocrine receptors, factors down-regulating inflammation, factors inducing apoptosis, enzymatic inhibitors and gene therapy [141]. This review attempts to summarize the great progress made over the past few years and thus to elucidate the role of PPARγ and its ligands in lung cancer development and treatment. The regulation of differentiation and the growth inhibition of cancer cells by PPARγ and its ligands have attracted more and more attention. Better understanding of the biological role of PPARγ and its associated pathways in lung cancer cells may allow the rational development of selective PPARγ modulators as well as completely novel therapies that target PPARγ downstream molecules critical for lung cancer cells growth.

Acknowledgements

The study was supported by a grant from the Research Grants Council of the Hong Kong Special Administrative Region (Project No. CUHK 4390/03M).

References

[1] Vineis, P., Alavanja, M., Buffler, P., Fontham, E., Franceschi, S., Gao, Y. T., Gupta, P. C., Hackshaw, A., Matos, E., Samet, J., Sitas, F., Smith, J., Stayner, L., Straif, K., Thun, M. J., Wichmann, H. E., Wu, A. H., Zaridze, D., Peto, R., and Doll, R. Tobacco and cancer: recent epidemiological evidence. *J. Natl. Cancer Inst.*, *96:* 99-106, 2004.

[2] Mulshine, J. L., Treston, A. M., Brown, P. H., Birrer, M. J., and Shaw, G. L. Initiators and promoters of lung cancer. *Chest, 103:* 4S-11S, 1993.

[3] Nicholson, S. A., Beasley, M. B., Brambilla, E., Hasleton, P. S., Colby, T. V., Sheppard, M. N., Falk, R., and Travis, W. D. Small cell lung carcinoma (SCLC): a clinicopathologic study of 100 cases with surgical specimens. *Am. J. Surg. Pathol., 26:* 1184-1197, 2002.

[4] Sekido, Y., Fong, K. M., and Minna, J. D. Progress in understanding the molecular pathogenesis of human lung cancer. *Biochim. Biophys. Acta*, *1378:* F21-59, 1998.

[5] Fong, K. M., Sekido, Y., and Minna, J. D. Molecular pathogenesis of lung cancer. *J. Thorac .Cardiovasc. Surg.*, *118:* 1136-1152, 1999.

[6] Mitsudomi, T., Viallet, J., Mulshine, J. L., Linnoila, R. I., Minna, J. D., and Gazdar, A. F. Mutations of ras genes distinguish a subset of non-small-cell lung cancer cell lines from small-cell lung cancer cell lines. *Oncogene, 6:* 1353-1362, 1991.

[7] Nemenoff, R. A. and Winn, R. A. Role of nuclear receptors in lung tumourigenesis. *Eur .J. Cancer, 41:* 2561-2568, 2005.

[8] Feng, G., Xu, X., Youssef, E. M., and Lotan, R. Diminished expression of S100A2, a putative tumor suppressor, at early stage of human lung carcinogenesis. *Cancer Res., 61:* 7999-8004, 2001.

[9] Gargiullo, P., Wingo, P.A., Coates, R.J., Thompson, T.D. *MMWR 51:* 49-53, 2002.

[10] Coudert, B., Anthoney, A., Fiedler, W., Droz, J. P., Dieras, V., Borner, M., Smyth, J. F., Morant, R., de Vries, M. J., Roelvink, M., and Fumoleau, P. Phase II trial with ISIS 5132 in patients with small-cell (SCLC) and non-small cell (NSCLC) lung cancer. A European Organization for Research and Treatment of Cancer (EORTC) Early Clinical Studies Group report. *Eur. J. Cancer*, *37:* 2194-2198, 2001.

[11] Perry, C. F., Stevens, M., Rabie, I., Yarker, M. E., Cochrane, J., Perry, E., Traficante, R., and Coman, W. Chemoprevention of head and neck cancer with retinoids: a negative result. *Arch. Otolaryngol. Head Neck. Surg., 131:* 198-203, 2005.

[12] Argiris, A. and Schiller, J. H. Can current treatments for advanced non-small-cell lung cancer be improved? *Jama, 292:* 499-500, 2004.

[13] Parkin, D. M., Pisani, P., Lopez, A. D., and Masuyer, E. At least one in seven cases of cancer is caused by smoking. Global estimates for 1985. *Int. J. Cancer, 59:* 494-504, 1994.

[14] Stellman, S. D., Takezaki, T., Wang, L., Chen, Y., Citron, M. L., Djordjevic, M. V., Harlap, S., Muscat, J. E., Neugut, A. I., Wynder, E. L., Ogawa, H., Tajima, K., and Aoki, K. Smoking and lung cancer risk in American and Japanese men: an international case-control study. *Cancer Epidemiol. Biomarkers Prev., 10:* 1193-1199, 2001.

[15] Wistuba, II, Bryant, D., Behrens, C., Milchgrub, S., Virmani, A. K., Ashfaq, R., Minna, J. D., and Gazdar, A. F. Comparison of features of human lung cancer cell lines and their corresponding tumors. *Clin. Cancer Res., 5:* 991-1000, 1999.

[16] Hecht, S. S. and Hoffmann, D. Tobacco-specific nitrosamines, an important group of carcinogens in tobacco and tobacco smoke. *Carcinogenesis, 9:* 875-884, 1988.

[17] Hoffmann, D., Rivenson, A., and Hecht, S. S. The biological significance of tobacco-specific N-nitrosamines: smoking and adenocarcinoma of the lung. *Crit. Rev. Toxicol., 26:* 199-211, 1996.

[18] Schuller, H. M. Mechanisms of smoking-related lung and pancreatic adenocarcinoma development. *Nat. Rev. Cancer, 2:* 455-463, 2002.

[19] Schuller, H. M., Plummer, H. K., 3rd, and Jull, B. A. Receptor-mediated effects of nicotine and its nitrosated derivative NNK on pulmonary neuroendocrine cells. *Anat. Rec. A Discov. Mol. Cell Evol. Biol., 270:* 51-58, 2003.

[20] Mangelsdorf, D. J., Thummel, C., Beato, M., Herrlich, P., Schutz, G., Umesono, K., Blumberg, B., Kastner, P., Mark, M., Chambon, P., and Evans, R. M. The nuclear receptor superfamily: the second decade. *Cell, 83:* 835-839, 1995.

[21] Gearing, K. L., Gottlicher, M., Teboul, M., Widmark, E., and Gustafsson, J. A. Interaction of the peroxisome-proliferator-activated receptor and retinoid X receptor. *Proc. Natl .Acad. Sci. U S A, 90:* 1440-1444, 1993.

[22] Vamecq, J. and Latruffe, N. Medical significance of peroxisome proliferator-activated receptors. *Lancet, 354:* 141-148, 1999.

[23] Desvergne, B. and Wahli, W. Peroxisome proliferator-activated receptors: nuclear control of metabolism. *Endocr. Rev., 20:* 649-688, 1999.

[24] Chang, T. H. and Szabo, E. Enhanced growth inhibition by combination differentiation therapy with ligands of peroxisome proliferator-activated receptor-gamma and inhibitors of histone deacetylase in adenocarcinoma of the lung. *Clin. Cancer Res., 8:* 1206-1212, 2002.

[25] Di-Poi, N., Tan, N. S., Michalik, L., Wahli, W., and Desvergne, B. Antiapoptotic role of PPARbeta in keratinocytes via transcriptional control of the Akt1 signaling pathway. *Mol. Cell., 10:* 721-733, 2002.

[26] Tontonoz, P., Hu, E., and Spiegelman, B. M. Stimulation of adipogenesis in fibroblasts by PPAR gamma 2, a lipid-activated transcription factor. *Cell, 79:* 1147-1156, 1994.

[27] Hu, E., Kim, J. B., Sarraf, P., and Spiegelman, B. M. Inhibition of adipogenesis through MAP kinase-mediated phosphorylation of PPARgamma. *Science, 274:* 2100-2103, 1996.

[28] Lambe, K. G. and Tugwood, J. D. A human peroxisome-proliferator-activated receptor-gamma is activated by inducers of adipogenesis, including thiazolidinedione drugs. Eur. *J. Biochem., 239:* 1-7, 1996.

[29] Elstner, E., Muller, C., Koshizuka, K., Williamson, E. A., Park, D., Asou, H., Shintaku, P., Said, J. W., Heber, D., and Koeffler, H. P. Ligands for peroxisome proliferator-activated receptorgamma and retinoic acid receptor inhibit growth and induce apoptosis of human breast cancer cells in vitro and in BNX mice. *Proc. Natl. Acad. Sci U S A, 95:* 8806-8811, 1998.

[30] Mueller, E., Sarraf, P., Tontonoz, P., Evans, R. M., Martin, K. J., Zhang, M., Fletcher, C., Singer, S., and Spiegelman, B. M. Terminal differentiation of human breast cancer through PPAR gamma. *Mol. Cell., 1:* 465-470, 1998.

[31] Keshamouni, V. G., Reddy, R. C., Arenberg, D. A., Joel, B., Thannickal, V. J., Kalemkerian, G. P., and Standiford, T. J. Peroxisome proliferator-activated receptor-gamma activation inhibits tumor progression in non-small-cell lung cancer. *Oncogene, 23:* 100-108, 2004.

[32] Masamune, A., Satoh, K., Sakai, Y., Yoshida, M., Satoh, A., and Shimosegawa, T. Ligands of peroxisome proliferator-activated receptor-gamma induce apoptosis in AR42J cells. *Pancreas, 24:* 130-138, 2002.

[33] Lee, T. W., Chen, G. G., Xu, H., Yip, J. H., Chak, E. C., Mok, T. S., and Yim, A. P. Differential expression of inducible nitric oxide synthase and peroxisome proliferator-activated *receptor gamma in non-small cell lung carcinoma.* European Journal of Cancer, *39:* 1296-1301, 2003.

[34] Dreyer, C., Krey, G., Keller, H., Givel, F., Helftenbein, G., and Wahli, W. Control of the peroxisomal beta-oxidation pathway by a novel family of nuclear hormone receptors. *Cell, 68:* 879-887, 1992.

[35] Chen, F., Law, S. W., and O'Malley, B. W. Identification of two mPPAR related receptors and evidence for the existence of five subfamily members. *Biochem. Biophys. Res. Commun., 196:* 671-677, 1993.

[36] Issemann, I. and Green, S. Activation of a member of the steroid hormone receptor superfamily by peroxisome proliferators. *Nature, 347:* 645-650, 1990.

[37] Werman, A., Hollenberg, A., Solanes, G., Bjorbaek, C., Vidal-Puig, A. J., and Flier, J. S. Ligand-independent activation domain in the N terminus of peroxisome proliferator-activated receptor gamma (PPARgamma). Differential activity of PPARgamma1 and -2 isoforms and influence of insulin. J. *Biol. Chem., 272:* 20230-20235, 1997.

[38] Kliewer, S. A., Umesono, K., Noonan, D. J., Heyman, R. A., and Evans, R. M. Convergence of 9-cis retinoic acid and peroxisome proliferator signalling pathways through heterodimer formation of their receptors. *Nature, 358:* 771-774, 1992.

[39] Berger, J. and Moller, D. E. The mechanisms of action of PPARs. *Annu. Rev. Med., 53:* 409-435, 2002.

[40] Fajas, L., Fruchart, J. C., and Auwerx, J. PPARgamma3 mRNA: a distinct PPARgamma mRNA subtype transcribed from an independent promoter. FEBS Lett, *438:* 55-60, 1998.

[41] Braissant, O., Foufelle, F., Scotto, C., Dauca, M., and Wahli, W. Differential expression of peroxisome proliferator-activated receptors (PPARs): tissue distribution

of PPAR-alpha, -beta, and -gamma in the adult rat. *Endocrinology, 137:* 354-366, 1996.

[42] Nagy, L., Tontonoz, P., Alvarez, J. G., Chen, H., and Evans, R. M. Oxidized LDL regulates macrophage gene expression through ligand activation of PPARgamma. *Cell, 93:* 229-240, 1998.

[43] Huang, J. T., Welch, J. S., Ricote, M., Binder, C. J., Willson, T. M., Kelly, C., Witztum, J. L., Funk, C. D., Conrad, D., and Glass, C. K. Interleukin-4-dependent production of PPAR-gamma ligands in macrophages by 12/15-lipoxygenase. *Nature, 400:* 378-382, 1999.

[44] Willson, T. M., Cobb, J. E., Cowan, D. J., Wiethe, R. W., Correa, I. D., Prakash, S. R., Beck, K. D., Moore, L. B., Kliewer, S. A., and Lehmann, J. M. The structure-activity relationship between peroxisome proliferator-activated receptor gamma agonism and the antihyperglycemic activity of thiazolidinediones. *J. Med. Chem., 39:* 665-668, 1996.

[45] Shibata, T., Matsui, K., Nagao, K., Shinkai, H., Yonemori, F., and Wakitani, K. Pharmacological profiles of a novel oral antidiabetic agent, JTT-501, an isoxazolidinedione derivative. *Eur. J. Pharmacol., 364:* 211-219, 1999.

[46] Suh, N., Wang, Y., Williams, C. R., Risingsong, R., Gilmer, T., Willson, T. M., and Sporn, M. B. A new ligand for the peroxisome proliferator-activated receptor-gamma (PPAR-gamma), GW7845, inhibits rat mammary carcinogenesis. *Cancer Res., 59:* 5671-5673, 1999.

[47] Hwang, B. Y., Lee, J. H., Nam, J. B., Kim, H. S., Hong, Y. S., and Lee, J. J. Two new furanoditerpenes from Saururus chinenesis and their effects on the activation of peroxisome proliferator-activated receptor gamma. *J. Nat. Prod., 65:* 616-617, 2002.

[48] Liang, Y. C., Tsai, S. H., Tsai, D. C., Lin-Shiau, S. Y., and Lin, J. K. Suppression of inducible cyclooxygenase and nitric oxide synthase through activation of peroxisome proliferator-activated receptor-gamma by flavonoids in mouse macrophages. *FEBS Lett, 496:* 12-18, 2001.

[49] Kuroda, M., Mimaki, Y., Sashida, Y., Mae, T., Kishida, H., Nishiyama, T., Tsukagawa, M., Konishi, E., Takahashi, K., Kawada, T., Nakagawa, K., and Kitahara, M. Phenolics with PPAR-gamma ligand-binding activity obtained from licorice (Glycyrrhiza uralensis roots) and ameliorative effects of glycyrin on genetically diabetic KK-A(y) mice. *Bioorg. Med. Chem. Lett., 13:* 4267-4272, 2003.

[50] Wang, Y., Porter, W. W., Suh, N., Honda, T., Gribble, G. W., Leesnitzer, L. M., Plunket, K. D., Mangelsdorf, D. J., Blanchard, S. G., Willson, T. M., and Sporn, M. B. A synthetic triterpenoid, 2-cyano-3,12-dioxooleana-1,9-dien-28-oic acid (CDDO), is a ligand for the peroxisome proliferator-activated receptor gamma. *Mol. Endocrinol., 14:* 1550-1556, 2000.

[51] Wright, H. M., Clish, C. B., Mikami, T., Hauser, S., Yanagi, K., Hiramatsu, R., Serhan, C. N., and Spiegelman, B. M. A synthetic antagonist for the peroxisome proliferator-activated receptor gamma inhibits adipocyte differentiation. *J. Biol. Chem., 275:* 1873-1877, 2000.

[52] Juge-Aubry, C., Pernin, A., Favez, T., Burger, A. G., Wahli, W., Meier, C. A., and Desvergne, B. DNA binding properties of peroxisome proliferator-activated receptor

subtypes on various natural peroxisome proliferator response elements. Importance of the 5'-flanking region. *J. Biol. Chem., 272:* 25252-25259, 1997.

[53] Lemberger, T., Desvergne, B., and Wahli, W. Peroxisome proliferator-activated receptors: a nuclear receptor signaling pathway in lipid physiology. *Annu. Rev. Cell Dev Biol,* Vol. 12, pp. 335-363. 1996.

[54] Xu, L., Glass, C. K., and Rosenfeld, M. G. Coactivator and corepressor complexes in nuclear receptor function. *Curr. Opin. Genet .Dev., 9:* 140-147, 1999.

[55] Zhu, Y., Qi, C., Calandra, C., Rao, M. S., and Reddy, J. K. Cloning and identification of mouse steroid receptor coactivator-1 (mSRC-1), as a coactivator of peroxisome proliferator-activated receptor gamma. *Gene Expr., 6:* 185-195, 1996.

[56] Zhu, Y., Qi, C., Jain, S., Rao, M. S., and Reddy, J. K. Isolation and characterization of PBP, a protein that interacts with peroxisome proliferator-activated receptor. *J. Biol. Chem., 272:* 25500-25506, 1997.

[57] Allred, C. D. and Kilgore, M. W. Selective activation of PPARgamma in breast, colon, and lung cancer cell lines. *Mol. Cell Endocrinol., 235:* 21-29, 2005.

[58] Li, M., Lee, T. W., Mok, T. S., Warner, T. D., Yim, A. P., and Chen, G. G. Activation of peroxisome proliferator-activated receptor-gamma by troglitazone (TGZ) inhibits human lung cell growth. *J. Cell Biochem., 96:* 760-774, 2005.

[59] Rochette-Egly, C. Nuclear receptors: integration of multiple signalling pathways through phosphorylation. *Cell Signal., 15:* 355-366, 2003.

[60] Camp, H. S., Tafuri, S. R., and Leff, T. c-Jun N-terminal kinase phosphorylates peroxisome proliferator-activated receptor-gamma1 and negatively regulates its transcriptional activity. *Endocrinology, 140:* 392-397, 1999.

[61] Gardner, O. S., Dewar, B. J., Earp, H. S., Samet, J. M., and Graves, L. M. Dependence of peroxisome proliferator-activated receptor ligand-induced mitogen-activated protein kinase signaling on epidermal growth factor receptor transactivation. *J. Biol. Chem., 278:* 46261-46269, 2003.

[62] Chana, R. S., Lewington, A. J., and Brunskill, N. J. Differential effects of peroxisome proliferator activated receptor-gamma (PPAR gamma) ligands in proximal tubular cells: thiazolidinediones are partial PPAR gamma agonists. Kidney Int, *65:* 2081-2090, 2004.

[63] Zhang, B., Berger, J., Zhou, G., Elbrecht, A., Biswas, S., White-Carrington, S., Szalkowski, D., and Moller, D. E. Insulin- and mitogen-activated protein kinase-mediated phosphorylation and activation of peroxisome proliferator-activated receptor gamma. *J. Biol. Chem., 271:* 31771-31774, 1996.

[64] Al-Rasheed, N. M., Chana, R. S., Baines, R. J., Willars, G. B., and Brunskill, N. J. Ligand-independent activation of peroxisome proliferator-activated receptor-gamma by insulin and C-peptide in kidney proximal tubular cells: dependent on phosphatidylinositol 3-kinase activity. *J. Biol. Chem., 279:* 49747-49754, 2004.

[65] Li, M. Y., Lee, T. W., Yim, A. P., and Chen, G. G. Apoptosis induced by troglitazone is both peroxisome proliferator-activated receptor-γ- and ERK-dependent in human non-small lung cancer cells. *J. Cell. Physiol., accepted*, 2006.

[66] Li, M. Y., Lee, T. W., Yim, A. P., and Chen, G. G. Function of PPARgamma and its ligands in lung cancer. Crit Rev Clin Lab Sci, *43:* 183-202, 2006.

[67] Camp, H. S. and Tafuri, S. R. Regulation of peroxisome proliferator-activated receptor gamma activity by mitogen-activated protein kinase. *J. Biol. Chem., 272:* 10811-10816, 1997.

[68] Adams, M., Reginato, M. J., Shao, D., Lazar, M. A., and Chatterjee, V. K. Transcriptional activation by peroxisome proliferator-activated receptor gamma is inhibited by phosphorylation at a consensus mitogen-activated protein kinase site. *J. Biol. Chem., 272:* 5128-5132, 1997.

[69] Han, J., Hajjar, D. P., Tauras, J. M., Feng, J., Gotto, A. M., Jr., and Nicholson, A. C. Transforming growth factor-beta1 (TGF-beta1) and TGF-beta2 decrease expression of CD36, the type B scavenger receptor, through mitogen-activated protein kinase phosphorylation of peroxisome proliferator-activated receptor-gamma. *J. Biol. Chem., 275:* 1241-1246, 2000.

[70] Tsubouchi, Y., Sano, H., Kawahito, Y., Mukai, S., Yamada, R., Kohno, M., Inoue, K., Hla, T., and Kondo, M. Inhibition of human lung cancer cell growth by the peroxisome proliferator-activated receptor-gamma agonists through induction of apoptosis. *Biochem. Biophys. Res. Commun., 270:* 400-405, 2000.

[71] Takahashi, N., Okumura, T., Motomura, W., Fujimoto, Y., Kawabata, I., and Kohgo, Y. Activation of PPARgamma inhibits cell growth and induces apoptosis in human gastric cancer cells. *FEBS Lett, 455:* 135-139, 1999.

[72] Koga, H., Sakisaka, S., Harada, M., Takagi, T., Hanada, S., Taniguchi, E., Kawaguchi, T., Sasatomi, K., Kimura, R., Hashimoto, O., Ueno, T., Yano, H., Kojiro, M., and Sata, M. Involvement of p21(WAF1/Cip1), p27(Kip1), and p18(INK4c) in troglitazone-induced cell-cycle arrest in human hepatoma cell lines. *Hepatology, 33:* 1087-1097, 2001.

[73] Sarraf, P., Mueller, E., Jones, D., King, F. J., DeAngelo, D. J., Partridge, J. B., Holden, S. A., Chen, L. B., Singer, S., Fletcher, C., and Spiegelman, B. M. Differentiation and reversal of malignant changes in colon cancer through PPARgamma. *Nat. Med., 4:* 1046-1052, 1998.

[74] Kubota, T., Koshizuka, K., Williamson, E. A., Asou, H., Said, J. W., Holden, S., Miyoshi, I., and Koeffler, H. P. Ligand for peroxisome proliferator-activated receptor gamma (troglitazone) has potent antitumor effect against human prostate cancer both in vitro and in vivo. *Cancer Res., 58:* 3344-3352, 1998.

[75] Elnemr, A., Ohta, T., Iwata, K., Ninomia, I., Fushida, S., Nishimura, G., Kitagawa, H., Kayahara, M., Yamamoto, M., Terada, T., and Miwa, K. PPARgamma ligand (thiazolidinedione) induces growth arrest and differentiation markers of human pancreatic cancer cells. *Int. J. Oncol., 17:* 1157-1164, 2000.

[76] Eibl, G., Wente, M. N., Reber, H. A., and Hines, O. J. Peroxisome proliferator-activated receptor gamma induces pancreatic cancer cell apoptosis. *Biochem. Biophys. Res. Commun., 287:* 522-529, 2001.

[77] Inoue, K., Kawahito, Y., Tsubouchi, Y., Kohno, M., Yoshimura, R., Yoshikawa, T., and Sano, H. Expression of peroxisome proliferator-activated receptor gamma in renal cell carcinoma and growth inhibition by its agonists. *Biochem. Biophys. Res. Commun., 287:* 727-732, 2001.

[78] Tontonoz, P., Singer, S., Forman, B. M., Sarraf, P., Fletcher, J. A., Fletcher, C. D., Brun, R. P., Mueller, E., Altiok, S., Oppenheim, H., Evans, R. M., and Spiegelman, B. M. Terminal differentiation of human liposarcoma cells induced by ligands for peroxisome proliferator-activated receptor gamma and the retinoid X receptor. *Proc. Natl. Acad. Sci. U S A, 94:* 237-241, 1997.

[79] Hirase, N., Yanase, T., Mu, Y., Muta, K., Umemura, T., Takayanagi, R., and Nawata, H. Thiazolidinedione induces apoptosis and monocytic differentiation in the promyelocytic leukemia cell line HL60. *Oncology, 57 Suppl 2:* 17-26, 1999.

[80] Fujimura, S., Suzumiya, J., Nakamura, K., and Ono, J. Effects of troglitazone on the growth and differentiation of hematopoietic cell lines. *Int. J. Oncol., 13:* 1263-1267, 1998.

[81] Nakashiro, K. I., Hayashi, Y., Kita, A., Tamatani, T., Chlenski, A., Usuda, N., Hattori, K., Reddy, J. K., and Oyasu, R. Role of peroxisome proliferator-activated receptor gamma and its ligands in non-neoplastic and neoplastic human urothelial cells. *Am. J. Pathol., 159:* 591-597, 2001.

[82] Nikitakis, N. G., Hebert, C., Lopes, M. A., Reynolds, M. A., and Sauk, J. J. PPARgamma-mediated antineoplastic effect of NSAID sulindac on human oral squamous carcinoma cells. *Int. J. Cancer, 98:* 817-823, 2002.

[83] Nikitakis, N. G., Hamburger, A. W., and Sauk, J. J. The nonsteroidal anti-inflammatory drug sulindac causes down-regulation of signal transducer and activator of transcription 3 in human oral squamous cell carcinoma cells. *Cancer Res., 62:* 1004-1007, 2002.

[84] Kato, M., Nagaya, T., Fujieda, M., Saito, K., Yoshida, J., and Seo, H. Expression of PPARgamma and its ligand-dependent growth inhibition in human brain tumor cell lines. *Jpn. J .Cancer Res., 93:* 660-666, 2002.

[85] Ohta, K., Endo, T., Haraguchi, K., Hershman, J. M., and Onaya, T. Ligands for peroxisome proliferator-activated receptor gamma inhibit growth and induce apoptosis of human papillary thyroid carcinoma cells. *J. Clin. Endocrino.l Metab., 86:* 2170-2177, 2001.

[86] Inoue, K., Kawahito, Y., Tsubouchi, Y., Yamada, R., Kohno, M., Hosokawa, Y., Katoh, D., Bishop-Bailey, D., Hla, T., and Sano, H. Expression of peroxisome proliferator-activated receptor (PPAR)-gamma in human lung cancer. *Anticancer Res., 21:* 2471-2476, 2001.

[87] Bren-Mattison, Y., Van Putten, V., Chan, D., Winn, R., Geraci, M. W., and Nemenoff, R. A. Peroxisome proliferator-activated receptor-gamma (PPAR(gamma)) inhibits tumorigenesis by reversing the undifferentiated phenotype of metastatic non-small-cell lung cancer cells (NSCLC). *Oncogene, 24:* 1412-1422, 2005.

[88] Wick, M., Hurteau, G., Dessev, C., Chan, D., Geraci, M. W., Winn, R. A., Heasley, L. E., and Nemenoff, R. A. Peroxisome proliferator-activated receptor-gamma is a target of nonsteroidal anti-inflammatory drugs mediating cyclooxygenase-independent inhibition of lung cancer cell growth. *Mol. Pharmacol., 62:* 1207-1214, 2002.

[89] Han, S. and Roman, J. Suppression of prostaglandin E2 receptor subtype EP2 by PPARgamma ligands inhibits human lung carcinoma cell growth. *Biochem. Biophys. Res. Commun., 314:* 1093-1099, 2004.

[90] Satoh, T., Toyoda, M., Hoshino, H., Monden, T., Yamada, M., Shimizu, H., Miyamoto, K., and Mori, M. Activation of peroxisome proliferator-activated receptor-gamma stimulates the growth arrest and DNA-damage inducible 153 gene in non-small cell lung carcinoma cells. *Oncogene, 21:* 2171-2180, 2002.

[91] Han, S. and Roman, J. Rosiglitazone suppresses human lung carcinoma cell growth through PPARgamma-dependent and PPARgamma-independent signal pathways. *Mol. Cancer Ther., 5:* 430-437, 2006.

[92] Pierce, G. B. and Speers, W. C. Tumors as caricatures of the process of tissue renewal: prospects for therapy by directing differentiation. *Cancer Res., 48:* 1996-2004, 1988.

[93] Jarrard, J. A., Linnoila, R. I., Lee, H., Steinberg, S. M., Witschi, H., and Szabo, E. MUC1 is a novel marker for the type II pneumocyte lineage during lung carcinogenesis. *Cancer Res., 58:* 5582-5589, 1998.

[94] Kwiatkowski, D. J. Predominant induction of gelsolin and actin-binding protein during myeloid differentiation. *J. Biol. Chem., 263:* 13857-13862, 1988.

[95] Dosaka-Akita, H., Hommura, F., Fujita, H., Kinoshita, I., Nishi, M., Morikawa, T., Katoh, H., Kawakami, Y., and Kuzumaki, N. Frequent loss of gelsolin expression in non-small cell lung cancers of heavy smokers. *Cancer Research, 58:* 322-327, 1998.

[96] Cultraro, C. M., Bino, T., and Segal, S. Function of the c-Myc antagonist Mad1 during a molecular switch from proliferation to differentiation. *Mol. Cell. Biol,. 17:* 2353-2359, 1997.

[97] Zhang, W., Grasso, L., McClain, C. D., Gambel, A. M., Cha, Y., Travali, S., Deisseroth, A. B., and Mercer, W. E. p53-independent induction of WAF1/CIP1 in human leukemia cells is correlated with growth arrest accompanying monocyte/macrophage differentiation. *Cancer Res., 55:* 668-674, 1995.

[98] Parker, S. B., Eichele, G., Zhang, P., Rawls, A., Sands, A. T., Bradley, A., Olson, E. N., Harper, J. W., and Elledge, S. J. p53-independent expression of p21Cip1 in muscle and other terminally differentiating cells. *Science, 267:* 1024-1027, 1995.

[99] Chang, T. H. and Szabo, E. Induction of differentiation and apoptosis by ligands of peroxisome proliferator-activated receptor gamma in non-small cell lung cancer. *Cancer Res., 60:* 1129-1138, 2000.

[100] Brown, P. D., Bloxidge, R. E., Stuart, N. S., Gatter, K. C., and Carmichael, J. Association between expression of activated 72-kilodalton gelatinase and tumor spread in non-small-cell lung carcinoma. *J. Natl .Cancer Inst, 85:* 574-578, 1993.

[101] White, E. S., Strom, S. R., Wys, N. L., and Arenberg, D. A. Non-small cell lung cancer cells induce monocytes to increase expression of angiogenic activity. *J. Immunol., 166:* 7549-7555, 2001.

[102] Strieter, R. M., Polverini, P. J., Kunkel, S. L., Arenberg, D. A., Burdick, M. D., Kasper, J., Dzuiba, J., Van Damme, J., Walz, A., Marriott, D., and et al. The functional role of the ELR motif in CXC chemokine-mediated angiogenesis. *J. Biol. Chem., 270:* 27348-27357, 1995.

[103] Keshamouni, V. G., Arenberg, D. A., Reddy, R. C., Newstead, M. J., Anthwal, S., and Standiford, T. J. PPAR-gamma activation inhibits angiogenesis by blocking ELR+CXC chemokine production in non-small cell lung cancer. *Neoplasia, 7:* 294-301, 2005.

[104] Takano, N., Kawakami, T., Kawa, Y., Asano, M., Watabe, H., Ito, M., Soma, Y., Kubota, Y., and Mizoguchi, M. Fibronectin combined with stem cell factor plays an important role in melanocyte proliferation, differentiation and migration in cultured mouse neural crest cells. *Pigment Cell Res., 15:* 192-200, 2002.

[105] Wilson, S. H., Ljubimov, A. V., Morla, A. O., Caballero, S., Shaw, L. C., Spoerri, P. E., Tarnuzzer, R. W., and Grant, M. B. Fibronectin fragments promote human retinal endothelial cell adhesion and proliferation and ERK activation through alpha5beta1 integrin and PI 3-kinase. *Invest. Ophthalmol. Vis. Sci., 44:* 1704-1715, 2003.

[106] Han, S., Sidell, N., Roser-Page, S., and Roman, J. Fibronectin stimulates human lung carcinoma cell growth by inducing cyclooxygenase-2 (COX-2) expression. *Int. J. Cancer, 111:* 322-331, 2004.

[107] Han, S., Ritzenthaler, J. D., Rivera, H. N., and Roman, J. Peroxisome proliferator-activated receptor-gamma ligands suppress fibronectin gene expression in human lung carcinoma cells: involvement of both CRE and Sp1. *Am. J. Physiol. Lung Cell Mol. Physiol., 289:* L419-428, 2005.

[108] Adachi, M., Taki, T., Higashiyama, M., Kohno, N., Inufusa, H., and Miyake, M. Significance of integrin alpha5 gene expression as a prognostic factor in node-negative non-small cell lung cancer. *Clin. Cancer Res., 6:* 96-101, 2000.

[109] Han, S., Rivera, H. N., and Roman, J. Peroxisome proliferator-activated receptor-gamma ligands inhibit alpha5 integrin gene transcription in non-small cell lung carcinoma cells. *Am. J. Respir. Cell Mol. Biol., 32:* 350-359, 2005.

[110] Richards, J. A. and Brueggemeier, R. W. Prostaglandin E2 regulates aromatase activity and expression in human adipose stromal cells via two distinct receptor subtypes. *J. Clin. Endocrinol. Metab., 88:* 2810-2816, 2003.

[111] Yano, T., Zissel, G., Muller-Qernheim, J., Jae Shin, S., Satoh, H., and Ichikawa, T. Prostaglandin E2 reinforces the activation of Ras signal pathway in lung adenocarcinoma cells via EP3. *FEBS Lett, 518:* 154-158, 2002.

[112] Dohadwala, M., Batra, R. K., Luo, J., Lin, Y., Krysan, K., Pold, M., Sharma, S., and Dubinett, S. M. Autocrine/paracrine prostaglandin E2 production by non-small cell lung cancer cells regulates matrix metalloproteinase-2 and CD44 in cyclooxygenase-2-dependent invasion. *J. Biol. Chem., 277:* 50828-50833, 2002.

[113] Green, D. R. and Reed, J. C. Mitochondria and apoptosis. *Science, 281:* 1309-1312, 1998.

[114] Gross, A., McDonnell, J. M., and Korsmeyer, S. J. BCL-2 family members and the mitochondria in apoptosis. *Genes Dev., 13:* 1899-1911, 1999.

[115] Li, P., Nijhawan, D., Budihardjo, I., Srinivasula, S. M., Ahmad, M., Alnemri, E. S., and Wang, X. Cytochrome c and dATP-dependent formation of Apaf-1/caspase-9 complex initiates an apoptotic protease cascade. *Cell, 91:* 479-489, 1997.

[116] Nicholson, D. W., Ali, A., Thornberry, N. A., Vaillancourt, J. P., Ding, C. K., Gallant, M., Gareau, Y., Griffin, P. R., Labelle, M., Lazebnik, Y. A., and et al. Identification and inhibition of the ICE/CED-3 protease necessary for mammalian apoptosis. *Nature, 376:* 37-43, 1995.

[117] Han, S., Sidell, N., Fisher, P. B., and Roman, J. Up-regulation of p21 gene expression by peroxisome proliferator-activated receptor gamma in human lung carcinoma cells. *Clin. Cancer Res., 10:* 1911-1919, 2004.

[118] Liu, Y., Lee, M. O., Wang, H. G., Li, Y., Hashimoto, Y., Klaus, M., Reed, J. C., and Zhang, X. Retinoic acid receptor beta mediates the growth-inhibitory effect of retinoic acid by promoting apoptosis in human breast cancer cells. *Mol. Cell Biol., 16:* 1138-1149, 1996.

[119] Lotan, R., Xu, X. C., Lippman, S. M., Ro, J. Y., Lee, J. S., Lee, J. J., and Hong, W. K. Suppression of retinoic acid receptor-beta in premalignant oral lesions and its up-regulation by isotretinoin. *N. Engl. J. Med., 332:* 1405-1410, 1995.

[120] Chambon, P. A decade of molecular biology of retinoic acid receptors. *Faseb. J., 10:* 940-954, 1996.

[121] James, S. Y., Lin, F., Kolluri, S. K., Dawson, M. I., and Zhang, X. K. Regulation of retinoic acid receptor beta expression by peroxisome proliferator-activated receptor gamma ligands in cancer cells. *Cancer Res., 63:* 3531-3538, 2003.

[122] Shankaranarayanan, P. and Nigam, S. IL-4 induces apoptosis in A549 lung adenocarcinoma cells: evidence for the pivotal role of 15-hydroxyeicosatetraenoic acid binding to activated peroxisome proliferator-activated receptor gamma transcription factor. *J. Immunol., 170:* 887-894, 2003.

[123] Xia, Z., Dickens, M., Raingeaud, J., Davis, R. J., and Greenberg, M. E. Opposing effects of ERK and JNK-p38 MAP kinases on apoptosis. *Science, 270:* 1326-1331, 1995.

[124] Nguyen, T. T., Tran, E., Nguyen, T. H., Do, P. T., Huynh, T. H., and Huynh, H. The role of activated MEK-ERK pathway in quercetin-induced growth inhibition and apoptosis in A549 lung cancer cells. *Carcinogenesis, 25:* 647-659, 2004.

[125] Schweyer, S., Soruri, A., Meschter, O., Heintze, A., Zschunke, F., Miosge, N., Thelen, P., Schlott, T., Radzun, H. J., and Fayyazi, A. Cisplatin-induced apoptosis in human malignant testicular germ cell lines depends on MEK/ERK activation. *Br. J. Cancer, 91:* 589-598, 2004.

[126] Yuan, Z., Taatjes, D. J., Mossman, B. T., and Heintz, N. H. The duration of nuclear extracellular signal-regulated kinase 1 and 2 signaling during cell cycle reentry distinguishes proliferation from apoptosis in response to asbestos. *Cancer Res., 64:* 6530-6536, 2004.

[127] Pettersson, F., Couture, M. C., Hanna, N., and Miller, W. H. Enhanced retinoid-induced apoptosis of MDA-MB-231 breast cancer cells by PKC inhibitors involves activation of ERK. *Oncogene, 23:* 7053-7066, 2004.

[128] Chen, C. H., Wang, W. J., Kuo, J. C., Tsai, H. C., Lin, J. R., Chang, Z. F., and Chen, R. H. Bidirectional signals transduced by DAPK-ERK interaction promote the apoptotic effect of DAPK. *EMBO Journal, 24:* 294-304, 2005.

[129] Lee, W. C., Choi, C. H., Cha, S. H., Oh, H. L., and Kim, Y. K. Role of ERK in hydrogen peroxide-induced cell death of human glioma cells. *Neurochem Res., 30:* 263-270, 2005.

[130] Tamura, Y., Simizu, S., and Osada, H. The phosphorylation status and anti-apoptotic activity of Bcl-2 are regulated by ERK and protein phosphatase 2A on the mitochondria. *FEBS Lett, 569:* 249-255, 2004.

[131] Roovers, K. and Assoian, R. K. Integrating the MAP kinase signal into the G1 phase cell cycle machinery. *Bioessays, 22:* 818-826, 2000.

[132] Brandlin, I., Hubner, S., Eiseler, T., Martinez-Moya, M., Horschinek, A., Hausser, A., Link, G., Rupp, S., Storz, P., Pfizenmaier, K., and Johannes, F. J. Protein kinase C (PKC)eta-mediated PKC mu activation modulates ERK and JNK signal pathways. *J. Biol .Chem., 277:* 6490-6496, 2002.

[133] Avis, I., Martinez, A., Tauler, J., Zudaire, E., Mayburd, A., Abu-Ghazaleh, R., Ondrey, F., and Mulshine, J. L. Inhibitors of the arachidonic acid pathway and peroxisome proliferator-activated receptor ligands have superadditive effects on lung cancer growth inhibition. *Cancer Res., 65:* 4181-4190, 2005.

[134] Andela, V. B., Altuwaijri, S., Wood, J., and Rosier, R. N. Inhibition of beta-oxidative respiration is a therapeutic window associated with the cancer chemo-preventive activity of PPARgamma agonists. *FEBS Lett, 579:* 1765-1769, 2005.

[135] Lee, S. Y., Hur, G. Y., Jung, K. H., Jung, H. C., Lee, S. Y., Kim, J. H., Shin, C., Shim, J. J., In, K. H., Kang, K. H., and Yoo, S. H. PPAR-gamma agonist increase gefitinib's antitumor activity through PTEN expression. *Lung Cancer, 51:* 297-301, 2006.

[136] Yao, C. J., Lai, G. M., Chan, C. F., Cheng, A. L., Yang, Y. Y., and Chuang, S. E. Dramatic synergistic anticancer effect of clinically achievable doses of lovastatin and troglitazone. *Int. J. Cancer, 118:* 773-779, 2006.

[137] Chawla, A., Barak, Y., Nagy, L., Liao, D., Tontonoz, P., and Evans, R. M. PPAR-gamma dependent and independent effects on macrophage-gene expression in lipid metabolism and inflammation. *Nat. Med., 7:* 48-52, 2001.

[138] Okano, H., Shiraki, K., Inoue, H., Yamanaka, Y., Kawakita, T., Saitou, Y., Yamaguchi, Y., Enokimura, N., Yamamoto, N., Sugimoto, K., Murata, K., and Nakano, T. 15-deoxy-delta-12-14-PGJ2 regulates apoptosis induction and nuclear factor-kappaB activation via a peroxisome proliferator-activated receptor-gamma-independent mechanism in hepatocellular carcinoma. *Lab. Invest., 83:* 1529-1539, 2003.

[139] Vaidya, S., Somers, E. P., Wright, S. D., Detmers, P. A., and Bansal, V. S. 15-Deoxy-Delta12,1412,14-prostaglandin J2 inhibits the beta2 integrin-dependent oxidative burst: involvement of a mechanism distinct from peroxisome proliferator-activated receptor gamma ligation. *J. Immunol., 163:* 6187-6192, 1999.

[140] Theocharisa, S., Margeli, A., and Kouraklis, G. Peroxisome proliferator activated receptor-gamma ligands as potent antineoplastic agents. Curr Med Chem Anticancer Agents, *3:* 239-251, 2003.

[141] Krzystyniak, K. L. Current strategies for anticancer chemoprevention and chemoprotection. *Acta Pol. Pharm., 59:* 473-478, 2002.

In: Oncogene Proteins: New Research
Editors: A. H. Malloy and E.C. Carson, pp.273-287
ISBN 978-1-60456-290-3

Chapter 11

Tumor Necrosis Factor and Carcinoma by Hepatitis B and C Virus Infection

Kazuya Shirato[1] *and Tetsuya Mizutani*[2*]
[1]Laboratory of Acute Respiratory Viral Diseases and Cytokines,
Department of Virology III,
[2]Laboratory of Special Pathogens, Department of Virology I,
National Institute of Infectious Diseases, Gakuen 4-7-1,
Musashimurayama, Tokyo 208-0011 Japan

Abstract

Hepatitis B virus (HBV) and hepatitis C virus (HCV) are global public health problems. The clinical courses of both HBV and HCV infection vary from acute hepatitis to chronic persistent infection that may progress to cirrhosis and carcinoma. Acute inflammation is a defense response and chronic inflammation can lead to cancer. Several proinflammatory gene products have been identified in both HBV- and HCV-infected patients. The expression of these genes is mainly regulated by the transcription factor NF-κB, which is constitutively active in most tumors and is induced by tumor promoters and carcinogenic viral proteins. Anti-inflammatory agents that suppress NF-κB may be useful in both the prevention and treatment of cancer. Tumor necrosis factor alpha (TNFα) is thought to be an important factor underlying the mechanisms of action of these types of viral hepatitis and carcinoma because TNFα plays an important role in the host immune response to HBV and HCV infection. Recent studies indicated that TNFα promoter polymorphisms are significantly associated with viral clearance. Neutralization of TNFα(anti-TNF therapy) has been shown to be associated with activation of HBV infection, but not HCV infection. However, TNF has been reported to induce clearance of HBV. One of the inhibitory mechanisms of anti-TNF therapy is that TNF blocks HBV replication by promoting destabilization of viral nucleocapsids. HCV proteins are able to trigger production of TNFα and modulate nuclear factor kappa B activation and

* Corresponding author. Telephone +81-42-561-0771, Facsimile +81-42-565-3315, Email tmizutan@nih.go.jp.

apoptosis stimulated by this cytokine. This review highlights the importance of TNFα production for hepatitis and carcinoma by HBV and HCV infection.

Abbreviations

CAH:	Chronic active hepatitis
HBV:	Human hepatitis B virus
HCC:	Hepatocellular carcinoma
HCV:	Human hepatitis C virus
IFN:	Interferon
IL:	Interleukin
MHV:	Mouse hepatitis virus
NF-κB:	Nuclear factor kappa B
NS:	Nonstructural protein
PBMC:	Peripheral blood mononuclear cell
pDC:	Plasmacytoid dendritic cells
TNFα:	Tumor necrosis factor alpha

Introduction

Hepatitis B virus (HBV) and hepatitis C virus (HCV) are global public health problems. The clinical courses of both HBV and HCV infection vary from acute hepatitis to chronic persistent infection that may progress to cirrhosis and carcinoma. Several inflammatory mediators are up-regulated during virus-induced hepatic diseases and these proinflammatory cytokines are regulated by the nuclear factor kappa B (NF-κB) pathway. Among these proinflammatory cytokines, tumor necrosis factor alpha (TNFα) is thought to be important for viral hepatitis progression or remission.

TNFα is a multifunctional cytokine, mainly produced by activated monocytes and macrophages, which is closely related to the development and regulation of the immune system and host defense. In addition, it was reported that TNFα plays an important role in acute and chronic liver injury induced by various stimuli, such as severe alcoholic hepatitis and hepatotoxic drugs.

This chapter presents an overview of the importance of TNFα expression during HBV- and HCV-induced hepatic injury.

Hepatitis B and C Viruses

1. Hepatitis B Virus

Human hepatitis B virus (HBV) is a member of the family *Hepadnaviridae* and has a circular DNA consisting of a longer complete negative-sense strand, with a piece missing at a nick site, and an incomplete positive-sense strand. HBV causes acute and chronic hepatocyte injury and is associated with liver cancer (Beasley *et al.*, 1981). Although geographic correlations have been reported, chronic infection by HBV is a major risk factor for hepatocellular carcinoma (HCC) (Block *et al.*, 2003) and occult HBV infection, which is characterized by HBV DNA persistence in HBV surface antigen-negative patients, and increased risk of HCC (Squadrito *et al.*, 2006). HBV vaccine consisting of purified HBV s antigen is now commercially available, and the incidence rates of chronic HBV infection have begun to decrease (Kane, 1995; Liu and Kao, 2007).

2. Hepatitis C Virus

Human hepatitis C virus (HCV) belongs to the family *Flaviviridae* and causes acute, and in many cases chronic, hepatitis. In 1989, the cloning of part of the genome of non-A and non-B hepatitis virus was first reported and this virus was HCV, a major cause of non-A and non-B hepatitis (Choo *et al.*, 1989; Kuo *et al.*, 1989). It was reported that the core (C) protein of HCV can transform HCV-infected cells (Kato, 2001; Ray and Ray, 2001) and chronic HCV infection appears to be a major factor involved in chronic hepatitis, liver cirrhosis, and hepatic cancer (Erhardt *et al.*, 2002; Kato, 2001). About 150 to 200 million people worldwide are infected with HCV. The prevalence rate of HCV is generally less than 3% in developed nations. It has been reported that the major reason for the spread of HCV in Japan and Italy is medical treatment, while that in the USA is drug abuse (Guadagnino *et al.*, 1997; Kelen *et al.*, 1992; Noguchi *et al.*, 1997).

Several commercial therapeutic agents based on IFNα are available and IFNα treatment in combination with ribavirin has been shown to reduce the HCV genome copy number, although the effect is dependent on the HCV genotype and various side effects have been reported (Fried *et al.*, 2002; Manns *et al.*, 2001; Manns, Wedemeyer, and Cornberg, 2006; McHutchison *et al.*, 1998; Poynard *et al.*, 1998).

Cytokine Induction During Hepatic Inflammation

Various inflammatory mediators, such as cytokines and chemokines, are secreted during hepatic inflammation and are correlated with disease progression or host protection.

In the murine model of hepatic inflammation induced by concanavalin A, D-galactosamine/lipopolysaccharide injection, TNFα, IFNγ, IL-2, IL-4, IL-10, and IL-12 are highly up-regulated and it has been shown that TNFα induces liver injury while IL-10 has a protective role (Sass *et al.*, 2002) . Mouse hepatitis virus (MHV) is a positive-strand RNA virus that causes a variety of clinical diseases, such as encephalomyelitis, respiratory diseases, and hepatitis in susceptible mice (De Albuquerque *et al.*, 2006; Lai and Cavanagh, 1997; Levy, Leibowitz, and Edgington, 1981; Li *et al.*, 1992). The MHV3 infection model is a commonly used murine model of virus-induced hepatitis. In MHV-3-induced fulminant hepatitis, it was reported that TNFα, IL-1, and TGFβ were up-regulated in susceptible mouse strains but not in resistant mice, and TNFα is considered one of the pathogenic determinants (Ning *et al.*, 1998; Pope *et al.*, 1995).

Serum levels of proinflammatory cytokines, such as IL-1β, IL-6, and TNFα have been reported to be up-regulated in patients with hepatitis A-, B-, and C-induced hepatitis (Torre *et al.*, 1994). In HBV-induced chronic active hepatitis (CAH), high levels of TNFα production were seen in plasma, and the TNFα level was reduced in advanced cirrhosis (Kiki *et al.*, 2006). Barnaba *et al.* also reported that cytotoxic T cells isolated from HBV-induced CAH patients secreted TNFα and IL-2 (Barnaba *et al.*, 1994). In hepatitis C virus infection, it was reported that IFNα levels were diminished by reducing the number of plasmacytoid dendritic cells (pDC) *via* the monocyte-derived TNFα and IL-10, which were induced by HCV core antigen stimulation, and neutralization of TNFα and IL-10 prevented HCV core-induced inhibition of IFNα production (Dolganiuc *et al.*, 2006). In addition, peripheral blood mononuclear cells (PBMC) from patients with chronic HCV infection cultured with HCV core antigen were shown to secrete IFNγ and IL-10 (Rigopoulou *et al.*, 2007).

As described above, TNFα is considered a key mediator of liver injury.

Relations between TNFα and Cancer and Liver Diseases

1. TNFα and Cancer

Inflammation is an important host reaction to various stimuli, such as injury or infection; however, inflammation may be either protective or immunopathological for the host. Cytokines are bioactive molecules that have signaling activities, and many cytokines are secreted during acute and chronic inflammation and build complicated networks. Chronic inflammation is a risk factor for various diseases, such as cardiovascular diseases, diabetes, arthritis, Alzheimer's disease, pulmonary diseases, autoimmune diseases, and cancer (Aggarwal, 2004; Aggarwal *et al.*, 2006). Some proinflammatory cytokines, such as TNFα, IL-1β, and IL-6, are associated with these inflammatory diseases, and these proinflammatory cytokines are regulated by the transcription factor NF-κB. Among the proinflammatory cytokines, it was reported that TNFα induced cellular transformation, proliferation, and tumor promotion (Balkwill and Coussens, 2004; Balkwill and Mantovani, 2001; Mantovani, 2005). TNFα is a multifunctional cytokine, mainly produced by activated monocytes and macrophages, which is intricately involved in the development and regulation of the immune system and host defense (Tracey and Cerami, 1993). TNFα was first identified as an anti-tumor cytokine (Aggarwal, 2003) and has cytolytic effects on tumor cells (Sugarman *et al.*, 1985) and blocks the growth of cancer cells (Shen *et al.*, 2002). On the other hand, TNFα is also secreted by various tumor cells (Aggarwal *et al.*, 2006), and TNFα can promote cancer progression (Balkwill, 2006; Mocellin *et al.*, 2005). In addition, TNFα can contribute to various steps of tumorigenesis, such as cellular transformation, promotion, proliferation, invasion, angiogenesis, and metastasis (Aggarwal *et al.*, 2006).

2. TNFα and Liver Diseases

TNFα is secreted at the early stages of many types of liver injury and TNFα induces other cytokines production and hepatic curing responses. TNFα promotes hepatocellular proliferation rather than cell death (Tilg and Diehl, 2000), and hepatectomy liver regeneration is induced through TNF signaling (Akerman *et al.*, 1992). However, it was reported that TNFα plays a detrimental role in acute and chronic liver injury induced by severe alcoholic hepatitis, human hepatitis C virus infection, and hepatotoxic drugs (Kusumoto *et al.*, 2006; Muto *et al.*, 1988; Rothe *et al.*, 1993; Tilg, Kaser, and Moschen, 2006; Winwood and Arthur, 1993). Plasma levels of TNFα and soluble TNF receptor are correlated with liver disease progression, and neutralization of TNFα successfully prevents liver injury in alcohol-fed rats (Iimuro *et al.*, 1997). In addition, TNF receptor p55-deficient mice (TNFRp55-/-) resisted lipopolysaccharide-, acetaminophen-, or dimethylnitrosamine-induced liver injury (Ishida *et al.*, 2004; Kitamura *et al.*, 2002; Tsuji *et al.*, 1997). On the other hand, TNFRp55-/- mice exhibited high susceptibility to *Listeria monocytogenes* infection (Pfeffer *et al.*, 1993; Rothe *et al.*, 1993), and Puro *et al.* reported that TNF blocked hepatitis B virus replication by promoting destabilization of viral nucleocapsids (Puro and Schneider, 2007). These reports suggest that TNFα plays an important role in liver injury or protection.

Apoptosis is thought to be one of the processes involved in TNFα-induced liver injury. Fas, which belongs to the TNF receptor family, can induce cell apoptosis *via* its ligand, FasL (Nagata and Golstein, 1995), and the up-regulation of Fas/FasL-induced apoptosis was seen in human liver failure caused by human hepatitis B and C infection (Galle *et al.*, 1995; Nagata and Golstein, 1995; Ryo *et al.*, 2000). TNFα induces cell apoptosis through TNFR (Bradham *et al.*, 1998; Nagata, 1997) and TNF is also able to induce Fas expression by up-regulating NF-κB expression (Streetz *et al.*, 2000). Therefore, it was suggested that TNFα contributes to the apoptosis seen in liver injury.

TNFα and HBV and HCV Infection

1. Role of TNFα in HBV and HCV Infection

The roles of TNFα in HBV and HCV infection are summarized in figure 1. Many reports have suggested that TNFα plays a protective role during HBV infection. Generally, plasma level of TNFα in patients with acute HBV infection are higher than those in non-HBV-infected individuals (Koulentaki *et al.*, 2004), and TNFα down-regulates the expression and replication of HBV in the liver (Chen *et al.*, 2005) and the expression of HBV antigen was shown to be inhibited by TNFα in HBV transgenic mice (Cavanaugh *et al.*, 1998). In addition, TNFα and IFNγ induced activated immunoeffector cells in lesions and TNFα and IFNγ can induce non-cytolytic HBV suppression (Kim *et al.*, 2003). Moreover, Puro *et al.* reported that TNFα destabilizes HBV nucleocapsids by inhibiting accumulation of transcription template without ubiquitination or methylation of protein (Puro and Schneider, 2007).

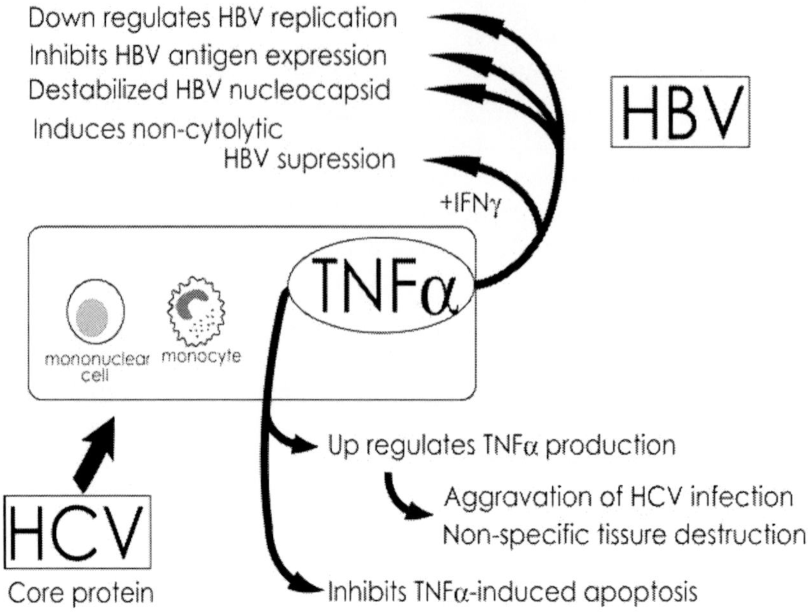

Figure 1. The role of TNFa in HBV and HCV infection.

On the other hand, Dunn *et al.* reported that natural killer cells may contribute to liver inflammation by increasing TNF-related apoptosis-inducing ligand-mediated hepatocellular death during active HBV infection (Dunn *et al.*, 2007).

In HCV infection, the role of TNFα is not understood in detail. HCV core protein, which is a 20-kDa structural protein, induces TNFα production in monocytes. It was also reported that HCV core protein inhibits TNF-induced apoptosis of human breast carcinoma cells (Ray *et al.*, 1998). In chronic hepatitis C patients, plasma levels of TNFα and TNF receptor are up-regulated and increased TNFα and TNF receptor levels lead to aggravation of HCV infection (Kallinowski *et al.*, 1998; Nelson *et al.*, 1997). TNFα is also secreted by infiltrating mononuclear cells in HCV infection, and these infiltrating lymphocytes are usually not specific for HCV antigen (Ando *et al.*, 1997; Kinnman, Andersson, and Hultcrantz, 2000). Therefore, TNFα is considered a critical mediator in nonspecific tissue destruction by infiltrates (Tilg, Kaser, and Moschen, 2006).

TNFα expression is regulated by the NF-κB signaling pathway (Aggarwal, 2004). It has been reported that the HCV core proteins can induce NF-κB pathway activation (Marusawa *et al.*, 1999; You, Chen, and Lee, 1999). Dolganiuc *et al.* reported that core protein or nonstructural protein (NS) 3 of HCV reduced cytoplasmic Iκ-Bα protein levels, which is the initiation step of the NF-κB signal cascade (Dolganiuc *et al.*, 2003). In addition, Choi *et al.* reported that NS5B protein of HCV inhibited TNFα-induced NF-κB activation and NS5B protein also activated TNFα-mediated JNK activity in hepatic cells (Choi *et al.*, 2006). Moreover, NS5A protein of HCV negatively regulates TNFα-induced NF-κB activation *via* TNF receptor-associated factor 2 signaling (Park *et al.*, 2002).

2. Association of TNFα Promoter Polymorphism with the Outcome of Hepatic Virus Infection

As described above, TNFα plays a critical role in the host immune response against HBV infection, and the capacity for TNFα production in an individual is a major genetic determinant of susceptibility to HBV. A number of recent reports described the relation between TNFα promoter polymorphism and the outcome of HBV infection (Cheong *et al.*, 2006; Du *et al.*, 2006; Kim *et al.*, 2003; Li *et al.*, 2005; Niro *et al.*, 2005). More than 10 SNPs have been reported in the promoter region of the TNFα gene (*e.g.*, –238G/A, –244A/G, –308G/A, –376A/G, –575A/G, –857C/T, –863C/A, –1031T/C, –1125G/C, –1196C/T), and these SNPs influence TNFα production (Baena *et al.*, 2002; Du *et al.*, 2006; Heesen *et al.*, 2004; Higuchi *et al.*, 1998; Jaber *et al.*, 2004; Negoro *et al.*, 1999). Among these SNPs, TNFα promoter haplotypes (–238/–308/–857/–863/–1031) are associated with the outcome of HBV infection (Cheong *et al.*, 2006; Du *et al.*, 2006; Kim *et al.*, 2003; Li *et al.*, 2005; Niro *et al.*, 2005). Several reports have suggested that the frequencies of –238GG, –308GG, and –857CC are significantly associated with persistent HBV infection, and the frequencies of –308AA or AG, –857TT, and –863CC are significantly associated with spontaneous clearance of HBV (Du *et al.*, 2006; Li *et al.*, 2005; Niro *et al.*, 2005). In addition, it was reported that TNFα haplotypes (–238G, –308G, –857C, –863A, –1031T) and (–238G, –308G, –857T, –863A, –1031T) are strongly associated with chronic HBV infection, while haplotypes (–238G, –308G, –857C, –863C, –1031T) and (–238G, –308G, –857C, –863A, –1031C) are associated with HBV clearance (Du *et al.*, 2006; Kim *et al.*, 2003).

On the other hand, although some reports suggested that TNF□ promoter polymorphisms are associated with response to IFNα therapy, the contribution of TNFα promoter polymorphism to HCV infection remains controversial. Dai *et al.* reported that the TNF-308A allele was associated with the failure of high-dose IFNα treatment in HCV patients, and the failure was particularly remarkable in HCV genotype 1b infection and in cases with high serum HCV RNA levels (Dai *et al.*, 2006a). They also reported that the TNF-308A allele appeared to be associated with variability in severity of fibrosis and viral load in chronic HCV infection (Dai *et al.*, 2006b). On the other hand, Rosen *et al.* and Yu *et al.* reported that TNF-308 polymorphism did not play a direct role in susceptibility or pathogenesis of HCV infection or response to antiviral IFNα therapy (Rosen *et al.*, 2002; Yu *et al.*, 2003). However, Kusumoto *et al.* reported that TNFα promoter variations were not associated with alterations in HCV clearance or alanine aminotransferase release but may be associated with hepatic fibrosis (Kusumoto *et al.*, 2006).

Anti-TNFα Therapy During HBV and HCV Infection

A number of anti-TNFα agents, such as infliximab, which is a human-mouse chimeric anti-TNFα monoclonal antibody, and etanercept, which is a soluble p75 TNFα receptor-conjugated Fc of IgG, are commercially available and have been used successfully in chronic inflammatory diseases, such as rheumatic disease and Crohn's disease. TNFα has been suggested to be involved in the pathogenesis of liver diseases. Therefore, treatments capable of neutralizing TNFα may be useful to ameliorate liver injury. In alcoholic steatohepatitis, anti-TNFα therapy combined with steroid medicine was useful in a rat model (Iimuro *et al.*, 1997) and was significantly effective in disease recovery in some clinical trials (Spahr *et al.*, 2002; Tilg *et al.*, 2003), although another report suggested that anti-TNFα therapy had detrimental effects in patient with severe alcoholic hepatitis (Naveau *et al.*, 2004).

In HBV infection, TNFα plays a protective role as described above and several reports suggested that anti-TNF□ therapy induces reactivation of HBV and fulminant hepatitis (Esteve *et al.*, 2004; Michel *et al.*, 2003; Ostuni *et al.*, 2003; Wendling *et al.*, 2005). Anti-TNFα treatment in HBV surface antigen carrier and Crohn's disease patients leads to severe reactivation of HBV (Millonig *et al.*, 2006; Tilg, Kaser, and Moschen, 2006). In addition, it was reported that HBV reactivation occurred in HBV antigen-negative patient with Crohn's disease treated with anti-TNFα therapy (Madonia *et al.*, 2007). Anti-TNFα therapy is commonly applied in the treatment of Crohn's disease. Therefore, it is important to check the HBV status of patients scheduled to receive anti-TNFα therapy.

On the other hand, as described above, serum levels of both TNFα and TNFα receptor were up-regulated in HCV-infected patients, and it has been suggested that TNFα enhances liver injury during HCV infection. However, the effects of anti-TNFα therapy in HCV infection remain unclear. Some reports indicated that rheumatic disease patients infected with HCV and receiving long-term treatment with anti-TNFα agents did not show reactivation of hepatitis C infection or aggravation of hepatic inflammation (Aslanidis *et al.*, 2007; Linardaki *et al.*, 2007; Parke and Reveille, 2004). Zein *et al.* reported that anti-TNFα therapy was directly effective against HCV infection. They reported that IFNα-2b and ribavirin treatment with etanercept significantly improved virological response in HCV patients (Zein, 2005) and this is the first evidence that an anti-TNFα strategy is beneficial for HCV infection. However, further studies are needed to confirm the effectiveness of anti-TNFα therapy in the treatment of HCV infection.

References

Aggarwal, B. B. (2003). Signalling pathways of the TNF superfamily: a double-edged sword. *Nat. Rev. Immunol.* 3(9), 745-56.

Aggarwal, B. B. (2004). Nuclear factor-kappaB: the enemy within. *Cancer Cell* 6(3), 203-8.

Aggarwal, B. B., Shishodia, S., Sandur, S. K., Pandey, M. K., and Sethi, G. (2006). Inflammation and cancer: how hot is the link? *Biochem. Pharmacol* .72(11), 1605-21.

Akerman, P., Cote, P., Yang, S. Q., McClain, C., Nelson, S., Bagby, G. J., and Diehl, A. M. (1992). Antibodies to tumor necrosis factor-alpha inhibit liver regeneration after partial hepatectomy. *Am. J. Physio.* 263(4 Pt 1), G579-85.

Ando, K., Hiroishi, K., Kaneko, T., Moriyama, T., Muto, Y., Kayagaki, N., Yagita, H., Okumura, K., and Imawari, M. (1997). Perforin, Fas/Fas ligand, and TNF-alpha pathways as specific and bystander killing mechanisms of hepatitis C virus-specific human CTL. *J. Immunol.* 158(11), 5283-91.

Aslanidis, S., Vassiliadis, T., Pyrpasopoulou, A., Douloumpakas, I., and Zamboulis, C. (2007). Inhibition of TNFalpha does not induce viral reactivation in patients with chronic hepatitis C infection: two cases. *Clin. Rheumatol.* 26(2), 261-4.

Baena, A., Leung, J. Y., Sullivan, A. D., Landires, I., Vasquez-Luna, N., Quinones-Berrocal, J., Fraser, P. A., Uko, G. P., Delgado, J. C., Clavijo, O. P., Thim, S., Meshnick, S. R., Nyirenda, T., Yunis, E. J., and Goldfeld, A. E. (2002). TNF-alpha promoter single nucleotide polymorphisms are markers of human ancestry. *Genes Immun.* 3(8), 482-7.

Balkwill, F. (2006). TNF-alpha in promotion and progression of cancer. *Cancer Metastasis Rev* .25(3), 409-16.

Balkwill, F., and Coussens, L. M. (2004). Cancer: an inflammatory link. *Nature* 431(7007), 405-6.

Balkwill, F., and Mantovani, A. (2001). Inflammation and cancer: back to Virchow? *Lancet* 357(9255), 539-45.

Barnaba, V., Franco, A., Paroli, M., Benvenuto, R., De Petrillo, G., Burgio, V. L., Santilio, I., Balsano, C., Bonavita, M. S., Cappelli, G., and *et al.* (1994). Selective expansion of cytotoxic T lymphocytes with a CD4+CD56+ surface phenotype and a T helper type 1 profile of cytokine secretion in the liver of patients chronically infected with Hepatitis B virus. *J. Immunol.* 152(6), 3074-87.

Beasley, R. P., Hwang, L. Y., Lin, C. C., and Chien, C. S. (1981). Hepatocellular carcinoma and hepatitis B virus. A prospective study of 22 707 men in Taiwan. *Lancet* 2(8256), 1129-33.

Block, T. M., Mehta, A. S., Fimmel, C. J., and Jordan, R. (2003). Molecular viral oncology of hepatocellular carcinoma. *Oncogene* 22(33), 5093-107.

Bradham, C. A., Plumpe, J., Manns, M. P., Brenner, D. A., and Trautwein, C. (1998). Mechanisms of hepatic toxicity. I. TNF-induced liver injury. *Am. J. Physiol.* 275(3 Pt 1), G387-92.

Cavanaugh, V. J., Guidotti, L. G., and Chisari, F. V. (1998). Inhibition of hepatitis B virus replication during adenovirus and cytomegalovirus infections in transgenic mice. *J. Virol.* 72(4), 2630-7.

Chen, Y., Wei, H., Gao, B., Hu, Z., Zheng, S., and Tian, Z. (2005). Activation and function of hepatic NK cells in hepatitis B infection: an underinvestigated innate immune response. *J. Viral. Hepat* .12(1), 38-45.

Cheong, J. Y., Cho, S. W., Hwang, I. L., Yoon, S. K., Lee, J. H., Park, C. S., Lee, J. E., Hahm, K. B., and Kim, J. H. (2006). Association between chronic hepatitis B virus infection and interleukin-10, tumor necrosis factor-alpha gene promoter polymorphisms. *J. Gastroenterol. Hepatol.* 21(7), 1163-9.

Choi, S. H., Park, K. J., Ahn, B. Y., Jung, G., Lai, M. M., and Hwang, S. B. (2006). Hepatitis C virus nonstructural 5B protein regulates tumor necrosis factor alpha signaling through effects on cellular IkappaB kinase. *Mol. Cell. Biol.* 26(8), 3048-59.

Choo, Q. L., Kuo, G., Weiner, A. J., Overby, L. R., Bradley, D. W., and Houghton, M. (1989). Isolation of a cDNA clone derived from a blood-borne non-A, non-B viral hepatitis genome. *Science* 244(4902), 359-62.

Dai, C. Y., Chuang, W. L., Chang, W. Y., Chen, S. C., Lee, L. P., Hsieh, M. Y., Hou, N. J., Lin, Z. Y., Huang, J. F., Wang, L. Y., and Yu, M. L. (2006a). Tumor necrosis factor-alpha promoter polymorphism at position -308 predicts response to combination therapy in hepatitis C virus infection. *J. Infect. Dis* .193(1), 98-101.

Dai, C. Y., Chuang, W. L., Lee, L. P., Chen, S. C., Hou, N. J., Lin, Z. Y., Hsieh, M. Y., Wang, L. Y., Chang, W. Y., and Yu, M. L. (2006b). Associations of tumour necrosis factor alpha promoter polymorphisms at position -308 and -238 with clinical characteristics of chronic hepatitis C. *J .Viral. Hepat.* 13(11), 770-4.

De Albuquerque, N., Baig, E., Ma, X., Zhang, J., He, W., Rowe, A., Habal, M., Liu, M., Shalev, I., Downey, G. P., Gorczynski, R., Butany, J., Leibowitz, J., Weiss, S. R., McGilvray, I. D., Phillips, M. J., Fish, E. N., and Levy, G. A. (2006). Murine hepatitis virus strain 1 produces a clinically relevant model of severe acute respiratory syndrome in A/J mice. *J. Virol.* 80(21), 10382-94.

Dolganiuc, A., Chang, S., Kodys, K., Mandrekar, P., Bakis, G., Cormier, M., and Szabo, G. (2006). Hepatitis C virus (HCV) core protein-induced, monocyte-mediated mechanisms of reduced IFN-alpha and plasmacytoid dendritic cell loss in chronic HCV infection. *J. Immunol* .177(10), 6758-68.

Dolganiuc, A., Kodys, K., Kopasz, A., Marshall, C., Do, T., Romics, L., Jr., Mandrekar, P., Zapp, M., and Szabo, G. (2003). Hepatitis C virus core and nonstructural protein 3 proteins induce pro- and anti-inflammatory cytokines and inhibit dendritic cell differentiation. *J. Immunol* .170(11), 5615-24.

Du, T., Guo, X. H., Zhu, X. L., Li, J. H., Lu, L. P., Gao, J. R., Gou, C. Y., Li, Z., Liu, Y., and Li, H. (2006). Association of TNF-alpha promoter polymorphisms with the outcomes of hepatitis B virus infection in Chinese Han population. *J. Viral Hepa.t* 13(9), 618-24.

Dunn, C., Brunetto, M., Reynolds, G., Christophides, T., Kennedy, P. T., Lampertico, P., Das, A., Lopes, A. R., Borrow, P., Williams, K., Humphreys, E., Afford, S., Adams, D. H., Bertoletti, A., and Maini, M. K. (2007). Cytokines induced during chronic hepatitis B virus infection promote a pathway for NK cell-mediated liver damage. *J. Exp. Med.* 204(3), 667-80.

Erhardt, A., Hassan, M., Heintges, T., and Haussinger, D. (2002). Hepatitis C virus core protein induces cell proliferation and activates ERK, JNK, and p38 MAP kinases together with the MAP kinase phosphatase MKP-1 in a HepG2 Tet-Off cell line. *Virology* 292(2), 272-84.

Esteve, M., Saro, C., Gonzalez-Huix, F., Suarez, F., Forne, M., and Viver, J. M. (2004). Chronic hepatitis B reactivation following infliximab therapy in Crohn's disease patients: need for primary prophylaxis. *Gut* 53(9), 1363-5.

Fried, M. W., Shiffman, M. L., Reddy, K. R., Smith, C., Marinos, G., Goncales, F. L., Jr., Haussinger, D., Diago, M., Carosi, G., Dhumeaux, D., Craxi, A., Lin, A., Hoffman, J., and Yu, J. (2002). Peginterferon alfa-2a plus ribavirin for chronic hepatitis C virus infection. *N. Engl. J. Med.* 347(13), 975-82.

Galle, P. R., Hofmann, W. J., Walczak, H., Schaller, H., Otto, G., Stremmel, W., Krammer, P. H., and Runkel, L. (1995). Involvement of the CD95 (APO-1/Fas) receptor and ligand in liver damage. *J Exp Med* 182(5), 1223-30.

Guadagnino, V., Stroffolini, T., Rapicetta, M., Costantino, A., Kondili, L. A., Menniti-Ippolito, F., Caroleo, B., Costa, C., Griffo, G., Loiacono, L., Pisani, V., Foca, A., and Piazza, M. (1997). Prevalence, risk factors, and genotype distribution of hepatitis C virus infection in the general population: a community-based survey in southern Italy. *Hepatology* 26(4), 1006-11.

Heesen, M., Kunz, D., Wessiepe, M., van der Poll, T., Zwinderman, A. H., and Blomeke, B. (2004). Rapid genotyping for tumor necrosis factor-alpha (TNF-alpha)-863C/A promoter polymorphism that determines TNF-alpha response. *Clin. Chem.* 50(1), 226-8.

Higuchi, T., Seki, N., Kamizono, S., Yamada, A., Kimura, A., Kato, H., and Itoh, K. (1998). Polymorphism of the 5'-flanking region of the human tumor necrosis factor (TNF)-alpha gene in Japanese. *Tissue Antigens* 51(6), 605-12.

Iimuro, Y., Gallucci, R. M., Luster, M. I., Kono, H., and Thurman, R. G. (1997). Antibodies to tumor necrosis factor alfa attenuate hepatic necrosis and inflammation caused by chronic exposure to ethanol in the rat. *Hepatology* 26(6), 1530-7.

Ishida, Y., Kondo, T., Tsuneyama, K., Lu, P., Takayasu, T., and Mukaida, N. (2004). The pathogenic roles of tumor necrosis factor receptor p55 in acetaminophen-induced liver injury in mice. *J. Leukoc. Biol* .75(1), 59-67.

Jaber, B. L., Rao, M., Guo, D., Balakrishnan, V. S., Perianayagam, M. C., Freeman, R. B., and Pereira, B. J. (2004). Cytokine gene promoter polymorphisms and mortality in acute renal failure. *Cytokine* 25(5), 212-9.

Kallinowski, B., Haseroth, K., Marinos, G., Hanck, C., Stremmel, W., Theilmann, L., Singer, M. V., and Rossol, S. (1998). Induction of tumour necrosis factor (TNF) receptor type p55 and p75 in patients with chronic hepatitis C virus (HCV) infection. *Clin. Exp. Immunol.* 111(2), 269-77.

Kane, M. (1995). Global programme for control of hepatitis B infection. *Vaccine* 13 Suppl 1, S47-9.

Kato, N. (2001). Molecular virology of hepatitis C virus. *Acta Med. Okayama* 55(3), 133-59.

Kelen, G. D., Green, G. B., Purcell, R. H., Chan, D. W., Qaqish, B. F., Sivertson, K. T., and Quinn, T. C. (1992). Hepatitis B and hepatitis C in emergency department patients. *N. Engl. J. Med.* 326(21), 1399-404.

Kiki, I., Yilmaz, O., Erdem, F., Gundogdu, M., Demircan, B., and Bilici, M. (2006). Tumour necrosis factor-alpha levels in hepatitis B virus-related chronic active hepatitis and liver cirrhosis and its relationship to Knodell and Child-Pugh scores. *Int. J. Clin. Pract.* 60(9), 1075-9.

Kim, Y. J., Lee, H. S., Yoon, J. H., Kim, C. Y., Park, M. H., Kim, L. H., Park, B. L., and Shin, H. D. (2003). Association of TNF-alpha promoter polymorphisms with the clearance of hepatitis B virus infection. *Hum. Mol. Genet.* 12(19), 2541-6.

Kinnman, N., Andersson, U., and Hultcrantz, R. (2000). *In situ* expression of transforming growth factor-beta1-3, latent transforming growth factor-beta binding protein and tumor necrosis factor-alpha in liver tissue from patients with chronic hepatitis C. *Scan.d J. Gastroenterol.*35(12), 1294-300.

Kitamura, K., Nakamoto, Y., Akiyama, M., Fujii, C., Kondo, T., Kobayashi, K., Kaneko, S., and Mukaida, N. (2002). Pathogenic roles of tumor necrosis factor receptor p55-mediated signals in dimethylnitrosamine-induced murine liver fibrosis. *Lab. Invest.* 82(5), 571-83.

Koulentaki, M., Notas, G., Petinaki, E., Valatas, V., Mouzas, I. A., Castanas, E., and Kouroumalis, E. A. (2004). Nitric oxide and pro-inflammatory cytokines in acute hepatitis B. *Eur. J. Intern. Med* .15(1), 35-38.

Kuo, G., Choo, Q. L., Alter, H. J., Gitnick, G. L., Redeker, A. G., Purcell, R. H., Miyamura, T., Dienstag, J. L., Alter, M. J., Stevens, C. E., and *et al.* (1989). An assay for circulating antibodies to a major etiologic virus of human non-A, non-B hepatitis. *Science* 244(4902), 362-4.

Kusumoto, K., Uto, H., Hayashi, K., Takahama, Y., Nakao, H., Suruki, R., Stuver, S. O., Ido, A., and Tsubouchi, H. (2006). Interleukin-10 or tumor necrosis factor-alpha polymorphisms and the natural course of hepatitis C virus infection in a hyperendemic area of Japan. *Cytokine* 34(1-2), 24-31.

Lai, M. M., and Cavanagh, D. (1997). The molecular biology of coronaviruses. *Adv Virus Res* 48, 1-100.

Levy, G. A., Leibowitz, J. L., and Edgington, T. S. (1981). Induction of monocyte procoagulant activity by murine hepatitis virus type 3 parallels disease susceptibility in mice. *J. Exp. Med.* 154(4), 1150-63.

Li, C., Fung, L. S., Chung, S., Crow, A., Myers-Mason, N., Phillips, M. J., Leibowitz, J. L., Cole, E., Ottaway, C. A., and Levy, G. (1992). Monoclonal antiprothrombinase (3D4.3) prevents mortality from murine hepatitis virus (MHV-3) infection. *J. Exp. Med.* 176(3), 689-97.

Li, H. Q., Li, Z., Liu, Y., Li, J. H., Dong, J. Q., Gao, J. R., Gou, C. Y., and Li, H. (2005). Association of polymorphism of tumor necrosis factor-alpha gene promoter region with outcome of hepatitis B virus infection. *World J. Gastroenterol* .11(33), 5213-7.

Linardaki, G., Katsarou, O., Ioannidou, P., Karafoulidou, A., and Boki, K. (2007). Effective etanercept treatment for psoriatic arthritis complicating concomitant human immunodeficiency virus and hepatitis C virus infection. *J. Rheumatol.* 34(6), 1353-5.

Liu, C. J., and Kao, J. H. (2007). Hepatitis B virus-related hepatocellular carcinoma: epidemiology and pathogenic role of viral factors. *J. Chin. Med. Assoc.* 70(4), 141-5.

Madonia, S., Orlando, A., Scimeca, D., Olivo, M., Rossi, F., and Cottone, M. (2007). Occult hepatitis B and infliximab-induced HBV reactivation. *Inflamm. Bowel Dis.* 13(4), 508-9.

Manns, M. P., McHutchison, J. G., Gordon, S. C., Rustgi, V. K., Shiffman, M., Reindollar, R., Goodman, Z. D., Koury, K., Ling, M., and Albrecht, J. K. (2001). Peginterferon alfa-2b plus ribavirin compared with interferon alfa-2b plus ribavirin for initial treatment of chronic hepatitis C: a randomised trial. *Lancet* 358(9286), 958-65.

Manns, M. P., Wedemeyer, H., and Cornberg, M. (2006). Treating viral hepatitis C: efficacy, side effects, and complications. *Gut* 55(9), 1350-9.

Mantovani, A. (2005). Cancer: inflammation by remote control. *Nature* 435(7043), 752-3.

Marusawa, H., Hijikata, M., Chiba, T., and Shimotohno, K. (1999). Hepatitis C virus core protein inhibits Fas- and tumor necrosis factor alpha-mediated apoptosis *via* NF-kappaB activation. *J. Virol* .73(6), 4713-20.

McHutchison, J. G., Gordon, S. C., Schiff, E. R., Shiffman, M. L., Lee, W. M., Rustgi, V. K., Goodman, Z. D., Ling, M. H., Cort, S., and Albrecht, J. K. (1998). Interferon alfa-2b alone or in combination with ribavirin as initial treatment for chronic hepatitis C. Hepatitis Interventional Therapy Group. *N. Engl. J. Med* .339(21), 1485-92.

Michel, M., Duvoux, C., Hezode, C., and Cherqui, D. (2003). Fulminant hepatitis after infliximab in a patient with hepatitis B virus treated for an adult onset still's disease. *J. Rheumatol.* 30(7), 1624-5.

Millonig, G., Kern, M., Ludwiczek, O., Nachbaur, K., and Vogel, W. (2006). Subfulminant hepatitis B after infliximab in Crohn's disease: need for HBV-screening? *World J. Gastroenterol.* 12(6), 974-6.

Mocellin, S., Rossi, C. R., Pilati, P., and Nitti, D. (2005). Tumor necrosis factor, cancer and anticancer therapy. *Cytokine Growth Factor Rev.* 16(1), 35-53.

Muto, Y., Nouri-Aria, K. T., Meager, A., Alexander, G. J., Eddleston, A. L., and Williams, R. (1988). Enhanced tumour necrosis factor and interleukin-1 in fulminant hepatic failure. *Lancet* 2(8602), 72-4.

Nagata, S. (1997). Apoptosis by death factor. *Cell* 88(3), 355-65.

Nagata, S., and Golstein, P. (1995). The Fas death factor. *Science* 267(5203), 1449-56.

Naveau, S., Chollet-Martin, S., Dharancy, S., Mathurin, P., Jouet, P., Piquet, M. A., Davion, T., Oberti, F., Broet, P., and Emilie, D. (2004). A double-blind randomized controlled trial of infliximab associated with prednisolone in acute alcoholic hepatitis. *Hepatology* 39(5), 1390-7.

Negoro, K., Kinouchi, Y., Hiwatashi, N., Takahashi, S., Takagi, S., Satoh, J., Shimosegawa, T., and Toyota, T. (1999). Crohn's disease is associated with novel polymorphisms in the 5'-flanking region of the tumor necrosis factor gene. *Gastroenterology* 117(5), 1062-8.

Nelson, D. R., Lim, H. L., Marousis, C. G., Fang, J. W., Davis, G. L., Shen, L., Urdea, M. S., Kolberg, J. A., and Lau, J. Y. (1997). Activation of tumor necrosis factor-alpha system in chronic hepatitis C virus infection. *Dig. Dis. Sci* .42(12), 2487-94.

Ning, Q., Brown, D., Parodo, J., Cattral, M., Gorczynski, R., Cole, E., Fung, L., Ding, J. W., Liu, M. F., Rotstein, O., Phillips, M. J., and Levy, G. (1998). Ribavirin inhibits viral-induced macrophage production of TNF, IL-1, the procoagulant fgl2 prothrombinase and preserves Th1 cytokine production but inhibits Th2 cytokine response. *J. Immunol* .160(7), 3487-93.

Niro, G. A., Fontana, R., Gioffreda, D., Valvano, M. R., Lacobellis, A., Facciorusso, D., and Andriulli, A. (2005). Tumor necrosis factor gene polymorphisms and clearance or progression of hepatitis B virus infection. *Liver Int* .25(6), 1175-81.

Noguchi, S., Sata, M., Suzuki, H., Mizokami, M., and Tanikawa, K. (1997). Routes of transmission of hepatitis C virus in an endemic rural area of Japan. Molecular epidemiologic study of hepatitis C virus infection. *Scand. J. Infect. Dis.* 29(1), 23-8.

Ostuni, P., Botsios, C., Punzi, L., Sfriso, P., and Todesco, S. (2003). Hepatitis B reactivation in a chronic hepatitis B surface antigen carrier with rheumatoid arthritis treated with infliximab and low dose methotrexate. *Ann. Rheum. Dis.* 62(7), 686-7.

Park, K. J., Choi, S. H., Lee, S. Y., Hwang, S. B., and Lai, M. M. (2002). Nonstructural 5A protein of hepatitis C virus modulates tumor necrosis factor alpha-stimulated nuclear factor kappa B activation. *J. Biol. Chem.* 277(15), 13122-8.

Parke, F. A., and Reveille, J. D. (2004). Anti-tumor necrosis factor agents for rheumatoid arthritis in the setting of chronic hepatitis C infection. *Arthritis Rheum.* 51(5), 800-4.

Pfeffer, K., Matsuyama, T., Kundig, T. M., Wakeham, A., Kishihara, K., Shahinian, A., Wiegmann, K., Ohashi, P. S., Kronke, M., and Mak, T. W. (1993). Mice deficient for the 55 kd tumor necrosis factor receptor are resistant to endotoxic shock, yet succumb to L. monocytogenes infection. *Cell* 73(3), 457-67.

Pope, M., Rotstein, O., Cole, E., Sinclair, S., Parr, R., Cruz, B., Fingerote, R., Chung, S., Gorczynski, R., Fung, L., and *et al.* (1995). Pattern of disease after murine hepatitis virus strain 3 infection correlates with macrophage activation and not viral replication. *J. Virol.* 69(9), 5252-60.

Poynard, T., Marcellin, P., Lee, S. S., Niederau, C., Minuk, G. S., Ideo, G., Bain, V., Heathcote, J., Zeuzem, S., Trepo, C., and Albrecht, J. (1998). Randomised trial of interferon alpha2b plus ribavirin for 48 weeks or for 24 weeks versus interferon alpha2b plus placebo for 48 weeks for treatment of chronic infection with hepatitis C virus. International Hepatitis Interventional Therapy Group (IHIT). *Lancet* 352(9138), 1426-32.

Puro, R., and Schneider, R. J. (2007). TNF Activates a Conserved Innate Antiviral Response to HBV that Destabilizes Nucleocapsids and Reduces Nuclear Viral DNA. *J Virol.*

Ray, R. B., Meyer, K., Steele, R., Shrivastava, A., Aggarwal, B. B., and Ray, R. (1998). Inhibition of tumor necrosis factor (TNF-alpha)-mediated apoptosis by hepatitis C virus core protein. *J. Biol. Chem.* 273(4), 2256-9.

Ray, R. B., and Ray, R. (2001). Hepatitis C virus core protein: intriguing properties and functional relevance. *FEMS Microbiol Lett* .202(2), 149-56.

Rigopoulou, E. I., Abbott, W. G., Williams, R., and Naoumov, N. V. (2007). Direct evidence for immunomodulatory properties of ribavirin on T-cell reactivity to hepatitis C virus. *Antiviral Res.* 75(1), 36-42.

Rosen, H. R., McHutchison, J. G., Conrad, A. J., Lentz, J. J., Marousek, G., Rose, S. L., Zaman, A., Taylor, K., and Chou, S. (2002). Tumor necrosis factor genetic polymorphisms and response to antiviral therapy in patients with chronic hepatitis C. *Am. J. Gastroenterol.* 97(3), 714-20.

Rothe, J., Lesslauer, W., Lotscher, H., Lang, Y., Koebel, P., Kontgen, F., Althage, A., Zinkernagel, R., Steinmetz, M., and Bluethmann, H. (1993). Mice lacking the tumour necrosis factor receptor 1 are resistant to TNF-mediated toxicity but highly susceptible to infection by Listeria monocytogenes. *Nature* 364(6440), 798-802.

Ryo, K., Kamogawa, Y., Ikeda, I., Yamauchi, K., Yonehara, S., Nagata, S., and Hayashi, N. (2000). Significance of Fas antigen-mediated apoptosis in human fulminant hepatic failure. *Am. J. Gastroenterol.* 95(8), 2047-55.

Sass, G., Heinlein, S., Agli, A., Bang, R., Schumann, J., and Tiegs, G. (2002). Cytokine expression in three mouse models of experimental hepatitis. *Cytokine* 19(3), 115-20.

Shen, W. H., Zhou, J. H., Broussard, S. R., Freund, G. G., Dantzer, R., and Kelley, K. W. (2002). Proinflammatory cytokines block growth of breast cancer cells by impairing signals from a growth factor receptor. *Cancer Res.* 62(16), 4746-56.

Spahr, L., Rubbia-Brandt, L., Frossard, J. L., Giostra, E., Rougemont, A. L., Pugin, J., Fischer, M., Egger, H., and Hadengue, A. (2002). Combination of steroids with infliximab or placebo in severe alcoholic hepatitis: a randomized controlled pilot study. *J. Hepatol.* 37(4), 448-55.

Squadrito, G., Pollicino, T., Cacciola, I., Caccamo, G., Villari, D., La Masa, T., Restuccia, T., Cucinotta, E., Scisca, C., Magazzu, D., and Raimondo, G. (2006). Occult hepatitis B virus infection is associated with the development of hepatocellular carcinoma in chronic hepatitis C patients. *Cancer* 106(6), 1326-30.

Streetz, K., Leifeld, L., Grundmann, D., Ramakers, J., Eckert, K., Spengler, U., Brenner, D., Manns, M., and Trautwein, C. (2000). Tumor necrosis factor alpha in the pathogenesis of human and murine fulminant hepatic failure. *Gastroenterology* 119(2), 446-60.

Sugarman, B. J., Aggarwal, B. B., Hass, P. E., Figari, I. S., Palladino, M. A., Jr., and Shepard, H. M. (1985). Recombinant human tumor necrosis factor-alpha: effects on proliferation of normal and transformed cells *in vitro*. *Science* 230(4728), 943-5.

Tilg, H., and Diehl, A. M. (2000). Cytokines in alcoholic and nonalcoholic steatohepatitis. *N. Engl. J. Med* .343(20), 1467-76.

Tilg, H., Jalan, R., Kaser, A., Davies, N. A., Offner, F. A., Hodges, S. J., Ludwiczek, O., Shawcross, D., Zoller, H., Alisa, A., Mookerjee, R. P., Graziadei, I., Datz, C., Trauner, M., Schuppan, D., Obrist, P., Vogel, W., and Williams, R. (2003). Anti-tumor necrosis factor-alpha monoclonal antibody therapy in severe alcoholic hepatitis. *J. Hepatol.* 38(4), 419-25.

Tilg, H., Kaser, A., and Moschen, A. R. (2006). How to modulate inflammatory cytokines in liver diseases. *Liver Int.* 26(9), 1029-39.

Torre, D., Zeroli, C., Giola, M., Ferrario, G., Fiori, G. P., Bonetta, G., and Tambini, R. (1994). Serum levels of interleukin-1 alpha, interleukin-1 beta, interleukin-6, and tumor necrosis factor in patients with acute viral hepatitis. *Clin. Infect. Dis.* 18(2), 194-8.

Tracey, K. J., and Cerami, A. (1993). Tumor necrosis factor, other cytokines and disease. *Annu Rev Cell Biol* 9, 317-43.

Tsuji, H., Harada, A., Mukaida, N., Nakanuma, Y., Bluethmann, H., Kaneko, S., Yamakawa, K., Nakamura, S. I., Kobayashi, K. I., and Matsushima, K. (1997). Tumor necrosis factor receptor p55 is essential for intrahepatic granuloma formation and hepatocellular apoptosis in a murine model of bacterium-induced fulminant hepatitis. *Infect. Immun.* 65(5), 1892-8.

Wendling, D., Auge, B., Bettinger, D., Lohse, A., Le Huede, G., Bresson-Hadni, S., Toussirot, E., Miguet, J. P., Herbein, G., and Di Martino, V. (2005). Reactivation of a latent precore mutant hepatitis B virus related chronic hepatitis during infliximab treatment for severe spondyloarthropathy. *Ann. Rheum. Dis.* 64(5), 788-9.

Winwood, P. J., and Arthur, M. J. (1993). Kupffer cells: their activation and role in animal models of liver injury and human liver disease. *Semin Liver Di.s* 13(1), 50-9.

You, L. R., Chen, C. M., and Lee, Y. H. (1999). Hepatitis C virus core protein enhances NF-kappaB signal pathway triggering by lymphotoxin-beta receptor ligand and tumor necrosis factor alpha. *J. Virol* .73(2), 1672-81.

Yu, M. L., Dai, C. Y., Chiu, C. C., Lee, L. P., Lin, Z. Y., Chen, S. C., Hsieh, M. Y., Wang, L. Y., Chen, C. J., Chuang, W. L., and Chang, W. Y. (2003). Tumor necrosis factor alpha promoter polymorphisms at position -308 in Taiwanese chronic hepatitis C patients treated with interferon-alpha. *Antiviral Re.s* 59(1), 35-40.

Zein, N. N. (2005). Etanercept as an adjuvant to interferon and ribavirin in treatment-naive patients with chronic hepatitis C virus infection: a phase 2 randomized, double-blind, placebo-controlled study. *J. Hepatol.* 42(3), 315-22.

In: Oncogene Proteins: New Research
Editors: A. H. Malloy and E.C. Carson, pp.289-309
ISBN 978-1-60456-290-3

Chapter 12

Estrogen and Apoptosis in Thyroid Cancer

Qiang Zeng, George G. Chen[6], Alexander C. Vlantis, and C. Andrew van Hasselt
Department of Surgery, The Chinese University of Hong Kong, Prince of Wales Hospital, Shatin, New Territories, Hong Kong, China

Abstract

Thyroid cancer is the commonest endocrine malignancy and accounts for the majority of endocrine cancer deaths each year. It is notable that the incidence of thyroid cancer is roughly three times more frequent in woman than in men, and further that the incidence of thyroid cancer decreases after menopause in females. This gender difference suggests that the growth and progression of thyroid tumors may be influenced by female sex hormones, particularly estrogen. Estrogen induction of cell viability plays a crucial role in the carcinogenesis of cancers such as breast, colon and prostate. Estrogen also plays a role in thyroid cancer. The action of estrogen on target sites in the human body is mediated through related but distinct estrogen receptors designated ERα and ERβ, both of which are known to be expressed in thyroid cancer. The literature and data from our laboratory has shown that estrogen can promote thyroid cancer cell survival through an ERα-non-genomic signaling pathway and can promote cell apoptosis through an ERβ-non-genomic signaling pathway. The non-genomic signaling pathways are different to the classical ER-mediated genomic signal transduction pathway. Estrogen has opposite effects on cell proliferation, growth and apoptosis, depending on the balance between ERα and ERβ in the cell. This review will address the way in which estrogen interacts with apoptotic molecules and how such an interaction contributes to the development of thyroid cancer.

6 Correspondence: George G Chen, Department of Surgery, The Chinese University of Hong Kong, Prince of Wales Hospital, Shatin, New Territories, Hong Kong. Tel: 852 2632 3934; Fax: 852 2645 0605; E-mail: gchen@cuhk.edu.hk.

1. Introduction

Thyroid cancer is the commonest endocrine malignancy and accounts for the majority of endocrine cancer deaths each year. The four types of thyroid cancers are papillary thyroid carcinoma and follicular thyroid carcinoma, both of which are classified as differentiated thyroid carcinomas, anaplastic thyroid carcinoma, also called undifferentiated thyroid carcinoma, and medullary thyroid carcinoma. Except for medullary thyroid carcinoma, which is derived from parafollicular C-cells, all other thyroid cancers originate from follicular epithelial cells. The incidence of thyroid cancer is eight times higher than cancers of other endocrine organs, excluding ovary and testis [1,2]. Each year in the United States of America, 18,400 new cases are reported [3], and 1,200 deaths occur in a population of approximately 200,000 individuals who have been diagnosed with thyroid cancer. The incidence of thyroid cancer is roughly three times more frequent in woman than in men. The use of oral contraceptives is thought to result in a moderately increased risk of developing thyroid cancer [4-6]. An increased risk is also reported in women who use estrogens for gynecological problems but not in postmenopausal women on low dose estrogen replacement therapy [4,7]. The difference in incidence between the genders suggests that the growth and progression of thyroid tumors is influenced by female sex hormones, particularly estrogen, which is implicated in the development and progression of numerous cancers, such as breast and ovary. The development of thyroid cancer is closely associated with an increased resistance of thyrocytes to apoptosis [8]. On the other hand, it is well known that estrogen regulates apoptosis [9]. Whether and how estrogen contributes to resistance to apoptosis in thyroid cancer cells and how estrogen helps thyroid cancer cells escape apoptosis is the focus of ongoing research. This chapter describes our current understanding and experimental data on the role of estrogen in the regulation of apoptosis in thyroid cancer cells.

2. Estrogen and Estrogen Receptors

2.1. Estrogen

Estrogens are a group of steroid compounds which function as the primary female sex hormones. While estrogens are present in men and women, their levels are significantly higher in women of reproductive age. They are produced by the adrenal cortex and ovary. They interact with multiple organ systems and play a pivotal role in the sentinel physiologic events that occur during a woman's life. Three estrogens occur naturally in the female [10]. In premenopausal women, 17β-estradiol (estradiol or E2), produced by the ovary, is the estrogen produced in the largest quantity and is the most potent as it has the highest affinity for estrogen receptors. In premenopausal women, circulating estradiol levels fluctuate from 40 to 200–400 pg/mL during the menstrual cycle [11]. After menopause, estradiol levels drop to less than 20 pg/mL [12]. The second endogenous estrogen is estrone, a less potent metabolite of estradiol. Estrone is produced from androstenedione, the immediate precursor of estrone, in adipose tissue. In postmenopausal women, the ovary ceases to produce estradiol while the adrenal gland continues to produce androstenedione, with the result that the level of estrone remains unchanged while the plasma level of estradiol falls significantly. The third endogenous estrogen is estriol (E3), also a metabolite of estradiol. Estriol is the principal estrogen produced by the placenta during pregnancy, and is found in smaller quantities than estradiol and estrone in nonpregnant women [11,12].

2.2. Estrogen Receptors

Estrogens induce cellular changes through several different mechanisms. Central to these mechanisms are the estrogen receptor (ER) proteins to which estrogens bind, ERα and ERβwhich are members of a large superfamily of nuclear receptors. The abundance and distribution of these receptors partly determines whether a ligand will have a particular effect. Using RT-PCR, Northern blot analysis, immunohistochemistry and *in situ* hybridization techniques, ERα and ERβ are found to be expressed in the breast, brain, cardiovascular system, urogenital tract, bone and thyroid tissues [13-16]. In some systems/organs, ERα and ERβ may be localized to different cell types. For example, in the ovary ER□ is largely present in thecal and interstitial cells whereas ERβ is predominantly found in granulosa cells [17,18]. In the prostate, ERβ is confined to epithelium whereas ERα is found in the stroma [19]. ERs are also expressed in human thyroid cancer cells. We have examined ERs in three different types of thyroid cancer cells: KAT5 (human thyroid papillary carcinoma cells), ARO (human anaplastic thyroid carcinoma cells) and FRO (human undifferentiated carcinoma cells). Both ERα and ERβare expressed in these three thyroid cancer cell lines [20] (Figure 1).

ERα and ERβ are encoded by separate genes, ESR1 and ESR2 respectively. ESR1 and ESR2 are located on different chromosomes, ERα ισ on chromosome 6q25.1 and ERβ on chromosome 14q22-24, and code for 595 and 530 amino acid receptors respectively. ERα and ERβalso have their own variants. The most common splice variants of ERα (Figure 2) are expressed in multiple tissues and arise from deletions of an internal exon, resulting in truncated proteins which lack segments of the DNA-binding domain (DBD) or ligand (hormone)-binding domain (LBD) of the receptor. Most variant isoforms possess little transcriptional activity, with the exception of ERα exon 5 deletion (ER△E5), which binds DNA but lacks most of the LBD, resulting in low levels of constitutive activity in some cell lines. ER△E2 lacks the DBD and the dimerization domain, while ER△E3 lacks part of the DBD. ER△E4, which lacks part of the LBD, does not bind DNA or hormones, and ER△E6 is missing part of the LBD and dimerization domain. ER△E7 lacks the activation function 2 (AF2) domain and part of the LBD. ER△E3, ER△E5, and ER△E7 variants demonstrate a dominant-negative effect on transcriptional activity mediated by wild-type ERs. Mammalian ERβ variants identified in humans (h), rats (r), mice (m), cows (b), sheep (o), and pigs (p) are shown in Figure 3. Wild-type ERβ (ERβ1) possesses both a DBD (C domain) and an LBD (E domain). ERβ2 codes for a variant that contains an additional 18 amino acids in the LBD, while ERβ1-δ3 lacks exon 3 and therefore a part of the DBD. ERβ2-δ3 contains both of these variations. ERβ1-δ5 lacks exon 5, while the C-terminal 61 amino acids are replaced by a unique sequence of 26 amino acids in $ERβ_{CX}$. ERβ4 is truncated at both the N and the C termini. In humans, variants lacking exon 2, exon 4, exon 6, and exon 7 also exist [21].

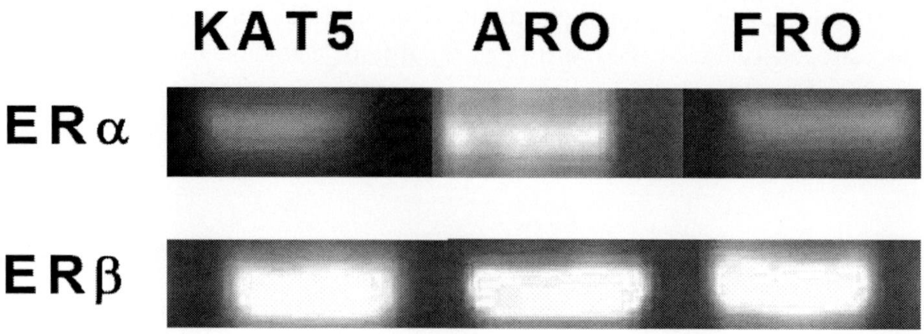

Figure 1. Expression of ERα and ERβ in thyroid cancer cells. Total RNA from KAT5, ARO and FRO cell lines was isolated and RT-PCR was performed to determine the expression of ERα and ERβ. The expected sizes for ERα and ERβ were 306bp and 229bp respectively.

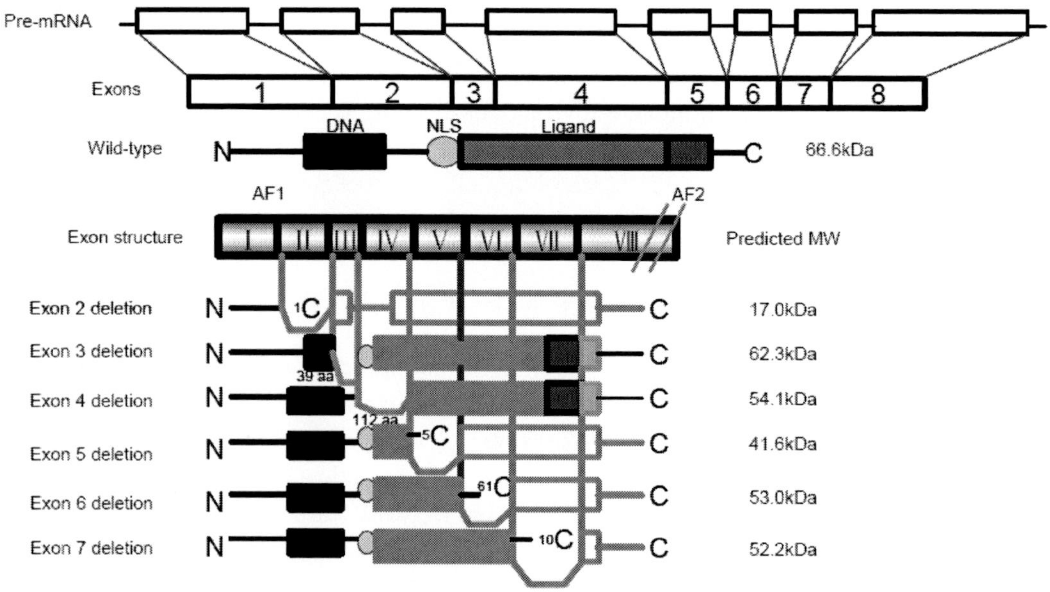

Figure 2. The splice variants of ERα. The most common splice variants of ERα are expressed in multiple tissues and arise from deletions of internal exons, resulting in truncated proteins lacking segments of the DNA-binding domain (DBD) or hormone-binding (ligand-binding) domain (LBD) of the receptor.

Because these splice variants are frequently co-expressed with their wild-type counterparts, the exact function and potential role of these variants in normal physiology and human disease remain to be elucidated. ERα and ERβ differ mostly in the N-terminal A/B and F domains, exhibiting a 15% and 18% identity respectively [13,22,23]. The ligand binding domain (E domain) is moderately conserved between both receptors and shows only a 59% amino acid identity. Despite the identity between ERα and ERβ in the ligand binding domain, the two receptors exhibit subtle differences in ligand binding specificity.

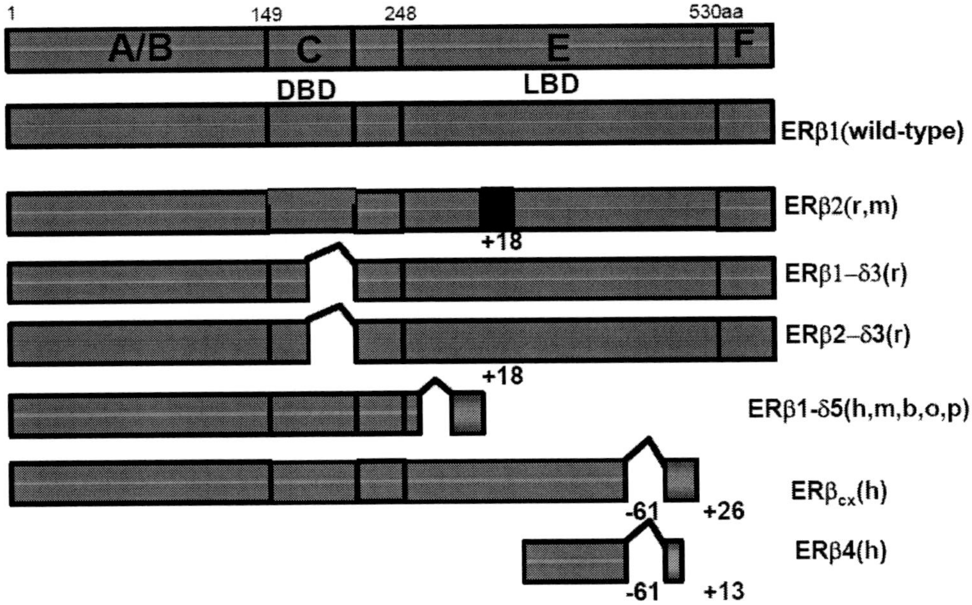

Figure 3. The splice variants of ERβ. Mammalian ERβ variants identified in humans (h), rats (r), mice (m), cows (b), sheep (o), and pigs (p).

A K_D for the ligand 16α-iodo-E2 of 0.6 nM for ERα and of 0.24 nM for ERβis similar to the range of estradiol binding to ERs (0.1–1 nM) in various systems [24]. Despite the slight differences in the affinity for the ligand 16α-iodo-estradiol of ERα and ERβ, both receptors are considered to have a similar affinity for estradiol. However, differences are observed for other ligands, such as anti-estrogens and phytoestrogens. A wide variety of structurally distinct compounds bind to ERs with differing affinities. Certain ligands act as ER agonists, and these include the natural ligand estradiol, as well as the synthetic estrogen diethylstilbestrol (DES). Certain phytoestrogens, which are environmental compounds produced by plants, can also be estrogenic. Genistein, present in soya beans and soy products, is a widely utilized phytoestrogen. Other compounds, such as ICI 182,780, are receptor antagonists. A final group of mixed agonists and antagonists are the selective ER modulators and examples include tamoxifen and raloxifene. The relative binding affinity of ERα for various ligands are: DES (468)>4-hydroxy tamoxifen(4-OHT) (178)> estradiol (100)>ICI 164,384(85)>genistein. In contrast, ERβdisplays the following relative binding affinities: 4-OHT (339)>DES (295)>ICI 164,384 (166)> estradiol (100)>genistein (36) [14]. Therefore, the ligand binding specificity and transcript tissue distribution of ERα and ERβ is different. These differences can result in functional consequences for a receptor subtype.

2.3. General Function of Estrogen and Estrogen Receptors

The classical ER-mediated gene transcription is stimulated by more than two distinct transactivation domains located in the amino-terminal A/B region (AF-1) and the carboxy-terminal E region of the receptor (AF-2) [25-27]. The AF-1 domain is hormone-independent, whereas the AF-2 domain is hormone-dependent [28]. Both AF-1 and AF-2 are probably required for maximal ER transcriptional activity [29]. The nuclear estrogen-ER complex binds to estrogen response element sequences in the promoter region of estrogen-responsive genes directly or indirectly, resulting in the recruitment of coregulatory proteins (co-activators or corepressors) to the promoter, in an increase or decrease in mRNA levels, in the production of associated proteins and finally in a physiological response. However, AF-1 and AF-2 can function independently with certain promoters [29]. Recently, it has been demonstrated that the activity of the ERβ AF-1 region is negligible compared with the ERαAF-1 region, while their AF-2 activities are comparable [30]. Thus, the activity of ERα may exceed that of ERβ on estrogen responsive element–containing genes that require both AF-1 and AF-2 transactivation domains. It is also known that anti-estrogens such as tamoxifen can function as pure antagonists on genes that require only the AF-2 domain for ER-mediated transcriptional activity. In contrast, in genes where the ER AF-2 domain is not required, transcription is driven by AF-1 only and tamoxifen can function as a partial agonist [29,31]. It is the capacity of many anti-estrogens to act differently on the two transactivation domains that, in part, make them clinically useful antagonists in the breast but agonists in the other organs such as the uterus, bone, and liver. This classical or 'genomic' mechanism typically occurs over the course of hours.

Besides this classical mechanism of direct DNA binding, the two ER subtypes can also function via other pathways [32,33]. It has been found that estrogen can act within seconds or minutes. The quick action, seconds, is termed 'nongenomic', and occurs either through the ER located in or adjacent to the plasma membrane, or through other non-ER plasma membrane–associated estrogen-binding proteins. The nongenomic action of estrogen results in cellular responses such as increased levels of Ca^{2+} or nitric oxide and the activation of kinases. The ER may be targeted to the plasma membrane by adaptor proteins such as caveolin-1 or Shc (an SH2 containing proto-oncogene). AP-1 response elements, for instance, are regulated indirectly through interactions between ERs and the AP-1 transcription factors c-fos and c-jun [34]. These transcription factors regulate genes involved in many cellular processes, including proliferation, differentiation, cell motility and apoptosis. Thus, the ER–AP-1 interaction can be pathologically important. It has been shown that tamoxifen can act as an agonist on genes under the control of an AP-1 response element when ERα and ERβ are expressed. Estradiol, in contrast, does not activate these genes when ERβ alone is expressed [35]. This finding may have important implications for the differential effects of selective ER modulators.

Our group has demonstrated that estradiol and PPT (a potent selective ERα agonist) can promote the expression of ERα in thyroid KAT5 cells. We showed that estradiol activates ERα and that PPT has a weak effect in thyroid ARO cells. The expression of ERβ in thyroid FRO cells gradually increased after treatment with estradiol and DPN (a potent selective ERβ agonist) (our unpublished data). Our data suggest that estradiol can selectively induce the expression of ERα in thyroid cells, which may not be the case in murine mammary epithelial cells [36]. Therefore, it is likely that different types of cells may respond differently to estradiol. The differential sensitivity of various cells to estradiol may be physiologically and pathologically significant.

3. Estrogen and Apoptosis in Thyroid Cancer

3.1. Differential Roles of ERα and ERβ in Apoptosis

Homeostasis in normal tissues is regulated by a balance between cell proliferation and apoptosis. Apoptosis is a physiological mechanism of cell death closely regulated by cytokines and hormones [37,38]. It plays an important role in organ and cancer development [39]. The morphological features of apoptosis were first described 30 years ago and are characterized by cell shrinkage, nuclear condensation, DNA fragmentation into oligonucleosome fragments and membrane blebbing [39]. The apoptotic process requires the activation of a family of cysteine proteases called caspases. The activation of caspases, under many different circumstances, irreversibly commits a cell to death. For example, perturbations in the cell cycle or cellular metabolism, growth factor availability and genotoxic agents that result in the loss of mitochondrial integrity can initiate cell death [41]. In the absence of mitochondrial dysfunction, an otherwise healthy cell can be committed to death through the engagement of cell surface death receptors. Therefore, cell death can be categorized according to the initiation, either cell surface death receptor engagement or mitochondrial dysfunction.

Estrogens have long been recognized as an important factor that contributes to the development of various human diseases and may mediate its effect through ERαand ERβ. The disruption of the estrogen pathway may serve as the basis for therapeutic interventions.

The role of ERα and ERβ in thyroid cancer is still unknown. It is well recognized that the incidence of human thyroid disorders is higher in females than in males [42-44]. Although thyroid stimulating hormone (TSH) plays an important role in regulating the growth and biological function of thyroid cells, the effect of age and sex on the incidence and prognosis of thyroid neoplasms cannot be explained by an increased level of serum TSH [2,45]. Epidemiological data as well as clinical and experimental evidence indicates that sex hormones play a role in the development of thyroid neoplasms. Thus, if sex hormones can affect thyroid tissue directly, thyroid cells should contain receptors for sex hormones [46,47]. Data from our laboratory have shown that human thyroid tumor cells are able to express both ERα and ERβ [20]. Other reports show that ERs are expressed in different parts of the thyrocyte and by different types of thyroid cells. Very few follicular cells are positive for ERαin the normal thyroid gland of both women and men. ERαimmunoreactivity is not detected in the nucleus of follicular cells in developing fetal thyroid glands. However, ERαimmunoreactivity is detectable in the nucleus of parenchymal cells in thyroid disorders [48]. The expression of ERα mRNA is notable in follicular adenomas and is marginally decreased in anaplastic thyroid carcinomas. ERαis also detectable in medullary carcinomas and similar levels have been seen in papillary and follicular carcinomas [49]. Estradiol has been shown to stimulate the proliferation of FRTL-5 rat cells, a well-established cell line derived from thyroid papillary carcinoma [50] and of HTC-TSHr cells of a human thyroid papillary carcinoma cell line [51]. Our research also supports the significant association between estrogen and cell growth in human papillary thyroid carcinoma KAT5 cells. These results suggest that higher estrogen concentrations in the serum may be responsible for the relatively higher incidence of papillary thyroid carcinoma in premenopausal women. In addition, a higher level of ERα labeling may also be involved in the tumorigenesis of papillary carcinoma in premenopausal women [48], but it needs further investigation for clarification. Similar results are found in breast cancer studies. As half of all breast cancers overexpress ERα, estradiol is implicated in the development of breast cancer based on data from both clinical and animal studies. The presence of elevated levels of ERα in benign breast epithelium appears to predispose a woman to an increased risk of breast cancer, suggesting a role for ERα in breast cancer initiation. Most resistant breast tumors remain ERα-positive and often respond to endocrine treatment, indicating the continued role of ERα in breast cancer cell proliferation [52].

ERβ nuclear immunoreactivity has been detected widely in both follicular, parenchymal and C-cells of normal thyroid glands as well as in the developing fetal thyroid gland. In addition, ERβimmunoreactivity has been detected in the nuclei of various thyroid lesions [48]. ERβmRNA expression levels are significantly higher in follicular carcinomas and ratios of ERβtoERαmRNA expression showed a significant decrease in follicular adenomas compared with normal thyroid tissue. Some results suggest that the estrogenic effect via ERβ plays a role in the maintenance of homeostasis of thyroid follicles, and possibly of thyroid hormone biosynthesis [48]. In our study, an increase in ERβ contributed to apoptosis in KAT5 cells. It is likely that ERβ inhibits human breast cancer MDA-MB-231 cell proliferation in a ligand-independent manner, whereas ER inhibition of proliferation is hormone dependent. ERβ is a negative modulator of breast cancer cell proliferation and invasion, supporting the concept that a loss of ERβ expression could be an event that leads to the development of breast cancer and that alterations in the ratio of ERα to ERβ may govern tumor development [53,54]. In ovarian cancers, although ERα is the main subtype expressed [55], ERβ might play a protective role against ERα mitogenic activity [56]. A decrease in ERβ mRNA is inversely correlated with tumor progression when compared to ERα mRNA [57]. The ratio of ERα mRNA to ERβ mRNA is markedly increased in ovarian cancer [57]. The loss of ERβ has been shown to be a marker of cell dedifferentiation in many studies. Molecules specifically activating ERβ or inducing ERβ re-expression may be beneficial by blocking tumor proliferation or invasion in ovarian cancers. Therefore, estrogen-induced gene expression, especially through ERβ, may constitute a new therapeutic target in cancers [56,58]. The levels of ERβ protein are reported to be lower in colon tumors than in normal colon tissue, and the loss of ERβ is associated with advanced stage colon cancer and tumor cell dedifferentiation which suggests a protective role for ERβ in colon tumorigenesis [59].

3.2. Bcl-2 Family

Bcl-2 family proteins play a central role in controlling mitochondrial-mediated apoptosis. In humans, more than 20 members of this family have been identified including proteins that suppress apoptosis (Bcl-2, Bcl-XL, Mcl-1, Bfl-1/A1, Bcl-W, Bcl-G) and proteins that promote apoptosis (Bax, Bak, Bok, Bad, Bid, Bik, Bim, Bcl-Xs, Krk, Mtd, Nip3, Nix, Noxa, Bcl-B) [60-63]. Bcl-2 proteins localize or translocate to the mitochondrial membrane and modulate apoptosis by permeabilization of the inner and/or outer membrane, leading to the release of cytochrome c or stabilization of the barrier function. The anti-apoptotic protein

Bcl-XL binds and inactivates the caspase adapter Apaf-1. This anti-apoptotic effect is balanced by pro-apoptotic Bcl-2 proteins that displace Bcl-xL from Apaf-1 [64]. The amount of individual Bcl-2 proteins in a cell depends on several variables including the lineage of the cell, presence of activated transcription factors, the presence of nonendogenous chemicals and estrogen responsiveness.

Expression of the anti-apoptotic protein Bcl-2 is increased by estradiol in estrogen-responsive MCF-7 cells, which were derived from the pleural effusion of a patient with metastatic breast cancer [65-67]. Environmental estrogens, including pesticides and alcohol, are thought to mimic this effect and have been shown to suppress TNF-induced apoptosis in MCF-7 cells but not in ER-negative cells [66]. Some studies suggest that high levels of Bcl-2 and low levels of Bax correlate with decreased protease activity in MCF-7 variants [68]. However, the imbalance between Bcl-2 and Bax may not entirely account for the intrinsic apoptotic resistance of MCF-7 cells.

It has been shown that estradiol can induce the reappearance of ER□ gene and protein expression on the ipsilateral side of an injured brain [69]. This gene is expressed at high levels during neonatal development when the cortex undergoes neurite outgrowth and synaptogenesis but decreases during later stages of development and in adulthood. Injury-induced re-appearance of the ERα gene and protein expression may indicate that the injured brain provides signals that communicate the need for expression of a receptor that is required for the re-initiation of 'developmental' events in the injured cortex [70,71]. Data obtained from ERα and ERβ knockout mice models have demonstrated that when the ERα is absent, estradiol no longer protects against middle cerebral artery occlusion-induced cell death. In contrast, estradiol continues to exert its protective effects in the absence of ERβ. These studies have clearly established that the classic estrogen receptor, ERα, which was once thought to be more important in reproductive functions than in nonreproductive functions, is the critical mechanistic link in the ability of low levels of estradiol to protect against delayed cell death [72]. Estradiol decreases Bcl-2 influenced cell death in ischemic brain injury. Estradiol prevents the injury-induced down-regulation of Bcl-2 expression on the ipsilateral side of the brain. These effects were specific to bcl-2, because expression of other members of the Bcl-2 family (Bax, Bcl-xL, Bcl-xS, Bad, and Bim) was unaffected in estradiol-treated rats [73].

Bcl-2 family members are known to be altered in thyroid cancer [74] and their levels are known to be regulated by estrogen in some cell systems [37]. Data from our laboratory showed that the anti-apoptotic protein Bcl-2 was up-regulated by estradiol and ERα agonist PPT but down-regulated by ERβ agonist DPN in KAT5 cells (Figure 4), suggesting that ERα induced Bcl-2 expression whereas ERβ reduced it. The role of estradiol and ERα in the expression of Bcl-2 was further supported by another experiment, in which the up-regulatory effect of estradiol was prevented by both the general ER antagonist ICI182780 and the ERα-selective antagonist MPP. In contrast to Bcl-2, the pro-apoptotic protein Bax was increased by DPN (Figure 5). However, estradiol and PPT did not have an obvious effect on Bax expression.

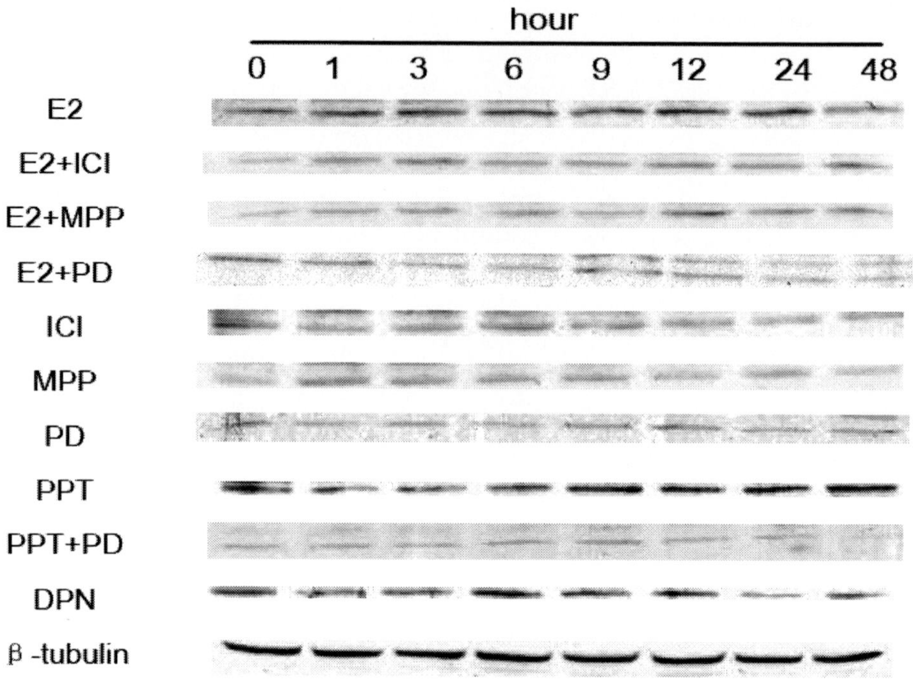

Figure 4. The expression of Bcl-2 in KAT5 cells. The expression of Bcl-2 protein (28 kDa) in response to 10^{-8}M E2 and its related agents was investigated by Western blot analysis in KAT5 cells. The representative results were shown. Compared with untreated cells, a significant increase was observed in the cell treated with E2 from 9h to 24h, and with 10^{-6}M PPT from 12h to 48h. In contrast, a significant decrease was observed in the cell treated with 10^{-6}M DPN from 24h to 48h.

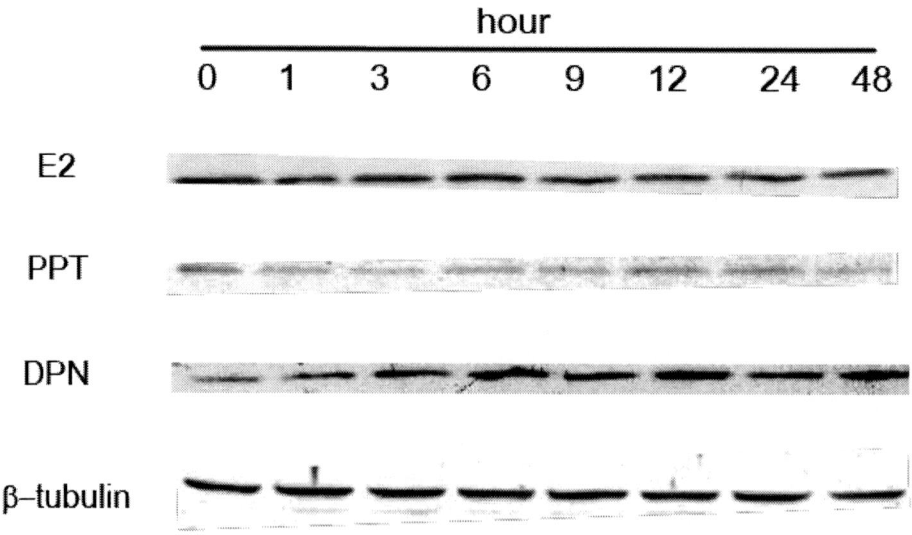

Figure 5. The expression of Bax in KAT5 cells. The expression of Bax protein (23kDa) in response to 10^{-8}M E2 and 10^{-6}M PPT and 10^{-6}M DPN was investigated by Western blot analysis in KAT5 cells. The representative results were shown. Compared with untreated cells, a significant increase was observed in the cell treated with DPN from 9h to 24h.

These results suggest that ERβ and not ERα promotes the level of Bax. It is well known that apoptosis regulated by Bcl-2 family proteins is closely associated with mitochondria, and their aberrant expression usually results in a change in the mitochondrial-mediated apoptotic pathway [20]. Therefore, it is possible that estradiol, ERα and ERβ regulate cell growth and death through a mitochondrial-mediated pathway.

3.3. Mitogen-Activated Protein Kinase Cascades

Stimulation of a variety of tyrosine kinase receptors leads to a rapid elevation of the enzymatic activity of a family of closely related serine-threonine kinases, known as MAP kinases. These kinases are able to convert extracellular stimuli to intracellular signals that control gene expression, and the products of these genes are able to regulate cell proliferation and differentiation [75,76]. MAP kinases have been classified into three subfamilies: ERKs, including ERK1 and ERK2 (also known as p44 MAPK and p42 MAPK respectively); SAPKs, also called c-jun N-terminus kinases (JNKs); and p38 kinase, the homologue of the Saccharomyces cerevisiae HOG1 gene. These MAPKs are members of separate modules and are regulated by distinct extracellular stimuli. For example, ERKs are activated by receptor tyrosine kinases and provide proliferation or differentiation signals. JNK and p38-type MAPKs are activated predominantly by stress stimuli and pathogenic insults but also by mitogens in some cell types. All 3 classes of MAPKs are involved in the regulation of distinct AP-1 components. C-Jun is regulated by JNK phosphorylation and also by ERK-mediated mechanisms in some cell types. c-Fos is a substrate for regulatory phosphorylation by ERK, and ATF-2 is regulated by JNK and p38 kinases. Due to this complex regulation of AP-1, the range of biological responses is broad. Recent studies have found that breast cancer frequently contains an increased proportion of cells with the activated form of MAP kinase. Activated MAP kinase provides an integrated measure of several biologic processes regulating cancer growth, which is considered to be a key molecule regulating breast cancer growth and apoptosis [76]. Therefore, it is worth examining how estrogen affects MAP kinase and how such an effect contributes to thyroid cancer cell proliferation and growth.

Several investigators have demonstrated that the functions of Bcl-2 family proteins are regulated by a number of protein kinases including ERK1/2 [77-79], although evidence for this in human thyroid cancer cells is lacking. It has been shown that a loss of pERK1/2 leads to a decrease in the expression of Bcl-2, and that the activation of the ERK pathway leads to the induction of Bcl-2 in murine B–cell lymphoma cells [77]. It has also been shown that activation of ERK1/2 protects gastric epithelial cells from apoptosis by maintaining Bcl-2 expression [78]. In rat hippocampal neuron cells, increased calcium influx caused by estradiol results in the activation of ERK, which subsequently enhances the level of Bcl-2 [79]. Data obtained in this study indicated that estradiol can induce the phosphorylation of ERK1/2. This finding of up-regulation of ERK1/2 phosphorylation by estradiol is further supported by the experiment in which the ER antagonist ICI182780 blocked this action of estradiol. We further explored the possible role of ER subtypes in the activation of ERK1/2 by estradiol (Figure 6).

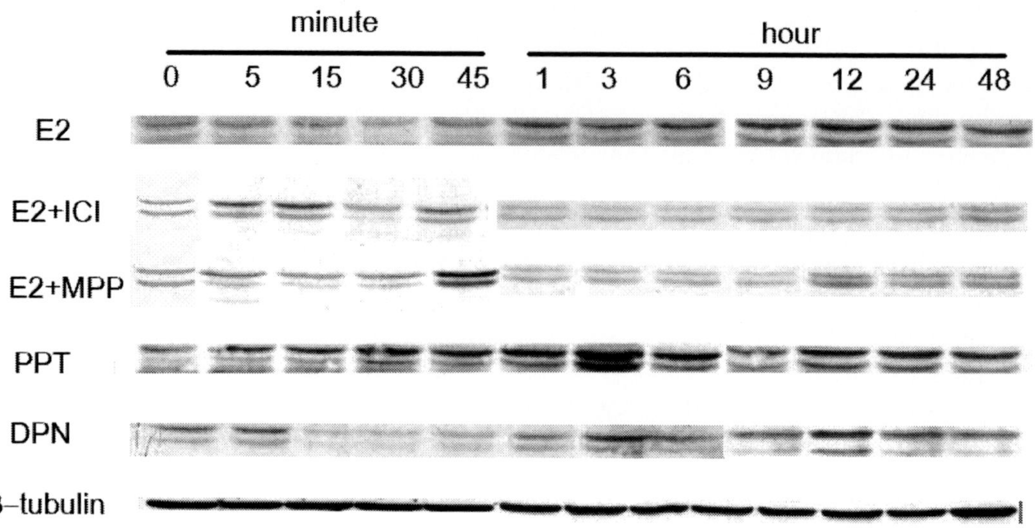

Figure 6. The expression of pERK1/2 in KAT5 cells. The expression of pERK1/2 protein (42/44 kDa) in response to 10^{-8}M E2 and its related agents was investigated by Western blot analysis in KAT5 cells. The representative results were shown. Compared with untreated cells, a significant increase was observed in the cell treated with E2 at 12h, and with 10^{-6}M PPT at 1h and 3h. However, a significant decrease was observed in the cell treated with 10^{-6}M DPN at 15min and 30min. The administration of either 10^{-6}M MPP or 10^{-6}M ICI could prevent the promoting effect of E2 on pERK1/2 protein expression.

The results suggest that the activation of ERK1/2 by estradiol was mainly due to the interaction between estradiol and ERα̃ The selective ERαantagonist MPP suppressed the effect of estradiol on the phosphorylation of ERK1/2 and its inhibition was similar to that by the general ER antagonist ICI182780. Therefore, our findings indicate that ERK1/2 activation by estradiol is positively associated with ERα in thyroid cancer cells.

Although we did not directly investigate the consequence of ERK1/2 activation on thyroid cancer cell proliferation and survival in this experiment, we showed that the inhibition of ERK1/2 by PD098059 could significantly reduce the level of Bcl-2 induced by both estradiol and the ERα agonist PPT, and that the increase in ERK1/2 occurred much earlier than the change in levels of Bcl-2 in thyroid cancer cells treated with estradiol and PPT (Figure 4).

These results suggest that the activation of ERK1/2 by either estradiol or ERα favors the expression of anti-apoptotic Bcl-2 and that the alteration in Bcl-2 expression is a downstream event in the estradiol /ERα-ERK1/2 pathway. These results are also consistent with our previous data which show that ERK1/2 activation protects thyroid cancer cells from apoptosis, as they were the same type of cells used in this study [80]. To the best of our knowledge, the activation of ERK1/2 by estradiol or ERα has not been reported in thyroid cancer cells. However, the relationship between ERK1/2 and estradiol or ERs has been investigated in several other cell systems including human breast cancer cells, human pheochromocytoma cells, rat hippocampal neuron cells and mouse male germ cells [81-84]. These studies all demonstrated that ERK1/2 activation provides a survival signal for cell proliferation and growth, and are thus in agreement with our findings in human thyroid cancer cells.

It is well known that estradiol can function through either genomic or non-genomic signaling pathways. Our data suggest that estradiol most likely activates ERK1/2 via a non-genomic signaling pathway. The increased phosphorylation of ERK1/2 occurs as early as 15 minutes after the treatment of cells with the ERα-selective agonist PPT, and an obvious elevation can be detected 1h after treatment. Similarly, the ERβ-selective agonist DPN also acts within 15 minutes to down-regulate the phosphorylation of ERK1/2 (Figure 6). These results agree with a recent study by Acconica F *et al.* [85] which showed that estradiol can promote Hela cell survival through ERα non-genomic signal activation of the ERK/MAPK pathway and cell death through ERβ non-genomic signal inhibition of the ERK/MAPK pathway. The study from our laboratory demonstrated a novel mechanism by which estradiol contributes to the proliferation and growth of thyroid cancer cells. Estradiol favors the expression of ERα over ERβ, and so causes an imbalance between an increased ERα and a decreased ERβ. This imbalance may change the behavior of thyroid cells, giving them the ability to proliferate and survive by enhancing ERK1/2 activity and subsequently the ratio of anti-apoptotic Bcl-2 to pro-apoptotic Bax.

As the activation of the MAP kinase pathway by various genetic alterations plays a pivotal role in thyroid tumor genesis and progression, efforts to target the inhibition of this pathway may lead to the arrest of cancer cell growth and induce apoptosis and thus be a novel and effective therapy for thyroid cancer [86]. AEE788, a novel dual specific EGFR and VEGFR inhibitor, can attenuate the activation status of the critical downstream targets involved in proliferation through MAPK, which agrees with a report by Chen KT, *et al.* [87] which showed that while total MAPK protein levels remained unaltered, phosphorylated MAPK levels were overexpressed in metastatic thyroid cancers compared with primary tumor samples.

3.4. NF-κB and Cellular Stress

Nuclear factorκBNF-κB) is a nuclear transcription factor that regulates the expression of a large number of genes which have an influence on apoptosis, viral replication, tumorigenesis, inflammation and immunological responses. In its inactive form, NF-κB is sequestered in the cytoplasm, bound to members of the IκB family of inhibitor proteins which include IκBα, IκBβ, IκBγ and IκBε. Various stimuli that activate NF-κB cause phosphorylation of IκB, which is followed by its ubiquitination and subsequent degradation. This results in the exposure of the nuclear localization signals (NLS) on NF-κB subunits and the subsequent translocation of the molecule into the nucleus. In the nucleus, NF-κB binds with a consensus sequence (5'GGGACTTTCC-3') of various genes and thus activates their transcription. NF-κB transcription factors are heterodimer and homodimer complexes of related proteins that contain a Rel homology domain involved in specific DNA binding, protein dimerization and nuclear import [88]. The Rel proteins, predominantly found in mammalian cells, consist of 2 transcriptionally inactive forms, NF-κB1 (p50) and NF-κB2 (p52), and NF-κB3 transcriptionally active subunits, RelA (p65), c-Rel and RelB [88].

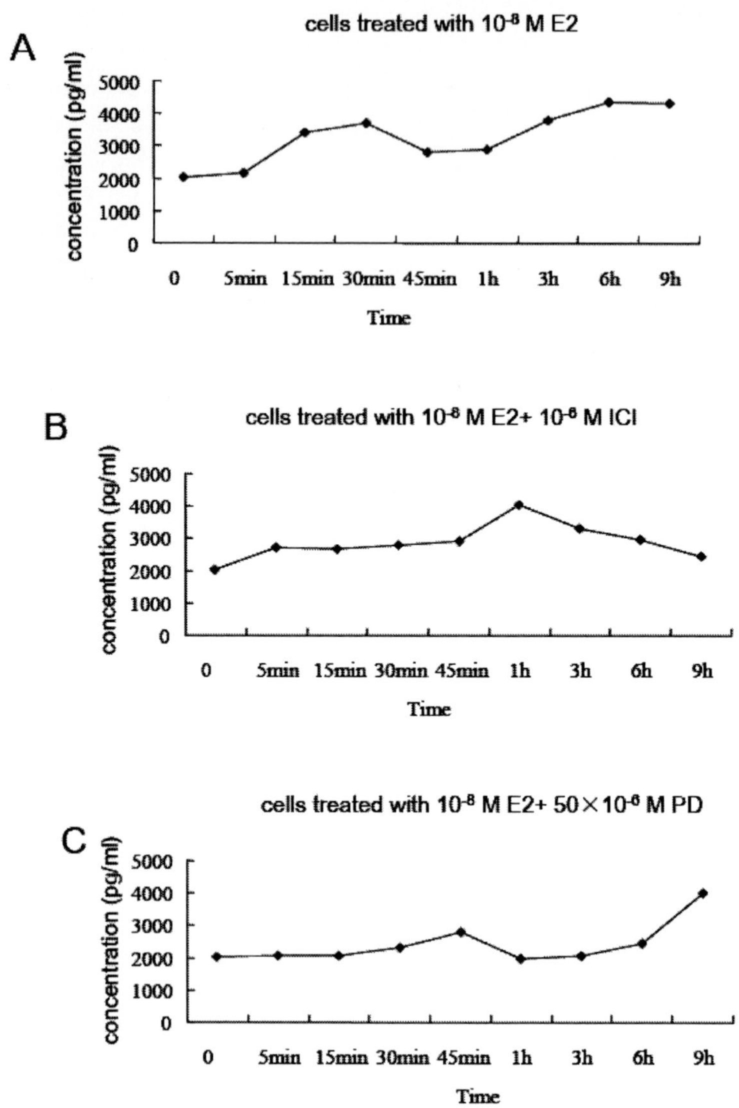

Figure 7. NF-*k*B activity in KAT5 cells. The colorimetric enzyme immunoassay kit was used to detect the activated NF-*k*B protein (p50) in the nucleus of KAT5 cells. The concentration of activated NF-*k*B was significantly increased in the cells treated by E2 and its activity reached the maximal level at 9 hours after the treatment. The administration of either 50 x 10^{-6}M PD098059, a specific inhibitor of activation of MAPK or 10^{-6}M ICI182780, a nonselective ER antagonist could prevent the promoting effect of E2 on the activity of NF-*k*B.

MAP kinases are known to activate NF-κB. Estradiol can influence MAPK through ERs as described in the previous section.

NF-κB may upregulate the expression of both manganese-superoxide dismutase (Mn-SOD) and glutathione peroxidase (GPx) genes through the interaction with their gene promoters which contain putative NF-κB-binding motifs [89]. On the other hand, an increased antioxidant level can inhibit the activity of NF-κB. For example, pyrrolidine dithiocarbamate (PDTC), an inhibitor of the IκB degradation, prevents NF-κB translocation into the nucleus.In contrast, increased stress/oxidant levels activate NF-κB [90]. Using two inhibitors, PDTC and UO126 (an inhibitor of the phosphorylation of these kinases), estradiol has been shown to upregulate the expression of Mn-SOD and GPx as follows. First, it interacts with the membrane ERs, then it activates MAP kinases and NF-κB and finally it upregulates antioxidant gene expression [89]. A high antioxidant level may make cancer cells less likely to undergo apoptosis [91], which agrees with the general function of NF-κB as a pro-survival signal in various types of cells [92]. Indeed, increased antioxidant levels have been documented in some cancers such as prostate and breast [93]. Data from our laboratory supports this concept that estradiol activates NF-κB via a pathway that requires the activation of MAKP in human thyroid cancer cells, since estradiol increases p50, a subunit of NF-κB. This promoting effect disappears in the presence of PD098059, a specific inhibitor of MAPK (Figure 7). However, it seems that ERs are not absolutely necessary for the action of estradiol because the activity of NF-κB can still be increased by estradiol in the presence of ICI182780, a nonselective ER antagonist, although to a lesser degree than by estradiol alone (Figure 8B). Furthermore, the enhanced activity of NF-κB by estradiol may also lead to an increased level of Bcl-2, as the expression of Bcl-2 is known to be positively regulated by NF-κB [94]. The fact that estradiol actually promotes the level of Bcl-2 in thyroid cancer cells as described in the previous section is strong evidence to support the positive role played by estradiol and NF-κB in the downregulation of apoptosis in human thyroid cancer cells. Further experiments are needed to clarify the role of ERs in the pathways described above.

4. Conclusion

It appears that estradiol can interact with distinct and compartmentalized pools of receptors, each of which may have unique effects on thyroid cancer cells. Under the influence of estradiol, cells can respond in two different ways [95]. Firstly, estradiol can interact with membrane ERαand ERβ which results in a rapid modification of proteins. Secondly, there is a slow response via nuclear receptors, which requires protein synthesis and function. In fact, both the rapid and slow responses can be exquisitely coordinated and be co-functional in physiological and pathological conditions. This combination of actions is thought to optimally regulate the proliferation and apoptosis of thyroid cancer cells. It is also possible that the large cytosolic pool of ERs, envisioned to be a reservoir prior to moving into the nucleus, can in fact act on cytosolic organelles. Further research into the function of ERα and ERβin mitochondria and their role in apoptosis in human thyroid cancer cells is needed.

Acknowledgments

The study was supported by a grant from the Research Grants Council of Hong Kong (Project No.: CUHK4556/05M).

References

[1] Fraker, D. L., Skarulis, M. and LiVolsi, V. (1997). In V. T. DeVita Jr., S. Hellman, and S. A. Rosenberg (Eds.), *Cancer: Principles and Practice of Oncology* (5th ed, pp. 1629-1652). Philadephia: Lippincott-Raven.

[2] Clark, O. H. (1981). TSH suppression in the management of thyroid nodules and thyroid cancer. *World J. Surg.*, *5*, 39-47.

[3] Greenlee, R. T., Hill-Harmon, M. B., Murray, T. and Thun, M. (2001). Cancer statistics, *CA. Cancer J. Clin.*, *51*, 15.

[4] Ron, E., Kleinerman, R. A., Boice Jr, J. D., Livolsi, V. A., Flannery, J. T. and Fraumeni, J. F. (1987). A population-based case-control study of thyroid cancer. *J. Natl. Cancer Inst.*, *79*, 1–12.

[5] Preston-Martin, S., Bernstein, L., Pike, M. C., Maldonado, A. A. and Henderson, B. E. (1987). Thyroid cancer among young women related to prior thyroid disease and pregnancy history. *Br. J. Cancer*, *55*, 191–195.

[6] McTiernan, A. M., Weiss, N. S. and Daling, J.R. (1984). Incidence of thyroid cancer in women in relation to previous exposure to radiation therapy and history of thyroid disease. *J. Natl. Cancer Inst.*, *73*, 575–581.

[7] Persson, I., Yuen, J., Bergkvist, L. and Schairer, C. (1996). Cancer incidence and mortality in women receiving estrogen and estrogen- progestin replacement therapy–long-term follow-up of a Swedish cohort. *Int. J. Cancer*, *67*, 327–332.

[8] Vlaeminck-Guillem, V., d'Herbomez-Boidein, M., Decoulx, M. and Wemeau, J. L. (2001). Apoptosis and the thyroid: the Fas/Fas L pathway. *Presse Med.*, *30*, 74-80.

[9] Telang, N. T., Katdare, M., Bradlow, H. L. and Osborne, M. P. (1997). Estradiol metabolism: an endocrine biomarker for modulation of human mammary carcinogenesis. *Environ. Health Perspect.*, *105*, 559-564.

[10] Speroff, L., Glass, R. H. and Kase, N. G. (1999). *Clinical gynecologic endocrinology and infertility* (6th ed). Philadelphia: Lippincott Williams and Wilkins.

[11] Ruggiero, R. J. and Likis, F. E. (2002). Estrogen: physiology, pharmacology, and formulations for replacement therapy. *J of Midwifery and Women's Health*, *47*, 130-138.

[12] Jones, K. P. (1992). Estrogens and progestins: what to use and how to use it. *Clin Obstet. Gynecol*, *32*, 871–883.

[13] Enmark, E., Pelto-Huikko, M., Grandien, K., Lagercrantz, S., Lagercrantz, J., Fried, G., Nordenskjold, M. and Gustafsson, J. A. (1997). Human estrogen receptor beta-gene structure, chromosomal localization, and expression pattern. *J. Clin. Endocrinol. Metab.*, *82*, 4258–4265.

[14] Kuiper, G. G., Carlsson, B., Grandien, K., Fujimori, K., Naganuma, H., Inoue, S., Kinouchi, Y., Kameyama, K., Takami, H., Shimosegawa, T. and Sasano, H. (1997). Comparison of the ligand binding specificity and transcript tissue distribution of estrogen receptors alpha and beta. *Endocrinology*, *138*, 863–870.

[15] Gustafsson, J. A. (1999). Estrogen receptor beta—a new dimension in estrogen mechanism of action. *J. Endocrinol.*, *163*, 379–383.

[16] Taylor, A. H. and Al-Azzawi, F. (2000). Immunolocalisation of oestrogen receptor beta in human tissues. *J. Mol. Endocrinol.*, *24*, 145–155.

[17] Hiroi, H., Inoue, S., Watanabe, T., Goto, W., Orimo, A., Momoeda, M., Tsutsumi, O., Taketani, Y. and Muramatsu, M. (1999). Differential immunolocalization of estrogen receptor alpha and beta in rat ovary and uterus. *J. Mol. Endocrinol.*, *22*, 37–44.

[18] Sar, M. and Welsch, F. (1999). Differential expression of estrogen receptor-beta and estrogen receptor-alpha in the rat ovary. *Endocrinology*, *140*, 963–971.

[19] Weihua, Z., Warner, M. and Gustafsson, J. A. (2002). Estrogen receptor beta in the prostate. *Mol. Cell Endocrinol.*, *193*, 1–5.

[20] Lee, M. L., Chen, G. G., Vlantis, A. C., Tse, G. M. K., Leung, B. C. H. and van Hasselt, C. A. (2005). Induction of thyroid papillary carcinoma cell proliferation by estrogen is associated with an altered expression of Bcl-xL. *Cancer J.*, *11*, 113-121.

[21] Deroo, B. J. and Korach, K. S. (2006). Estrogen receptors and human disease. *J. Clin. Invest.*, *116*, 561-570.

[22] Menasce, L. P., White, G. R., Harrison, C. J. and Boyle, J. M. (1993). Localization of the estrogen receptor locus (ESR) to chromosome 6q25.1 by FISH and a simple post-FISH banding technique. *Genomics*, *17*, 263–265.

[23] Kelley, S. T. and Thackray, V. G. (1999). Phylogenetic analyses reveal ancient duplication of estrogen receptor isoforms, *J. Mol. Evol.*, *49*, 609–614.

[24] Kuiper, G. G., Lemmen, J. G., Carlsson, B., Corton, C. J., Safe, S. H., van der Saag, P. T., van der Burg, B., and Gustafsson, J-A. (1998). Interaction of estrogenic chemicals and phytoestrogens with estrogen receptor beta. *Endocrinology*, *139*, 4252–4263.

[25] Kumar, V., Green, S., Staub, A. and Chambon, P. (1986). Localisation of the oestradiol-binding and putative DNA-binding domains of the human oestrogen receptor. *EMBO J.*, *5*, 2231-2236.

[26] Kumar, V., Green, S., Stack, G., Berry, M., Jin, J. R. and Chambon, P. (1987). Functional domains of the human estrogen receptor. *Cell*, *51*, 941-951.

[27] Tora, L., White, J., Brou, C., Tasset, D., Webster, N., Scheer, E. and Chambon, P. (1989). The human estrogen receptor has two independent nonacidic transcriptional activation functions. *Cell*, *59*, 477-487.

[28] Webster, N. J. G., Green, S., Tasset, D., Ponglikitmongkol, M. and Chambon, P. (1989). The transcription activation function located in the hormone-binding domain of the human oestrogen receptor is not encoded by a single exon. *EMBO J.*, *8*, 1441-1446.

[29] Tzukerman, M. T., Esty, A., Santisomere, D., Danielian, P., Parker, M. G., Stein, R. B., Pike, J. W. and McDonnell, D. P. (1994). Human estrogen receptor transactivational capacity is determined by both cellular and promoter context and mediated by two functionally distinct intramolecular regions. *Molecular Endocrinology*, *8*, 21-30.

[30] Cowley, S. and Parker, M. (1999). A comparison of transcriptional activation by ER alpha and ER beta. *J. Steroid Biochem. Mol. Biol.*, *69*: 165-175.

[31] McDonnell, D. P., Clemm, D. L., Hermann, T., Goldman, M. E. and Pike, J. W. (1995). Analysis of estrogen receptor function in vitro reveals distinct classes of antiestrogens. *Mol Endocrinol*, *9*, 659-669.

[32] Pettersson, K., Grandien, K., Kuiper, G. and Gustafsson, J. A. (1997). Mouse estrogen receptor beta forms estrogen response element-binding heterodimers with estrogen receptor alpha. Mol *Endocrinol.*, *11*, 1486-1496.

[33] Cowley, S. M., Hoare, S., Mosselman, S., and Parker, M. G. (1997). Estrogen receptors alpha and beta form heterodimers on DNA. *J. Biol. Chem.*, *272*: 19858-19862.

[34] Webb, P., Lopez, G. N., Uht, R. M. and Kushner, P, J. (1995). Tamoxifen activation of the estrogen receptor/AP-1 pathway: Potential origin for the cell-specific estrogen-like effects of antiestrogens. *Mol. Endocrinol.*, *9*, 443-456.

[35] Paech, K., Webb, P., Kuiper, G. G., Nilsson, S., Gustafsson, J. A., Kushner, P. J. and Scanlan, T. S. (1997). Differential ligand activation of estrogen receptors ERalpha and ERbeta at AP1 sites. *Science*, *277*, 1508-1510.

[36] Helguero, L. A., Faulds, M. H., Gustafsson, J. A. and Haldosen, L.A. (2005). Estrogen receptors alfa (ERa) and beta (ERb) differentially regulate proliferation and apoptosis of the normal murine mammary epithelial cell line HC11. *Oncogene*, *24*, 6605–6616.

[37] Song, R. X. and Santen, R. J. (2003). Apoptotic action of estrogen. *Apoptosis*, *8*, 55-60.

[38] Thompson, C. B. (1995). Apoptosis in the pathogenesis and treatment of disease. *Science*, *267*, 1456–1462.

[39] Steller, H. (1995). Mechanisms and genes of cellular suicide. *Science*, *267*, 1445–1449.

[40] Kerr, J. F., Wyllie, A. H. and Currie, A. R. (1972). Apoptosis: A basic biological phenomenon with wide-ranging implications in tissue kinetics. *Br. J. Cancer*, *26*, 239–257.

[41] Talapatra, S. and Thompson, C. B. (2001). Growth factor signaling in cell survival: Implications for cancer treatment. *J. Pharmacol. Exp. Ther.*, *298*, 873–878.

[42] Cady B, Sedgwick CE, Meissner WA, Marvin S. Wool, M. S., Salzman, F. A. and Werber, J. (1979). Risk factor analysis in differentiated thyroid cancer. *Cancer*, *43*, 810-820.

[43] Crile, G., Pontius, K. I. and Hawk, W. A. (1985). Factors influencing the survival of patients with follicular carcinoma of the thyroid gland. *Surg. Gynecol. Obstet.*, *160*, 409-413.

[44] Fransilla, K. O. (1975). Prognosis in the thyroid carcinoma. *Cancer*, *36*, 1138-1146.

[45] Clark, O. H., Gerend, P. L., Cote, T. C. and Nissenson, R. A. (1981). Thyrotropin binding and adenylate cyclase stimulation in the thyroid neoplasms. *Surgery*, *90*, 252-261.

[46] Christianson, D., Roti, E., Vagenakis, A. G. and Braverman, L. E. (1981). The sex-related difference in serum thyrotropin concentration is androgen mediated. *Endocrinology*, *108*, 529-535.

[47] Farbota, L., Hofmann, C., Oslapas, R. and Paloyan, E. (1987). Sex horman modulation of serum TSH levels. *Surgery*, *102*, 1081-1087.

[48] Kawabata, W., Suzuki, T., Moriya, T., Fujimori, K., Naganuma, H., Inoue, S., Kinouchi, Y., Kameyama, K., Takami, H., Shimosegawa, T., Sasano, H. (2003). Estrogen receptors (alpha and beta) and 17beta-hydroxysteroid dehydrogenase type 1 and 2 in thyroid disorders: possible in situ estrogen synthesis and actions. *Mod. Pathol,*. *6*, 437-444.

[49] Egawa, C., Miyoshi, Y., Iwao, K., Shiba, E. and Noguchi, S. (2001). Quantitative analysis of estrogen receptor-alpha and -beta messenger RNA expression in normal and malignant thyroid tissues by real-time polymerase chain reaction. *Oncology*, *61*, 293-298.

[50] Furlanetto, T. W., Nguyen, L. Q. and Jameson, J. L. (1999). Estradiol increases proliferation and down-regulates the sodium/iodide symporter gene in FRTL-5 cells. *Endocrinology*, *140*, 5705-5711.

[51] Manole, D., Schildknecht, B. and Gosnell, B. (2001). Estrogen promotes growth of human thyroid tumor cells by different molecular mechanisms. *J. Clin. Endocrinol. Metab.*, *86*, 1072-1077.

[52] Ali, S., Coombes, R. C. (2000). Estrogen receptor alpha in human breast cancer: occurrence and significance. *J. Mammary Gland Biol. Neoplasia*, *5*, 271-281.

[53] Leygue, E., Dotzlaw, H., Watson, P. H. and Murphy, L. C. (1998). Altered estrogen receptor alpha and beta messenger RNA expression during human breast tumorigenesis. *Cancer Res.*, *58*, 3197-3201.

[54] Lazennec, G., Bresson, D., Lucas A., Chauveau, C. and Vignon, F. (2001). ERβ inhibits proliferation and invasion of breast cancer cells. *Endocr*, *142*, 4120-4130.

[55] Rao, B. R. and Slotman, B. J. (1991). Endocrine factors in common epithelial ovarian cancer. *Endocr. Rev.*, *12*, 14–26.

[56] Mosselman, S., Polman, J. and Dijkema, R. (1996). ER beta: identification and characterization of a novel human estrogen receptor. *FEBS Lett*, *392*, 49–53.

[57] Cunat, S., Hoffmann, P. and Pujol, P. (2004). Estrogens and epithelial ovarian cancer. *Gynecol. Oncol.*, *94*, 25-32.

[58] Kuiper, G. G. and Gustafsson, J. A. (1997). The novel estrogen receptor-beta subtype: potential role in the cell- and promoter-specific actions of estrogens and anti-estrogens. *FEBS Lett*, *410*, 87–90.

[59] Foley, E. F., Jazaeri, A. A., Shupnik, M. A., Jazaeri, O. and Rice, L. W. (2000). Selective loss of estrogen receptor beta in malignant human colon. *Cancer Res.*, *60*, 245-248.

[60] Antonsson, B. and Martinou, J-C. (2000). The Bcl-2 protein family. *Exp. Cell Res.*, *256*, 50-57.

[61] Guo, B., Godzik, A. and Reed, J. C. (2001). Bcl-G, a novel pro-apoptotic member of the Bcl-2 family. *J. Biol. Chem.*, *276*, 2780-2785.

[62] Ke, N., Godzik, A., and Reed, J. C. (2001). Bcl-B, a novel Bcl-2 family member that differentially binds and regulates Bax and Bak. *J. Biol. Chem.*, *276*, 12481-12484.

[63] Oda, E., Ohki, R., Murasawa, H., Nemoto, J., Shibue, T., Yamashita, T., Tokino, T., Taniguchi, T. and Tanaka, N. (2000). Noxa, a BH3-only member of the Bcl-2 family and candidate mediator of p53-induced apoptosis. *Science*, *288*, 1053-1058.

[64] Gross, A., McDonnell, J. M. and Korsmeyer, S. J. (1999). Bcl-2 family members and the mitochondria in apoptosis. *Genes Dev.*, *13*, 1899–1911.

[65] Leung, L. K. and Wang, T. T. Y. (1999). Paradoxical regulation of Bcl-2 family proteins by 17ß-oestradiol in human breast cancer cells MCF-7. *Br. J. Cancer*, *81*, 387–392.

[66] Burow, M. E., Tang, Y., Collins-Burow, B. M., Krajewski, S, Reed, J. C., McLachlan, J. A. and Beckman, B. S. (1999). Effects of environmental estrogens on tumor necrosis factor -mediated apoptosis in MCF-7 cells. *Carcinogenesis*, *20*, 2057–2061.

[67] Vakkala, M., Lahteenmaki, K., Raunio, H., Paakko, P. and Soini, Y. (1999). Apoptosis during breast carcinoma progression. *Clin. Cancer Res.*, *5*, 319–324.

[68] Burow, M. E., Weldon, C. B., Tang, Y., Navar, G. L., Krajewski, S., Reed, J. C., Hammond, T. G., Clejan, S. and Beckman, B. S. (1998). Differences in susceptibility to tumor necrosis factor -induced apoptosis among MCF-7 breast cancer cell variants. *Cancer Res.*, *58*, 4940–4946.

[69] Dubal, D. B., Shughrue, P. J., Wilson, M. E., Merchenthaler, I. and Wise, P. M. (1999). Estradiol modulates Bcl-2 in cerebral ischemia: a potential role for estrogen receptors. *J. Neurosci.*, *19*, 6385–6393

[70] Dubal, D. B., Zhu, B., Yu, J., Rau, S. W., Shughrue, P. J., Merchenthaler, I., Kindy, M. S. and Wise, P. M. (2001). Estrogen receptor-α, not-β, is a critical link in estradiol-mediated protection against brain injury. *Proc. Natl. Acad .Sc.i USA*, *98*, 1952–1957.

[71] Wilson, M. E., Dubal, D. B. and Wise, P. M. (2000). Estradiol protects against injury-induced cell death in cortical explant cultures: a role for estrogen receptors. *Brain Res*, *873*, 235–242.

[72] Rau SW, Dubal DB, Böttner M and Wise, P. M. (2003). Estradiol differentially regulates c-fos after focal cerebral ischemia. *J Neurosci*, *23*, 10487–10494.

[73] Wise, P. M., Dubal, D. B., Rau, S. W., Brown, C. M. and Suzuki, S. (2005). Are estrogens protective or risk factors in brain injury and neurodegeneration? Reevaluation after the Women's health initiative. *Endocr. Rev.*, *26*, 308-312.

[74] Kossmehl, P., Shakibaei, M., Cogoli, A., Infanger, M., Curcio, F., Schonberger, J., Eilles, C., Bauer, J., Pickenhahn, H., Schulze-Tanzil, G., Paul, M. and Grimm, D. (2003). Weightlessness induced apoptosis in normal thyroid cells and papillary thyroid carcinoma cells via extrinsic and intrinsic pathways. *Endocrinology*, *144*, 4172-4179.

[75] Cano, E. and Mahadevan, L. C. (1995). Parallel signal processing among mammalian MAPKs. *Trends Biochem. Sci.*, *20*, 117–122.

[76] Cohen, P. (1997). The search for physiological substrates of MAP and SAP kinases in mammalian cells. *Trends Biochem Sci*, *7*, 353–361.

[77] Kurland, J. F., Voehringer, D. W. and Meyn, R. E. (2003). The MEK/ERK pathway acts upstream of NF kappa B1 (p50) homodimer activity and Bcl-2 expression in a murine B-cell lymphoma cell line. MEK inhibition restores radiation-induced apoptosis. *J. Biol. Chem.*, *278*, 32465-32470.

[78] Choi, I. J., Kim, J. S., Kim, J. M., Jung, H. C. and Song, I. S. (2003). Effect of inhibition of extracellular signal-regulated kinase 1 and 2 pathway on apoptosis and bcl-2 expression in Helicobacter pylori-infected AGS cells. *Infection and Immunity*, *71*, 830-837

[79] Trisciuoglio, D., Iervolino, A., Zupi, G. and Del Bufalo, D. (2005). Involvement of PI3K and MAPK signaling in bcl-2-induced vascular endothelial growth factor expression in melanoma cells. *Mol. Biol. Cell*, *16*, 4153-4162.

[80] Chen, G. G., Liu, Z. M., Vlantis, A. C., Tse, G. M. K., Leung, B. C. H. and van Hasselt, C. A. (2004). Heme oxygenase-1 protects against apoptosis induced by tumor necrosis

factor-alpha and cycloheximide in papillary thyroid carcinoma cells. *J. Cell Biochem.*, *92*, 1246-1256.

[81] Wu, T. W., Wang, J. M., Chen, S. and Brinton, R. D. (2005). 17Beta-estradiol induced Ca2+ influx via L-type calcium channels activates the Src/ERK/cyclic-AMP response element binding protein signal pathway and BCL-2 expression in rat hippocampal neurons: a potential initiation mechanism for estrogen-induced neuroprotection. *Neuroscience*, *135*, 59-72

[82] Zivadinovic, D. and Watson, C. S. (2005). Membrane estrogen receptor-alpha levels predict estrogen-induced ERK1/2 activation in MCF-7 cells. *Breast Cancer Res.*, *7*, R130-144.

[83] Alexaki, V. I., Charalampopoulos, I., Kampa, M., Nifli, A. P., Hatzoglou, A., Gravanis, A. and Castanas, E. (2006). Activation of membrane estrogen receptors induce pro-survival kinases. *J. Steroid Biochem. Mol. Biol.*, *98*, 97-110.

[84] Vicini, E., Loiarro, M., Di Agostino, S., Corallini, S., Capolunghi, F., Carsetti, R., Chieffi, P., Geremia, R., Stefanini, M. and Sette, C., (2006). 17-beta-estradiol elicits genomic and non-genomic responses in mouse male germ cells. *J. Cell Physiol.*, *206*, 238-245.

[85] Acconcia, F., Totta, P., Ogawa, S., Cardillo, I., Inoue, S., Leone, S., Trentalance, A., Muramatsu, M. and M Marino, M. (2005). Survival versus apoptotic 17beta-estradiol effect: role of ER alpha and ER beta activated non-genomic signaling. *J. Cell Physiol.*, *203*, 193-201.

[86] Xing, M. (2005). BRAF mutation in thyroid cancer. *Endocr Relat Cancer*, *12*, 245-262.

[87] Chen, K. T., Lin, J. D., Chao, T. C., Hsueh, C., Chang, C. A., Weng, H. F. and Chan, E. C. (2001). Identifying differentially expressed genes associated with metastasis of follicular thyroid cancer by cDNA expression array. *Thyroid*, *11*, 41–46.

[88] Baeuerle, P. A. and Baltimore, D. (1996). NF-B: ten years after. *Cell*, *87*, 13-20.

[89] Vina, J., Borras, C., Gambini, J., Sastre, J. and Pallardo, F. V. (2005). Why females live longer than males? Importance of the upregulation of longevity-associated genes by oestrogenic compounds. *FEBS Lett*, *2579*, 2541-2545.

[90] Pande, V. and Ramos, M. J. (2003). Nuclear factor kappa B: a potential target for anti-HIV chemotherapy. *Curr. Med. Chem.*, *10*, 1603-1615.

[91] Smyth, P. P. (2003). Role of iodine in antioxidant defence in thyroid and breast disease. *Biofactors*, *19*, 121-130.

[92] Karin, M. (2006). Nuclear factor-kappaB in cancer development and progression. *Nature*, *441*, 431-436.

[93] Bubici, C., Papa, S., Pham, C. G., Zazzeroni, F. and Franzoso, G. (2006). The NF-kappaB-mediated control of ROS and JNK signaling. *Histol. Histopathol.*, *21*, 69-80.

[94] Pozo-Guisado, E., Merino, J. M., Mulero-Navarro, S., Lorenzo-Benayas, M. J., Centeno, F., Alvarez-Barrientos, A. and Salguero, P. M. F. (2005). Resveratrol-induced apoptosis in MCF-7 human breast cancer cells involves a caspase-independent mechanism with downregulation of Bcl-2 and NF-kappaB. *Int. J. Cancer*, *115*, 74-84.

[95] Jacob, J., Sebastian, K. S., Devassy, S., Priyadarsini, L., Farook, M. F., Shameem, A., Mathew, D., Sreeja, S. and Thampan, R. V. (2006). Membrane estrogen receptors: genomic actions and post transcriptional regulation. *Mol. Cell Endocrinol.*, *246*, 34-41.

Index

A

B

C

D

E

F

G

H

I

M

N

T

U

V

W

X

Y

Z